STUDIES ON JAPANESE OSTRACODA

STUDIES ON JAPANESE OSTRACODA

Edited by
Tetsuro HANAI

UNIVERSITY OF TOKYO PRESS

The University Museum, The University of Tokyo, Bulletin No. 20, 1982

UTP 3044–67694–5149

Printed in Japan

ISBN 0-86008-314-4

CONTENTS

INTRODUCTORY NOTE ON THE STUDIES ON JAPANESE OSTRACODA

Tetsuro Hanai
Geological Institute, Faculty of Science, The University of Tokyo

I

In 1977, I mentioned the existence of two general frames of mind in the study of a particular taxonomic group. In one, the intent is to understand, more or less idiographically, the taxonomic group. In the other, constituents of the taxonomic group may be treated as experimental animals, simply because they are more useful than other groups of organisms for the more or less nomothetical study of causal relations in biology. In reality, however, a biologist's mind seems to oscillate between the two above attitudes, perhaps because in order to successfully utilize a particular taxonomic group in experimental studies one must know it as well as those specializing in the taxonomy of that group. Balance between the two attitudes seems to be the only way to attain proper and self-controllable biological knowledge and avoid monstrous deviations.

When the checklist of Ostracoda from Japan and Southeast Asia was compiled, we recognized that our idiographic knowledge of Japanese Ostracoda was more scanty than we had thought and that the place where we should send the first expedition to investigate ostracod faunas was Japan itself. Therefore, utilization of Ostracoda as experimental animals was still far away.

The first step in a series of idiographic studies of Japanese Ostracoda was commenced by building upon taxonomic knowledge that had been summarized in the checklist and directing effort in the following two directions. One was to determine the spatial and temporal distribution of Ostracoda in relation to the surrounding environment with hope "to elucidate paleontologically significant historical events which may be complicated but which may be interpreted in terms of biological theories" (Hanai, 1977, p. 85). The other was to study individuals in terms of the causal relation between structure of calcareous exoskeleton, which remains in the fossil record, and internal cells, which are responsible for secretion of the hard tissue but are easily destroyed after the death of the animal. The underlying hope is that certain groups of Ostracoda may become "marine *Drosophila*" in the not-too-distant future. Some progress has been made in both directions, and a few definite steps may have been made toward the understanding of Japanese Ostracoda.

Among the studies of the time and space distribution of ostracod faunas, the following questions are of special interest. How and how far will the transgressions and regressions, though the events are small in scale and local in nature, affect the environment and hence the change of ostracod fauna? How imperfect are fossiliferous sediments and consequently fossil records in terms of representing time, and how fragmental is their preservation in space they once occupied—that is, the problem of "imperfection of the fossil record"? What is the rate and way in which faunal compositions change over geologically short time intervals—namely, the "tempo and mode" of species migration and of new element addition?

II

Before discussing our research, a few terms commonly used in our papers will be explained. Observation of the Petri dish culture of species of different genera, which occur together in one bottom sample, shows that one species ignores the presence of the other. Thus, the coexistence of ostracod species in one sample may not imply any direct biological interrelation between the species. Study of the distribution pattern of some shallow water Ostracoda shows that no two species share exactly the same area of distribution, though some species quite often have overlapping distribution ranges, suggesting similar but different environmental requirements of the species. However, these differences in environment are often very small and not at all detectable in evidence preserved in the sediment. Further, areas actually occupied by ostracod species do not always coincide with the area where those species can potentially live. Thus, a sedimentary unit which is occupied or potentially occupied by an assemblage of living Ostracoda characteristic of a particular environment becomes a practical unit for research applicable to paleontology, and the sum of bilogical features of that unit exhibited by Ostracoda may be called ostracod biofacies. Biofacies are named by words describing a particular environment, for example, bay mouth biofacies, brackish water inlet biofacies, etc. Here again the constituent species of one biofacies do no necessarily have any direct or even indirect relationship with any other, except for the accidental relation arising from the preference for a similar environment.

Paleoecologists have discussed *ad nauseam* the preburial and postburial relation between the biocoenosis and the thanatocoenosis, pointing out that a group of fossils that occur together in a particular lithologic unit of a certain stratigraphic level, called fossil assemblage, may not directly indicate the particular environment in which the constituent species lived. Therefore, fossil assemblages of Ostracoda are named not by using environmental terms but by using characteristic constituent species, for example, *Spinileberis quadriaculeata—Nipponocythere bicarninata* assemblage, etc. The assemblage is a combination of ostracod species, which lived in several adjacent environments. However, if fossil evidence is adequate and proper knowledge of correlation between the distribution of living Ostracoda and of dead carapaces, applicable to that particular case, is available, biofacies—that is, an assemblage of Ostracoda characteristic of a particular environment when alive—may be detectable with high certainty through investigation of fossil assemblage. It is evident that fossil assemblages range from those in which biofacies of the past are detectable to those in which allochthonous occurrence of ostracod carapaces completely masked the original biofacies. In the former extreme case, the term biofacies is even applicable directly to the biofacies of the past inferred from the fossil assemblage. However, Frydl, in his paper on Holocene

Ostracoda, marked the biofacies of the past by prefixing the term biofacies with the first letter of the characteristic species in a fossil assemblage, for example, S biofacies for Holocene *Spinileberis quadriaculeata* biofacies, and he showed in detail the relation between inferred biofacies of the past and the present-day biofacies descended from them. In the latter case, Yajima employed a large-scale environmental term which includes several adjacent biofacies once probably characterized by distinct assemblages of living Ostracoda, but undistinguishable at present, to cover safely all the variations within a set of fossil assemblages, for example, subtidal sand assemblage, etc.

III

Even though the knowledge of change of population size and distribution of ostracod species due to transgression and regression is important from a paleontological point of view, there is a shortage of concrete data. Because the sediments of the Jomon transgression include material which can be dated by the ^{14}C method, it is possible to obtain their age and thus trace lateral shifts of ostracod species on a minute time scale. However, Jomon transgression sediments are generally not exposed and are thus difficult to examine in detail. In the recently uplifted southern part of the Boso Peninsula, these sediments form marine terraces where they are accessible to direct observation. These terraces are the result of both eustatic sea level changes and tectonic uplift of the area, and they and the sediments forming them can be used to infer the mode and rate of relative sea level changes, and hence the changes of fauna due to transgression and regression.

Frydl's work shows that Ostracoda of the *Spinileberis*, *Keijella* and *Keijella-Nipponocythere* assemblages, which occupied the inner bay areas of deeply incised coves during the transgressive phase, died out during the regressive phase. Only Ostracoda of the *Neonesidea-Schizocythere* assemblage and of further seaward located assemblages were able to retreat during the regressive phase. This demonstrates that even small-scale transgression and regression can have a profound influence on population size and the size of area occupied by ostracod species constituting the innermost bay assemblages. In the case of a large bay (i.e., Tateyama Bay) or slower regression, *Keijella* and *Nipponocythere* could retreat and continue to occupy the central bay mud biofacies. In general, the effect of a regression does not entirely depend on the scale of the regression itself, but also considerably on the scale of the preceding transgression.

A field survey of the Hamana-ko Bay made by Ikeya and Hanai illustrates the condition of a drowned valley after a long period of nearly constant sea level. At present the mouth of the valley is almost closed by a sand bar, and the enclosed bay is being filled from the entrance inwards with well-sorted coastal sand supplied by tidal current. Continuation of one condition may often emphasize or exaggerate some phenomena and minimize or efface others. In drowned valleys with a wide bay mouth and thus good marine water circulation, the area occupied by *Spinileberis* lies close to that occupied by *Keijella* or *Nipponocythere*, as shown by Frydl. In Hamana-ko Bay, however, the area with abundant *Spinileberis* is in direct contact with the Main channel biofaces, which are characterized by abundant *Hemicytherura* and *Semicytherura* living on the sandy bottom under direct influence of tidal current. Further, it seems probable that even if the present nearly standstill condition of the sea level continues, the area of *Spinileberis* may finally be effaced. *Keijella* and *Nipponocythere* have already disappeared from the present Hamana-ko Bay, though they are abundant in the Pleistocene sediments in the same area, which were probably deposited in a bay with wide bay mouth.

All of these facts suggest that the extreme change in selection pressure caused by transgression and regression exerts the strongest influence on ostracod species inhabiting the innermost parts of bays, which are often influenced by freshwater inflow, and the change may provide the trigger of a mechanism which develops unique elements of the subsequent transgressive fauna. For this reason Quaternary *Spinileberis quadriaculeata* and *S. furuyaensis*, and *Cytheromorpha acupunctata* represent suitable material for studies dealing with the relation between speciation and morphological variation as a function of time. Influence of selection pressure on offshore species, living in areas with good marine water circulation, seems relatively weak.

Further, when the change of population size and the size of the area occupied by a particular species of Ostracoda caused by a small-scale regression is extrapolated to a large-scale regression, such as Oligocene—Lower Miocene regression, the effect of the regression on some species appears to be drastic. *Schizocythere* with Paleogene-type ornamentation, for example, was completely replaced, on a worldwide scale, by the Neogene type during the period of the Oligocene—Lower Miocene regression (Hanai, 1970).

IV

The relation between phenomena occurring in the time span called ecological time and those occurring in geological time is one of the most important problems of paleontology. Upper Pleistocene sediments of northern Boso Peninsula under discussion were deposited between approximately 263,000 (marker tephra GoP) and 132,000 (marker tephra KlP) years B. P. and contain abundant fossils. They are thus particularly well suited for investigation of the faunal changes occurring in the time interval lying between ^{14}C (1,000 years) and biostratigraphical (1,000,000 years) time scales.

The Upper Pleistocene sediments of the area mentioned above were deposited during a period characterized by two geologic phenomena. One is volcanic activity, which produces volcanic ash, thus inserting exact time planes across various types of sediments in several horizons and further providing material that can be dated by the fission track method. The other is glacial and interglacial eustatic sea level change, which produces cyclic change of the environment and thus repetitious faunal sequences.

Yajima's study illustrates faunal changes in offshore areas of a large bay (i.e., Paleo-Tokyo Bay), complementing studies of faunal changes occurring in the nearshore environment at the transgression-regression front. Five sedimentary cycles (even though incomplete) separated by diastems can be recognized in the sediments deposited over a period of approximately 130,000 years.

In the last of the five cycles, which was formed by the Shimosueyoshi transgression, sedimentary sequence (Toyonari Member) is represented by a nearly complete cycle in the southern area, with nonmarine sediments in its lower and upper sequences and marine sediments of high sea level phase in its middle sequence. In the northern area (Kioroshi Member), depositional sequence starts with nearshore marine sediment of the high sea level phase, which directly overlies the lower cycle. Periods of non-deposition or erosion represented by the diastems generally get longer landwards and probably extended over a considerable time span, comparable in length to the periods of rising and stable high sea level. The evidence of events which occurred during the lowest sea level is not preserved in sediments of the Paleo-Tokyo Bay. The nearshore sediments of the regressive sea may be hidden somewhere between the Bay and the site of the deep sea cores.

The imperfection of the fossil record in sediments of a regressive sea of this magnitude both in time and area has so far been the "fatal imperfection" which can only be overcome by inference based on extrapolations from the result of small regressions and experimental simulation.

V

Composition of ostracod fauna is controlled mainly by the nature of local substratum and the salinity, oxygen content and other characteristics of sea water. On a 10,000-year scale these factors were influenced by the transgression, regression and sedimentary filling-up during sea level standstill. Thus the transgressive and regressive shift of sediments, which form the substratum, seems to prescribe the mode of the shift of ostracod fauna as a function of time, and it may be the basic controlling mechanism.

When considering events on a 100,000-year scale, in addition to the above changes, large-scale modification of topography and resulting alteration of sea water circulation must also be taken into account. This modification will, through change of paleogeography and of current circulation, give populations opportunities to interbreed and may aid some ostraced species, aided by their unchanging ecological requirements and homeostatic mechanisms, to maintain phenotypic *status quo* for a long time on a biostratigraphical time scale.

Paleo-Tokyo Bay was initially open to the east. Each of the three sedimentary cycles represented by the Yabu and Kamiizumi Members of the Yabu Formation and the Kiyokawa Formation has a middle horizon which consists of sediments deposited at the time of the highest sea level. The presence of warm water pteropods and bivalves indicates that higher sea level resulted in the opening of the Paleo-Tokyo Bay to the south as well as to the east, though only temporarily. Ostracod assemblages are, however, dominated by warm temperate species, suggesting bottom water staying in the Bay. The Shimosueyoshi transgression again connected the Paleo-Tokyo Bay to the south during the period of highest sea level, and the sea retreated toward the east as well as toward the south to form the present Tokyo Bay. It is highly possible that many ostracod species which inhabited the shallow water of the Paleo-Tokyo Bay followed the retreating sea to the inner part of the present Tokyo Bay.

VI

Ostracod faunas are also affected by changes in temperature. During the interglacial period of rising sea level, shallow thermophilic species extended their geographical range of the area of reproduction along the coast toward the north. The extension was controlled by a stepping-stonelike distribution pattern of substratum favorable to a given species. Ostracod migration over these 'stepping-stones" covering distances of several hundred kilometers along the coast may have occurred in a relatively short period of time of about 100 years or less. In the case of the Jomon transgression, at least one species, *Ambocythere japonica*, common in subtropical areas, appeared in shallow water sediments of the southern part of the Boso Peninsula deposited during the period of high sea level. It is interesting to note that thermophilic species, for example, *Trachyleberis niitsumai*, *Neocytheretta* sp., are represented in the area of their northernmost extent by abundant instars, but adult carapaces have been found only rarely or not at all, suggesting that the reproductive area of these species in the northern marginal area of the species range may be relatively small in comparison with the area of instar distribution.

Analogously, during the glacial period, cryophilic species might have extended their distribution toward the south along the coast of the retreating sea. It is, however, quite rare to find evidence of their migration, as it probably lies somewhere between the present shore and the site of the offshore drillings. The rareness may be accentuated by the fact that the nearshore environment predominated along the transgressive sea became a minor element along the shore of the retreating or retreated regressive sea. Cryophilic species, which migrated from the north and lived in the retreated sea of the glacial stage, entered the Paleo-Tokyo Bay with the rising sea of the interglacial transgression. Examples are given by *Finmarchinella* (*Finmarchinella*) *uranipponica*, *Howeina camptocytheroidea*, *H. higashimeyaensis* and *Robertsonites reticuliforma*. Cryophilic species are relatively abundant in the sedimentary cycle represented by the Kamiiwahashi Formation, in which warm water influence is not detectable even at the time of the highest sea level. In the sediments constituting the remaining four cycles, presence of warm water pteropods and bivalves indicates that the increased sea level resulted in the opening of the Paleo-Tokyo Bay also to the south. However, the increased sea level also caused deepening of the bay, and no marked influence of the warm water current is apparent in relatively deep ostracod faunas.

Elements may originate either as unique elements, which developed in response to extreme change in selection pressure during transgression and regression, or as marginal isolates of cryophilic species. In either case they are added to shallow water fauna during transgressive migration. It is interesting to note that the cryophilic species, which migrated from the north, are without exception smaller than the closely related northern forms.

The metaphysical domain lying between the paleontological facts and the biological mechanisms will not necessarily be overcome either by waiting passively for the new advancement of favorable biological theories or by speculations based on reallocation of insufficient evidence. In order to reduce the domain, tracing the factual tracks of faunal change and perceiving the missing details of the imperfect fossil records may be one approach, and a study of living animals viewed from the paleontological angle and finding out actively the biological mechanisms explaining paleontologically interesting phenomena may be another approach.

VII

Prior to the selection of experimental animals for study at the individual level in terms of the relation between the structure of calcareous exoskeleton and internal cell, ten years elapsed in collecting materials to find the habitat of species suitable for the study and, in examining materials for their viability, determining how long one can keep specimens alive and developing techniques to control culturing conditions. Hanai (1977) pointed out four basic conditions of Ostracoda specimens suitable for a culturing experiment: 1) ubiquitous and obtainable both in any season and in any place, 2) tolerant and standing up to environmental fluctuations produced unintentionally by culturing, 3) easy to feed, and 4) easy to reproduce. Because long-term culturing was not necessary for this study, conditions 3 and 4 were ruled out. Among the species which meet conditions 1 and 2 and are thus easy to culture by the simple Petri dish technique and easy to keep for a time in laboratory, *Keijella bisanensis*, a large species with simple but definite surface ornamentation, was selected. The movements of individuals are observable in a Petri dish with the naked eye, and thus are easy to manipulate.

Many of the methods of foraminiferal culture, which were explained briefly in Ikeya et al. (1976), were an application of the methods developed actually for ostracod culture in our laboratory. For collecting ostracod specimens, Ockelmann-type sledge sampler modified for semiquantitative sampling is used. From a specific station, mud samples which contain approximately 500 to 2,000 individuals of *K. bisanensis* are collected by a single sampling covering a bottom surface of approximately $15\times 1{,}000\ \text{cm}^2$.

Mud samples are placed in a shallow pail and left in the laboratory, keeping the temperature of the sea water as it was on the sea bottom. Within about ten minutes, small organisms, including *K. bisanensis*, crawl out from the disturbed mud toward the mud surface. After clay and mud particles settle on the bottom of the pail and the sea water becomes clear, surface layer mud 1 mm in thickness is sucked up and moved into small Petri dishes together with the soft bottom suspension and sea water using a pipette with an opening of about 2.5 mm in diameter. *K. bisanensis* crawls just below the sediment water interface, but its way of movement is detectable through the characteristic movement of the mud surface. Phototaxis of *K. bisanensis* can be utilized to concentrate individuals. Annelids, small crustacea other than *K. bisanensis*, molluscan larvae, etc., which consume oxygen while alive and produce poisonous substances after death, are removed. Thus *K. bisanensis* are kept alive for a maximum of 4 months or so in Petri dishes placed in an incubator at about 15°C. Replacement of half of the water in the Petri dishes with fresh sea water every 4 days facilitates culturing.

Observation of the life cycle of this species showed that experiments on development of eggs and on ecdysis ought to be made during the period from October to December and from November to April of the next year, respectively. In the other periods of a year, both the appearance of eggs in the female specimens and the occurrence of young instars become very rare. Details of the observations on the local variation and seasonal fluctuation of population structure in a deme are summarized by Abe (unpublished thesis, University of Tokyo, 1981).

VIII

Studies on the reticulation pattern of ostracod carapace were actually started with Pokorný's demonstration (1964, 1969) of individual meshes and ridges which grouped meshes into areas. The purpose of his study was simply to classify forms which differ from each other only in minor sculptural features. Liebau (1969, 1971, 1975a and b), following the method used by Pokorný, utilized the constancy in number and arrangement of the mesh pattern appearing on the carapace surface to trace phylogeny and to establish classification of Trachyleberididae s. l. Independent of the studies on reticulation pattern, the distribution pattern of normal pore canals has been noted as similar in all specimens of one species, following a definite pattern of distribution (Triebel, 1941; Morkhoven, 1962; Plusquellec and Sandberg, 1969). Hanai (1970) illustrated a remarkably conservative and almost identical pattern of pore distribution within one species, by superimposing the distribution pattern of one specimen on that of another.

These two lines of studies were accelerated by the development of the technique of scanning electron microscopy. It provided us with extreme details of the sculpture and structure of the ostracod carapace and stimulated us to introduce a diversity of minute terminology into the description of the sculpture of ostracod carapace. To describe and to name details of carapace morphology without seeking their biological meaning was largely a matter of refinement of laboratory technique, but it was an unavoidable step of study and soon combined the idea of the constancy of reticulation pattern with the idea of the constancy of distribution of pore canals—this time, especially of pore cones by giving reference points of pore cones on the reticulation walls.

The second step of study seems to have developed in two directions. Benson (1972, 1974, 1975) approached the surface sculpture and form of carapace from its mechanically functional skeletal structure, through constructing abstract models of carapace sculpture and form. Liebau (1977, 1978) tried to distinguish evolutionary levels in an evolutional trend of cytheracean carapace ornamentations. In the primitive stage of ornamental genetics, ornamental elements vary within a species in number and arrangement, and genetic changes affect the entire ornamentation or the inexactly defined part of it. In the most advanced stage of ornamental genetics, ornamental elements are constant in number and arrangement within a species, and genetic changes may even affect a single element. Liebau also recognized an intermediate stage between the two stages. Liebau's work seemed to have proved that practically everything on the ostracod carapace is genetically controlled (Sohn, 1975).

There is, however, another approach from the viewpoint of developmental biology. The short-circuit connection between precisely placed ornamentation, including meshes and pores, and their genetic control may result in the oversight of the importance of the study of developmental mechanisms. Thus Hanai (1977) emphasized the necessity of developmental study of the spacial pattern of specifically placed pore canal openings on the ostracod carapace, adopting the term "organules" coined by Lawrence

(1966) originally for specialized structures such as tactile bristles and secretory glands in integument of insects in the latter's paper on the pattern formation.

Studies on the carapace surface of Japanese Paijenborchellini (Hanai, 1970) suggest that the diversity of ornamentation of the tribe is explainable through modification of mesh pattern into two directions: one from reticulate to ridged (ex. *Hanaiborchella miurensis* to *H. triangularis*) and the other from reticulate to spinose (ex. *H. miurensis* to *H. spinosa*). The fact suggests that the mesh pattern of a certain size, excluding second-order reticulation, may form the basis of carapace ornamentation, and ridges and spines may be modifications of walls.

The smooth area of the carapace surface of genus *Neomonocertatina*, which is surrounded by strong ridges, reveals the lining of mesh structure under transmitted light. Further scanning electron microscopy demonstrates that the mesh pattern of ornamentation is observable on the adequately etched surface of a certain species, which have been described as having a smooth surface. Although the boundary between meshes does not always turn into the ridges of carapace surface, the ridges of surface ornamentation, if present, always coincide with the boundary between the meshes. This further suggests that the mesh pattern always underlies the cytheracean carapaces, homologously with the cuticular polygones of certain insects and other arthropods, which correspond to the underlying epidermal cells.

Pore canals of trachyleberidids usually open on the ridges of the mesh, either at the junction of the ridges or on the ridge between the junctions. When a pore canal opens on the floor of the mesh, it is located close to the ridge and the ridge quite often extends and is connected to the pore (Liebau, 1978). It may present an interesting problem to relate the precise locations of pore canal openings to the precise location of meshes and further to the lining of epidermal cells, approaching them from the pattern formation of meshes and pores.

The promise of the study depends, in general, on the selection of experimental species. Among the few species that I have already examined for the ease of keeping them for a while in the laboratory, a trachyleberidid, *Keijella bisanensis*, was selected for close examination, because the species is large in size and easy to manipulate and is ornamented by simple uniform meshes of similar size surrounded by ridges with pore canal openings.

Okada's work explained the constancy of mesh pattern. Cell junctions between outer epidermal cells, which adjoin each other, are strengthened by development of apical desmosomes and septate junctions. Fibrille de soutien (Rome, 1947) or supporting fibers (Kesling, 1951) consisting of clustered columns of microtubles connected to basement membranes of both outer and inner lamella cuticules by conical hemidesmosomes, and running perpendicularly to shell lamella through the outer and inner epidermal cells. On the boundary between inner and outer epidermal cells, where the columns meet, microtubles are connected by means of intermediate junctions. Thus, in addition to the cell junction of the apical desmosomes and septate junction, supporting fibers probably serve as fixed points maintaining the position and form of outer and inner epidermal cells. Further, just after the ecdysis, ridges being folded into a T-shaped cuticule in cross-section and forming the mesh are always underlain by apical desmosomes and septate junctions, suggesting that ridges correspond to the boundaries between outer epidermal cells, and therefore each reticule quite likely corresponds to a cell in this species. The area on the carapace where the number of reticule increases by division of a reticule after ecdysis from A-2 to A-1 is already known through Okada's work. The effects of actual cell division on the formation of new ridges are being investigated in our laboratory by inhibiting cell division using colchicine.

IX

Knowledge of the structure and function of the pores with setae on the cytheracean carapace has depended mostly on the classical works of G. W. Müller (1894) and Rome (1944, 1947) until comparatively recently. Their observations on the pore canals and setae were made solely by the light microscope, and inference of their function was based on the limited knowledge at that time of the physiology of the sensory mechanism. Yet Müller already noted the coexistence of two types of setae that are different in their function on a single valve: one is long and thick and is interpreted as receptor for direct touch by the solid object, and the other is fine and short and may be sensitive to the delicate touch-like movement of the surrounding water or even of sound oscillations (Hartmann, 1966). Rome (1944, 1947) in his classical works pointed out the similarity of ostracod sensilla to those of the insect and showed the presence of scolopidia in Ostracoda.

Simple pores without setae or of unknown nature have also been found on the outer surface of the cytheracean carapace. Van Morkhoven (1962) predicted that they are openings for hypodermal glands. Sylvester-Bradley and Benson (1971) called pore canals of this category tegumental ducts, a term employed to describe those of the other orders of crustacea. Scanning electron microscope observation, made by Hanai, Ikeya, Okada and Nishida, on *Keijella bisanensis* revealed the presence of probable tegumental ducts, each of which has a peculiarly shaped cap. The duct itself is a simple cylindrical tube 2 μm in diameter which opens to a large hollow space under the cap. The cap is surrounded by a circular furrow and consists of approximately 30 long, often bifurcate arms which project from the periphery of the pore opening centripetally towards the center of the hole, making a slightly swollen dome. Adjacent arms are often fused distally. Because the general shape of this cap reminded us of the function of the device to protect against the backward flow of dirt seen in the hole of a Japanese-style water closet (*Benjo* in Japanese) in a coach of the Shinkansen Line, we have called this type of pore Ben-type. Ben-type pores open always at precisely the same position in both male and female—on the floor close to the ridge of the mesh.

Certainly, pore canals were ideal subjects for scanning electron microscopy (Sylvester-Bradley and Benson, 1971). Studies on pore with setae have advanced rapidly since the invention of the scanning electron microscope (SEM). In illustrating SEM photographs, Sandberg and Hay (1967) materialized Triebel's (1941, 1950, 1956) prediction that "blind pores" of the sieve plate actually open to the exterior. The range of variations which have been found in the normal pore structure and in the nature of setae is so wide that it seemed difficult to put this diversity in order. However, a few generalizations seem to emerge in the course of SEM illustration and description of these variations.

In the early stage of SEM illustration, Sandberg and Plusquellec (1969) showed, in certain species of the Thaerocytherinae and Campylocytherinae, coexistence of the two types of pores located separately but side by side on one valve. One is a sieve plate with short, thin, furcate seta, or even without an opening for seta. Sieve pores are more or less regularly disposed and open on the level of the carapace surface or its extension. The other is a simple pore with a long, stout seta, or a deeply sunken sieve plate with irregularly perforate sieve pores and radiation of buttress-like supports for the long stout seta from a large central opening. Coexistence of two types of setal pore on one valve is now popularly known from the species of many other subfamilies (i.e. Cytherinae, Hemicytherinae).

The sieve plate without a setal pore may be merely an extreme case of the reduction of a short, thin sensory seta of the sieve plate. Recently, Liebau (1978) summarized the pore canal morphologies according to their historical development: the prototype of sieve pores moved from the muri into the floor of the mesh during the Lower Cretaceous, and the reductions in large central pore canals are traceable back to the Upper Cretaceous.

However, a question still remains on the relation between the simple pore and the deeply sunken sieve plate with a large central opening for the sensory seta. Puri (1974) illustrates, in the simple pore of *Reticulocythereis* sp., a structure called a circular reinforcement of the basal part of the seta, located between the simple pore and a structure similar to the deeply sunken sieve plate. Our reconnaissance SEM observation of the simple pores of *Keijella bisanensis* found that the circular reinforcement appears as a bellows-like structure at the base of the seta and may serve to preserve the roundness of the transverse section of the seta when it bends sharply. Below this is an anchoring structure, the deeply sunken sieve plate. It consists actually of a cluster of test-tube shaped tubes of various length which open distally but are blind proximally. In *Cythere omotenipponica* as well as in many species of cytherine and hemicytherine Ostracoda, the bellows-like structure is found at the base of the seta emerging from the typical sieve plate, suggesting its common occurrence among cytheracean Ostracoda.

When one follows Rome's idea of comparing sensory organs of the ostracod carapace with the sensilla of the insect, e.g. sensillum trichodeum, it may be predictable that the simple pore with a narrow lip is surrounded by the product of the epidermis, while the deeply sunken sieve plates and the setae are products of the outer enveloping (tormogen) and intermediate enveloping (trichogen) cells respectively. Further, the sheath may be generated by the internal and the glia cells.

Since the pore canals and setae which emerge from them are the external cuticle apparatus of the sensilla, it is quite natural to study the external features of pore canals, always taking the nature of the setae into consideration. The recent development of methods of fixation and sublimation for SEM observation have supplied us with details of the setae. Sandberg (1970) illustrated the sensory seta on the sieve pore of *Aurila conradi*. It is interesting to note that the seta of this species is dendritic with a stout upright stem which terminates like a tube. Thus, it may also be interesting to find out how far the dendrite with ciliary structure is generated distally into the tube-like seta by a bipolar sense cell, and to compare the sensilla with a certain kind of insect contact chemoreceptor because the receptor always makes contact with the sea water. *Cythere omotenipponica* seems to receive sensory information on delicate changes in the chemical condition of sea water even with its carapace tightly closed. It is well known among Japanese collectors that the trap for collecting *Cypridina hilgendorifii* is designed utilizing the animals' response to "olfactory" stimuli, though the location of the sensilla is not know yet.

Variations in the canal and seta related to position on the carapace have also been encountered. Our SEM observation of *Cythere omotenipponica* revealed that at least four types of setae are distinguishable on the carapace of this species (Hanai, Abe, Tabuki and Kamiya, in preparation). The first type is distributed in the marginal area along the free margin. Seta is stout in its lower half, tapering rapidly and easily twisted in its upper half, and terminating somewhat like a tube. The pore of this type of setae has irregular decoration around its opening. The second type is distributed also along the free margin but more or less on the inside area of the zone of the first type of pore canals. Seta of this type is bifurcate near its base, having branches of similar size and extending widely apart parallel to the free margin. Branched setae taper gradually into a sharp point. Pores of this type are simple and small with no decoration. The third and fourth types are distributed widely in the central and dorsal areas and corresponded to the two types of setae or two types of pore canals which coexist on one carapace and have been described elsewhere. The third type is a long staut seta tapering gradually and terminating with a pointed end. The pore has a wide and clearly rimmed lip. Pore canals of the fourth type correspond to the sieve-type pore canals. The fourth type of seta seems to include two forms. One is a stout seta without branches, and the other is a stout seta with one slender branch near its base. The setal pore of the former form seems to occupy a margin of the sieve plate, whereas that of the latter form emerges from the central area of the sieve. The nature of the stout seta of both forms seems similar to the first type seta.

The pores with the third type seta are constant in number, totaling 26 before and after ecdysis from the later stage instars to the adult, keeping a precise pattern of distribution on the carapace, while the other types of setal pore seem to increase in number with every ecdysis. In *Keijella bisanensis*, the Ben-type pores remain constant in number and location on the carpace, while the pores with setae increase in number through ecdysis. Thus it may be presumable that there exist two categories of pores on the cytheracean carapaces; one remains constant in number and the other increases in number through ecdysis. Since the third type seta of *Cythere omotenipponica* is a good example of the sensillum trichodeum, Ben-type pores of *K. bisanensis*, which are quite similar in number and distribution to those of the third type setal pores of *C. omotenipponica*, are likely to be a neuro-secretory modification of the sensillum.

Diversities of the pore canal structure which have been encountered in relation to position on the carapace are concordant with the continuous nature of the ostracod carapace across the hinge margin. Further, when one adopts the idea underlying the term "marginal infold" instead of using the term "calcified portion of the inner lamella," a homologous relation may be expected even between the structural elements of normal pore and radial pore canals. Work begun in our laboratory with these programs may provide a good starting point for arguments on these problems.

X

In the systematic description, only the synonymies after the publication of checklist (Hanai, et al., 1977) are listed. As stated by Neale (1965, p. 258), accurate synonymy is a *sine qua non* as the basis for biological study. Species are thus identified prudently, and there is a general agreement among authors in regard to the result of the species identification. However, a slight difference of opinion exists on the generic assignment of certain species, but this problem will not be considered here. This is simply because an element of arbitariness is involved in the generic assignment. Where emotive views can be used, as is exemplified by the case where a trivial name is chosen for a new species, is rare in the branches of natural science. The name of the new species without any explanation on etymology in Yajima's description is derived after the heroines of *Genji Monogatari* (*The Tale of Genji*), a classic novel completed by Murasaki Shikibu in ca. 1020 A.D.

The following abbreviations are used in the section of systematic description in all articles:

UMUT: University Museum, University of Tokyo.

IGSU: Institute of Geoscience, Shizuoka University.

CA: Cenozoic Arthropoda. O: Ostracoda.

S: Sample. Sa: Sample number. Ho: Sampling horizon. F: Formation. M: Member.

Sp: Specimen measured. C: Carapace. LV: Left valve. RV: Right valve.

A-1: Instar of adult minus one stage.

Me: Measurements. L: Length. H: Height. W: Width. N: Number of specimens measured. $\bar{X}$: Arithmetic mean (mm). Sd: Standard deviation (mm). V: Coefficient of variability. OR: Observed range (mm).

The following expressions are used to show the abundance of each species.

Abundant: The species occurs in more than 50% of sample collected from a given member of formation and occupies more than 10% of total individuals.

Common: The species occurs in more than 50% of samples and occupies less than 10% of total individuals.

Rare: The species occurs in less than 50% of samples and occupies less than 10% of total individuals.

All the types and illustrated specimens are deposited in the collection of either the University Museum, University of Tokyo or Institute of Geosciences, Shizuoka University.

Three articles were prepared by P. Frydl, M. Yajima and Y. Okada as a part of their doctoral dissertations based on studies made in our laboratory.

This introductory note is a result of a study supported in part by the grant-in-aid for co-operative research (project no. 434042) and the grant-in-aid for special project research (project no. 56117004) of the Ministry of Education, Science and Culture, the Government of Japan.

ECOLOGY OF RECENT OSTRACODS IN THE HAMANA-KO REGION, THE PACIFIC COAST OF JAPAN

Noriyuki Ikeya* and Tetsuro Hanai**
**Institute of Geoscience, Faculty of Science, Shizuoka University*
***Geological Institute, Faculty of Science, The University of Tokyo*

Abstract: The ostracods from 70 stations in the Hamana-ko region, on the Pacific Coast of central Japan, comprise 70 species belonging to 45 genera, among them, 44 species belonging to 31 genera are extant. The ostracod distribution in Hamana-ko Bay is dependent on characteristics of the water, the nature of substrate, and the bottom topography of the bay. Based on the characteristic species, 6 major biofacies and 7 sub-biofacies are recognized as follows: (I) fresh water pond and river biofacies; (II) brackish water inlet biofacies; (III) bay coast and sand bank biofacies; (IV) inner bay biofacies (dy sub-biofacies, sandy bottom sub-biofacies, silty bottom sub-biofacies, silty sand bottom sub-biofacies); (V) channel and extension channel biofacies (main channel sub-biofacies, extension channel sub-biofacies); (VI) bay mouth and open sea biofacies (bay mouth sub-biofacies, open coast sub-biofacies, offshore sub-biofacies). The standing crop size of ostracods are examined quantitatively in 27 localities. The species diversity and ratio of living specimens to the total number of specimens collected are calculated. A new genus, *Neopellucistoma* (type-species, *Neopellucistoma inflatum* n. sp.) and new species, *Pontocythere sekiguchii*, *Pontocythere minuta*, *Neocytherideis punctata*, *Cornucoquimba rugosa*, *Bythocythere maisakensis*, *Bythoceratina elongata*, *Semicytherura elongata*, *Phlyctocythere hamanensis* and *Neopellucistoma inflatum* are described.

Introduction

As paleontological investigations place increasing emphasis on lateral and vertical biofacies variation of ancient sediments, a need for more critical examination of Recent sedimentary processes and faunal distribution of ostracods is becoming apparent. In order to clarify paleontology and sedimentary environments of Pleistocene marine, brackish, and fresh water sediments distributed along the Pacific Coast of central Japan, a basic examination of Recent sediments from similar environments has been undertaken. Recent sediments suitable for such examination are found in the Hamana-ko Bay which occupies the central area of distribution of these Pleistocene sediments and contains a wide variety of sediments, some of them directly comparable to those of the Pleistocene. The Hamana-ko Bay, situated in the western part of Shizuoka Prefecture, is the largest brackish water embayment (polyhaline) on the Pacific Coast of Japan. The bay covers an area of about 79 km^2 and connects with the Pacific Ocean (Enshu-nada Sea) through a narrow entrance approximately 200 m wide. The relationship between the various sediments and the corresponding sedimentary environments of the Hamana-ko Bay was analyzed by Ikeya and Handa (1972); the faunal analysis of foraminifers was given in Ikeya (1977).

Table 1

List of station numbers, locations, depths, dates, methods of sampling, and bottom characters.

Station No.	N. Lat.	E. Long	Depth (m)	Date of sampling	Sampler**	Bottom*** character
1*	34°47′23″	137°37′54″	5.2	Aug. 8, 1971	C, G, N	M.S.sl.
2	34°47′16″	137°38′42″	2.0	,,	,,	W.S.fs.
3	34°46′56″	137°38′31″	2.5	,,	,,	M.S.sl.
4	34°46′52″	137°37′42″	6.8	,,	,,	,,
5	34°46′34″	137°36′54″	10.0	,,	,,	,,
6	34°45′52″	137°37′48″	2.4	,,	,,	,,
7	34°45′37″	137°37′27″	5.1	,,	,,	,,
8	34°45′42″	137°36′51″	4.7	Aug. 9, 1971	,,	W.S.fs.
9	34°46′30″	137°35′48″	9.1	,,	,,	M.S.sl.
10	34°46′51″	137°35′27″	5.1	,,	,,	P.S.fs.
11	34°46′32″	137°34′30″	5.6	,,	,,	,,
12*	34°46′00″	137°35′33″	12.0	,,	,,	M.S.sl.
13	34°45′21″	137°35′55″	10.2	,,	,,	,,
14	34°45′39″	137°34′47″	11.4	,,	,,	,,
15	34°45′33″	137°34′00″	7.7	,,	,,	P.S.gs.
16	34°45′39″	137°33′12″	8.3	,,	,,	M.S.sl.
17	34°46′08″	137°33′04″	6.4	,,	,,	,,
18*	34°46′47″	137°33′33″	7.2	,,	,,	,,
19	34°47′52″	137°33′20″	1.8	,,	,,	,,
20	34°45′44″	137°32′08″	3.7	,,	,,	P.S.gs.
21*	34°45′20″	137°31′33″	6.3	,,	,,	M.S.sl.
22	34°44′55″	137°32′11″	4.7	,,	,,	,,
23	34°44′52″	137°32′52″	8.2	Aug. 10, 1971	,,	,,
24	34°44′52″	137°33′53″	10.0	,,	,,	,,
25	34°44′55″	137°35′24″	5.5	,,	,,	W.S.fs.
26	34°44′19″	137°35′40″	2.1	,,	,,	P.S.gs.
27	34°44′08″	137°35′27″	3.3	,,	,,	W.S.fs.
28*	34°44′10″	137°34′07″	5.2	,,	,,	,,
29	34°44′00″	137°33′02″	3.7	,,	,,	,,
30*	34°43′28″	137°32′51″	3.4	,,	,,	,,
31	34°43′03″	137°32′49″	2.2	,,	,,	M.S.sl.
32	34°43′07″	137°34′03″	3.1	,,	,,	W.S.fs.
33	34°43′13″	137°34′53″	3.2	,,	,,	W.S.ms.
34	34°42′39″	137°33′56″	2.0	,,	,,	W.S.fs
35*	34°42′26″	137°34′24″	2.3	,,	,,	W.S.ms.
36	34°41′44″	137°34′41″	2.1	,,	,,	P.S.gs.
37	34°41′32″	137°35′18″	2.4	,,	,,	,,
38*	34°40′56″	137°35′30″	4.3	Aug. 11, 1971	,,	W.S.ms.
39*	34°40′37″	137°36′12″	3.1	,,	,,	,,
40*	34°41′39″	137°36′51″	2.1	,,	,,	,,
41	34°42′19″	137°35′55″	2.0	,,	,,	,,
42*	34°42′23″	137°36′39″	4.8	,,	,,	M.S.sl.
43	34°42′48″	137°37′01″	2.5	,,	,,	,,

(Continue)

Station No.	N. Lat.	E. Long.	Depth (m)	Date of sampling	Sampler**	Bottom*** character
44*	34°43′26″	137°37′22″	2.3	Aug. 11, 1971	C, G, N	M.S.sl.
45	34°44′47″	137°38′23″	3.5	,,	,,	,,
46	34°41′37″	137°37′18″	1.8	Nov. 20, 1974	C, N	,,
47	34°42′15″	137°36′57″	2.1	,,	,,	,,
48	34°42′35″	137°36′49″	4.4	,,	,,	,,
49	34°43′15″	137°37′00″	4.2	,,	,,	,,
50	34°44′03″	137°38′00″	3.8	,,	,,	,,
51	34°39′56″	137°35′12″	13.6	Sept. 15, 1975	N	W.S.fs.
52	34°40′22″	137°34′48″	5.6	,,	,,	W.S.ms.
53	34°39′59″	137°34′09″	13.2	,,	,,	W.S.fs.
54	34°40′19″	137°33′22″	7.3	,,	,,	W.S.ms.
55	34°40′00″	137°32′55″	12.4	,,	,,	,,
56	34°40′18″	137°32′03″	5.9	,,	,,	,,
57	34°39′59″	137°38′15″	7.6	,,	,,	W.S.fs.
58	34°41′14″	137°37′44″	5.1	,,	,,	W.S.ms.
59	34°39′51″	137°37′18″	13.2	,,	,,	W.S.fs.
60	34°40′14″	137°36′44″	7.0	,,	,,	W.S.ms.
61	34°40′16″	137°36′04″	9.7	,,	,,	W.S.cs.
62	34°43′16″	137°31′59″	0.7	Sept. 14, 1975	,,	P.S.fs.
63	34°43′28″	137°31′49″	1.8	,,	,,	clay
64	34°43′26″	137°31′45″	1.4	,,	,,	,,
65	34°44′15″	137°32′26″	—	,,	—	soil
66	34°44′39″	137°32′22″	0.8	,,	N	P.S.cs.
67	34°45′03″	137°31′05″	0.6	,,	,,	P.S.fs.
68	34°45′22″	137°31′01″	0.5	,,	,,	P.S.ms.
69	34°46′42″	137°32′45″	0.9	,,	,,	P.S.fs.
70	34°47′29″	137°33′04″	1.0	,,	,,	P.S.gs.
71	34°47′55″	137°33′13″	0.8	,,	,,	,,
72	34°47′29″	137°34′32″	1.3	,,	,,	P.S.cs.
73	34°46′55″	137°36′33″	2.0	,,	,,	,,
74	34°41′40″	137°36′14″	1.0	Sept. 16, 1975	,,	P.S.gs.
75	34°42′27″	137°35′07″	1.2	,,	,,	,,
76	34°43′07″	137°36′13″	1.2	,,	,,	clay
77	34°43′11″	137°36′12″	—	,,	—	soil
78	34°44′06″	137°36′35″	—	,,	—	,,
79	34°44′11″	137°36′39″	1.0	,,	N	clay
80	34°44′03″	137°36′56″	1.2	,,	,,	P.S.gs.
81	34°45′40″	137°37′48″	0.8	,,	,,	,,
82	34°46′19″	137°37′42″	0.9	,,	,,	,,
83	34°47′27″	137°38′58″	0.5	,,	,,	P.S.ms.
84	34°48′06″	137°39′28″	0.4	,,	,,	P.S.fs.
85	34°47′45″	137°37′41″	1.2	,,	,,	P.S.gs.
86	34°47′08″	137°37′02″	1.0	,,	,,	,,
87	34°45′21″	137°39′01″	1.2	,,	,,	,,

(Continue)

Station No.	N. Lat.	E. Long.	Depth (m)	Date of sampling	Sampler**	Bottom*** character
88	34°44′07″	137°38′34″	1.4	,,	,,	,,
89	34°41′19″	137°37′56″	1.2	Sept. 17, 1975	,,	M.S.sl.
90	34°41′37″	137°40′18″	1.6	,,	,,	,,
91	34°42′06″	137°41′22″	1.8	,,	,,	clay
92	34°42′07″	137°41′30″	2.4	,,	,,	,,
93	34°42′02″	137°41′09″	0.5	,,	,,	,,

* Fixed Station for Meteorological Observation.

** C: Phleger-type core sampler, G: Ekman-Birge-type grab sampler, N: Ockelmann-type bottom net sampler.

*** W.S.: Well-sorted, M.S.: Moderately sorted, P.S.: Poorly sorted, gs: gravelly sand, cs: coarse sand, ms: medium sand, fs: fine sand, sl: silt, soil: bottom surface of rice field.

This paper is a part of a reconnaissance study intended to examine the relationship between the environment and ecology of benthic organisms. Although the standing crop size of ostracods is very small in comparison with that of foraminifers, ostracods occur in a wide range of marine, brackish, and fresh-water environments; thus, ostracods have already been shown to be very useful as indicators for analysis of the sedimentary environments of ancient sediments. The object of the present study is to determine the factors that influence the distribution of Recent ostracod species and to evaluate the value of ostracods as indicators for the interpretation of ancient environments.

Materials and methods

Samples used in this study were collected from 93 stations during the following four survey periods: (1) August 8–11, 1971, stations 1–45 from the inner bay; (2) November 20, 1973, stations 46–50 from the northeastern inlet (Inohana-ko); (3) September 14–17, 1975, stations 62–93 along the bay coast and ponds; (4) September 15, 1975, stations 51–61 from the open sea (Enshu-nada). These stations were designed to cover effectively the whole area at intervals of approximately 1.5 km.

The samples from stations of surveys 1 and 2 in the bay area were collected by means of the Phleger-type core sampler (Phleger, 1951, 1960) and a modified Ockelmann-type bottom net sampler. Surveys were made by the research boat *Kamome-maru* of the Hamana-ko Branch, Shizuoka Fisheries Experimental Station. The result of studies on sediments and foraminifers of these samples has been given in previous papers (Ikeya and Handa, 1972; Ikeya, 1977). The collections of samples during surveys 3 and 4 were made by a Ockelmann-type bottom net sampler operating from a charter boat in survey 4 and from land in survey 3.

Sampling data (water depth and bottom character) are listed in table 1, and the locations of the sampling stations are shown in text-figure 1. Twelve among the 93 stations coincide with fixed stations where regular meteorological observations have been made monthly by the Hamana-ko Branch, Shizuoka Fisheries Experimental

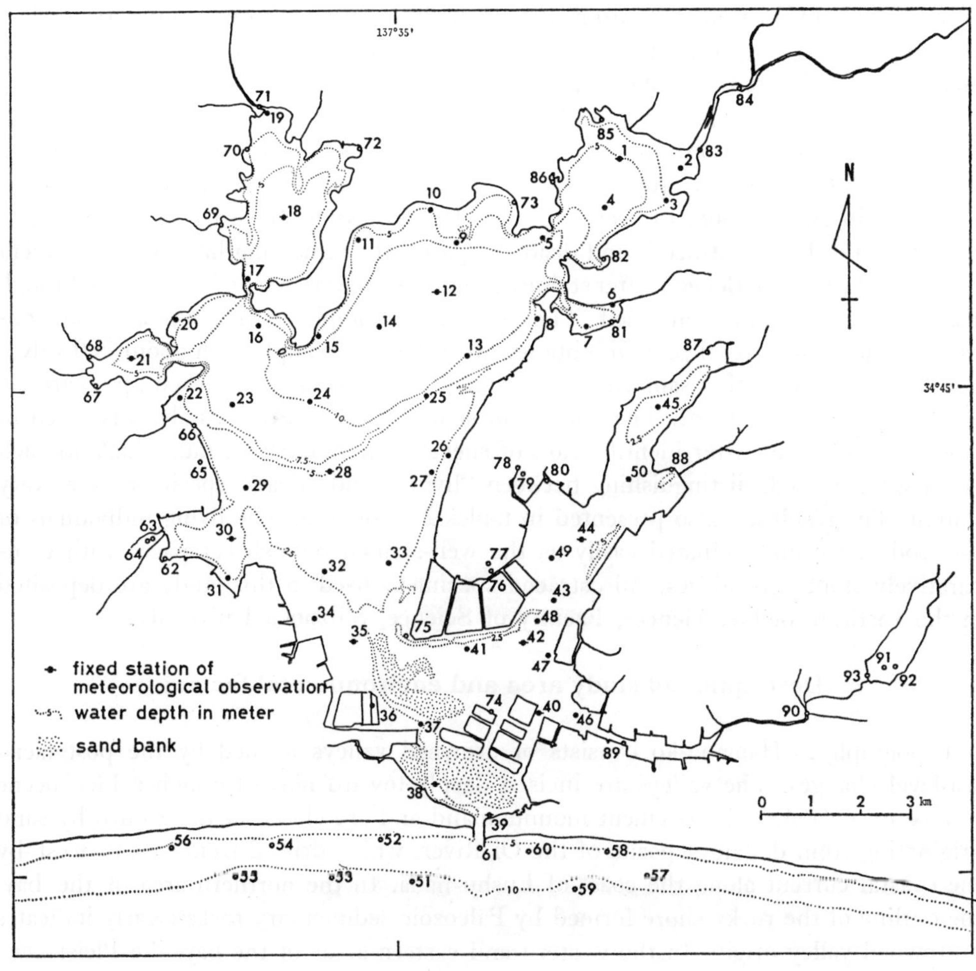

Text-fig. 1

The physiography and locations of the sampling stations in the Hamana-ko region. Numerals indicate station numbers.

Station, since 1949.

Quantitative samples of surface sediments were taken using a Phleger-type core sampler with a plastic inner tube. The samples used for quantitative analysis of ostracod fauna are composed of a few centimeters of the bottom water with suspended sediment in the tube and the uppermost 1 cm of undisturbed core of sediment, which is 3.5 cm in diameter. The standing crop size of ostracods is very small in comparison with that of foraminifers. Thus, the volume of sediments obtained with a core sampler designed for foraminiferal research is usually too small to analyze species diversity of ostracods. Supplementary sampling was made by an Ockelmann-type bottom net sampler at the localities where the core sampling was made. At stations with coarse sediments, only the Ockelmann-type bottom net sampler was used for sampling. The remodeled Ockelmann-type bottom net sampler is 15 cm wide and is designed to be drawn for about 3 m on the bottom to collect the upper-most one centimeter or less of surface sediments, including suspension at the water-sediment interface. A few liters of sediment are thus obtained.

These samples were placed in a plastic bottle and fixed immediately with 5% neutralized formalin solution. In the laboratory all samples were washed with water through a 200-mesh sieve (opening of 0.074 mm). Half of the washed sediment is preserved in 70% alcohol to keep ostracods for dissection. In order to facilitate discrimination of living ostracods and dead carapaces, the remaining half of the sediment was stained with rose Bengal. After staining, the samples were washed over again through a 200-mesh sieve and then dried. The bottom net samples with abundant ostracods were divided into several parts to obtain a workable size sample with about 200 valves of ostracods. In the other samples, all specimens of ostracods were picked up regardless of the volume of sediment. The volume of sediment used in this analysis is given in tables 3 through 5. After identification of species, the number of individuals in each taxon was counted, distinguishing between "living" and "dead" specimens for every station. The result are also presented in tables 2 through 5. The living individuals of ostracods were discriminated easily as the well-stained reddish carapaces with comparatively stout appendages. All ostracod specimens used in this study are deposited in the Institute of Geosciences, Faculty of Science, Shizuoka University.

Description of study area and environmental factors

Topography: Hamana-ko consists of drowned valleys formed by the postglacial sea-level changes. The valleys are incised deeply toward north through a Pleistocene terrace into a Paleozoic basement mountainland and are closed at the mouth by sand originating from deltaic deposits of the Oi River, which drifted from east to west by the coastal current along the coast of Enshu-nada. In the northern area of the bay, the outline of the rocky shore formed by Paleozoic sedimentary rocks clearly indicates a drowned valley origin. In the western and eastern areas of the bay, the Pleistocene formation occurs as terraces with flat upper depositional surface. In the southern area of the bay, a coastal lowland of alluvial deposits develops along the Pacific Coast. Thus, the coastline of the bay is characterized by much indented topography with rocky inlets in the northern area and a straight shoreline in west-to-east direction along the Enshu-nada.

Drowned valleys have been filled up with sand transported toward the north from the mouth of the bay. A sandbank of less than 4 m depth occupying the southern half of the bay impedes water circulation between the open sea and the embayment; a moderately deep area of stagnant water with a maximum depth of 16.1 m occupies the northern half of the bay. To circulate sea water toward the inner part of the bay, three artificial channels (maximum 4 m deep) have been excavated. The small adjoining inlets corresponding to the heads of the drowned valleys still retain their narrow entrances, which open toward the main bay, and are deeper (13–16 m deep) than the main bay. The continental shelf along the Enshu-nada is wide and occupies a flat bottomed area —2 km wide and less than 15 m deep—along the coast. The topography of this area is shown in text-figure 1.

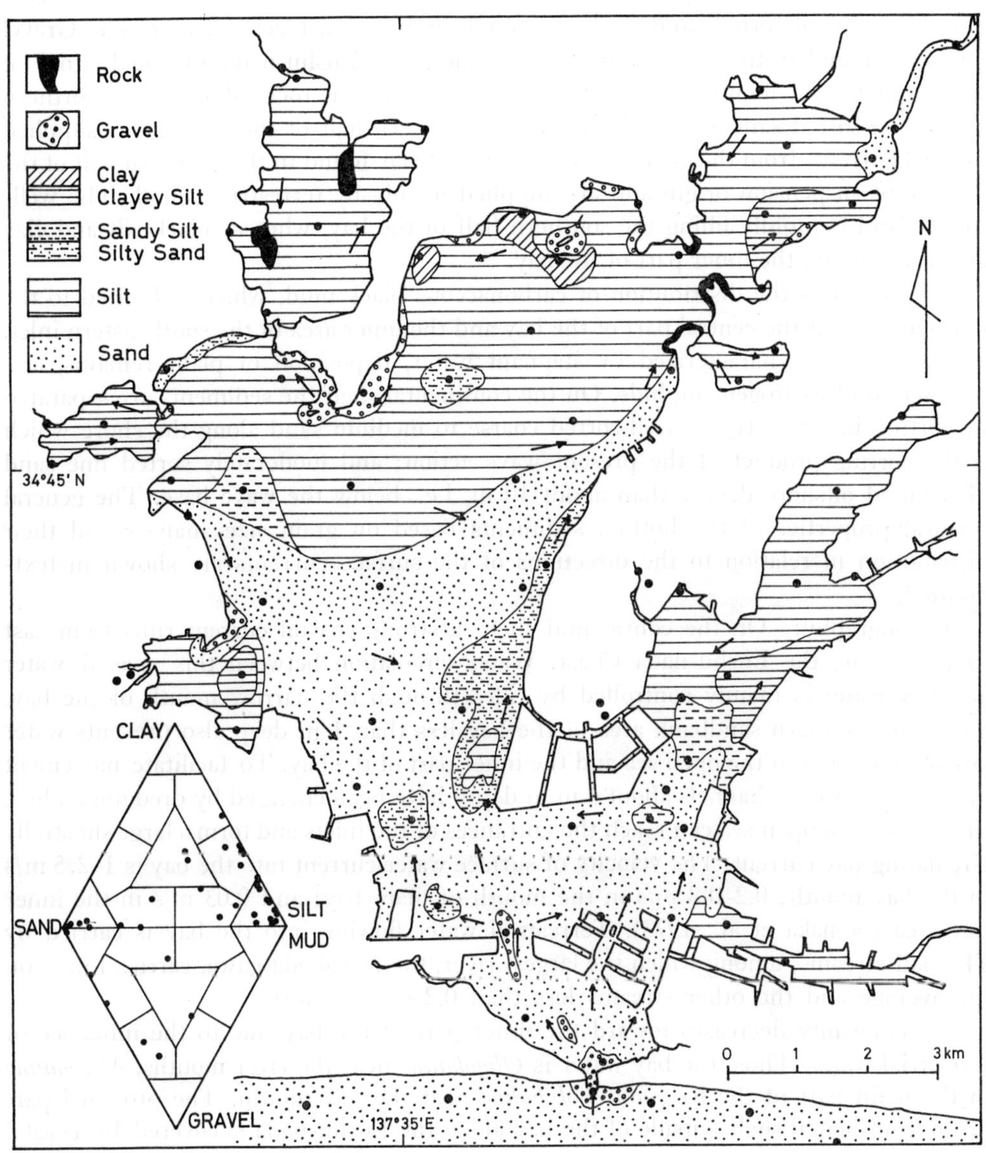

Text-fig. 2

Distribution and general textural properties of the bottom surface sediments based on grain size analysis. Arrows indicate the direction of surface currents.

Sediments: Three major types of sediment can be distinguished: poorly sorted gravel sand, well-sorted sand, and moderately well-sorted black silt and clay. Gravel sands are found in the shallow coastal areas along the Enshu-nada, on the bottom of the channel at the mouth of the bay, and in areas along the rocky shore in the northern area. Well-sorted sand is distributed in the southern half of the bay and is supplied by tidal currents from the open sea. Black silt and clay found in the northern half of the bay are of suspension origin and are supplied mainly by terrestrial runoff. The well-sorted sand is rapidly filling the southern half of the bay, whereas black silt and clay are slowly filling the inner part of the bay.

Noteworthy is the distribution of carbonaceous black mud, which is limited to the northern area of the central part of the bay and the inner area of the southeastern inlet. These areas are characterized by stagnant water, deposition of plant remains, and production of hydrogen sulphide. On the continental shelf the sediments are separated by sorting into two types: well-sorted coarse to medium sand along the shore which is the sorting product of the present wave action; and moderately sorted fine sand distributed offshore deeper than about 10 m, i.e., below the wave base. The general textural properties of the bottom sediments based on grain size analysis and their distribution in relation to the directions of the surface currents are shown in text-figure 2.

Oceanography: On the continental shelf, a strong coastal current runs from east to west along the Enshu-nada Coast. Water circulation between this coastal water and bay water is mainly controlled by tides through the narrow mouth of the bay. The wide southern sandbank area of the bay less than 1 m deep also prevents water circulation between the open sea and the inner part of the bay. To facilitate movement of sea water, three channels about 4 m in depth have been enlarged by dredging. Thus, the water from open sea can reach the entrances of the inlets and form a large sinistrally circulating bay current. The velocity of surface water current into the bay is 1–2.5 m/s in the bay mouth; 0.2–0.8 m/s in the mouth of main bay; and 0.03 m/s in the inner bay area (Nonaka et al., 1973). Terrestrial water flowing into the bay is carried by 14 small streams, among which the largest river, the Miyakoda-gawa, carries 1 m^3/s on the average and the other streams less than 0.2 m^3/s of flux.

The chlorinity decreases toward the inner part of the bay due to the influence of terrestrial water. Thus, the bay water is *Oligohaline* near the river mouths, *Mesohaline* in the main part of the bay, and *Polyhaline* near the bay mouth. The proximal part of the continental shelf outside of the bay along the Enshu-nada is covered by coastal water.

Vertical changes of temperature and salinity are indiscernible in the southern shallow area, which suggests the influence of a strong inflow of sea water. Yet, on the contrary, the water temperature and salinity of the northern part of the bay are characterized by wide seasonal variations. In winter, the mixing of the bay water takes place through a downward movement of the cooled upper water mass of high density into the lower water mass of low density. The difference between surface and bottom water temperature is not significant. In summer, the surface layer has a high temperature and low density, whereas the bottom layer has a low temperature and high salinity. Thus, the water becomes stagnant and stratified into two layers considerably different in

Table 2

Number of individual specimens of each species per unit volume, based on the samples collected by Phleger-type core sampler. (Cardinal numerals show the total number of specimens; italics indicate the number of living specimens.)

Species / Station number	8	10	15	19	20	25	26	27	28	29	32	33	34	35	36	38	39	40	41	42	43	44	46	47	48	49	50
Neonesidea oligodentata (KAJIYAMA)														*1* 2													
Propontocypris (P.) cf. *attenuata* (BRADY)																*1* 2											
Cyprididae n. sp.				*3* 5																		*1* 1	2	*1* 2			
Clithrocytheridea? japonica (ISHIZAKI)	2						2								4	1		1	*1* 2				2		2		
Pontocythere miurensis (HANAI)														1	1	1		1									
P. subjaponica (HANAI)																		1									
Callistocythere japonica HANAI																*1* 1											
Cythere lutea omotenipponica HANAI																			1								
Spinileberis quadriaculeata (BRADY)								*1* 1				*2* 5	2			*1* 1				*2* 2	2	*28* 90			*15* 20	*4* 10	*4* 5
Aurila hataii ISHIZAKI															*1* 1	1											
A. uranouchiensis ISHIZAKI															1												
Mutilus assimilis (KAJIYAMA)			*3* 29			*2* 2	*18* 102		*16* 94					*9* 17	2	*1* 4											
Proteoconcha tomokoae (ISHIZAKI)														1													
Trachyleberis sp.																					1						
Bythoceratina hanaii ISHIZAKI														1													
Hemicytherura tricarinata HANAI	*1* 6		6			*1* 3	*3* 36	*1* 1	1		2			*5* 10		*1* 1					*6* 6	*6* 11		5	*1* 1		
Semicytherura? miurensis (HANAI)	*4* 23	1				*2* 4	6	*3* 4		*8* 15	*2* 3	*1* 2		*1* 1		1			*1* 1						2		
Loxoconcha (L.) pulchra ISHIZAKI																3					1			1			
L. (L.) uranouchiensis ISHIZAKI					1									2	2												
Phlyctocythere hamanensis n. sp.														*3* 4													
Cytheromorpha acupunctata (BRADY)												*1* 1			1						*1* 1	2		3	*30* 48	*2* 2	1
Xestoleberis hanaii ISHIZAKI					1																						
X. setouchiensis OKUBO			1											*13* 40	1	*12* 18				*1* 1							
Paradoxostoma japonicum SCHORNIKOV							*2* 3																				
P. ovulare KAJIYAMA																1											
Cytherois zosterae SCHORNIKOV	1		*4* 7		2		*1* 2							*4* 7		*2* 2							*1* 1	*1* 1			
Sclerochilus mukaishimensis OKUBO														*3* 5													
Cytheroma? sp.																	*1* 1	*1* 1									
Total living specimens	5	-	7	3	-	5	24	5	16	8	2	4	-	39	1	19	1	1	2	3	7	35	1	2	46	6	4
Total specimens	32	1	43	5	4	9	151	6	95	15	5	8	2	91	13	37	1	4	4	3	11	104	5	12	73	12	6

Table 3

Number of individual specimens of each species collected from stations in the inner bay area. (Cardinal numerals show the total number of specimens; italics indicate the number of living specimens.)

Species \ Station number	2	3	5	8	10	12	13	15	19	20	21	22	23	25	26	27	28	29	32	33	34	35	38	39	40	41	42	43	44	45	46	47	48	49	50
Neonesidea oligodentata (KAJIYAMA)																							*3* 3	*5* 5											
Darwinula sp.					1																														
Propontocypris (P.) cf. *attenuata* (BRADY)																		1					1				1								
Cyprididae n. sp.	*21* 27	*9* 61							*439* 443			1								*1* 1					*57* 65	*1* 1	14	8	*2* 11	*2* 2	*9* 11				
Physocypria sp.																					1														
Candonocypris assimilis SARS				1								105												2			6								
Cypridopsis vidua (O. F. MÜLLER)												*2* 3																							
Potamocypris producta (SARS)												*3* 12													*1* 1										
Clithrocytheridea? japonica (ISHIZAKI)				2			1								3								12	2	*3* 3	*20* 94	*4* 63	4			*2* 24	*1* 1			
Pontocythere miurensis (HANAI)							1												1				1	3		1	1								
P. subjaponica (HANAI)																							*1* 10	7			8								
Callistocythere japonica HANAI																							*4* 5	*1* 3			1								
C. sp.																								1		1	1								
Cythere lutea omotenipponica HANAI																1							1				1					2			
Hanaiborchella triangularis (HANAI)																								1											
Spinileberis quadriaculeata (BRADY)						1	*1* 20						5			*17* 20		*1* 1							*1* 1		*15* 44	*15* 65	*270* 546		*1* 1	*13* 17	*14* 15	*69* 104	*65* 115
Aurila hataii ISHIZAKI																							*20* 22		*1* 1		1								
Mutilus assimilis (KAJIYAMA)				1		1	*1* 40	*3* 75			2	*1* 3	11	*5* 31	*54* 241	*1* 4	*1* 48					1	*2* 6	*1* 4	*17* 20		3	2							
Cornucoquimba rugosa n. sp.																							1	1	1										
Proteoconcha tomokoae (ISHIZAKI)																											1								
Trachyleberis scabrocuneata (BRADY)																							1												
T. sp.																											2								
Neocytheretta sp.																									1										
Hemicytherura tricarinata HANAI				*2* 5	8	*3* 16	*12* 96	27		*1* 1		*1* 1	*2* 12	1	*41* 84	*17* 24	*1* 3	*1* 8	*4* 5				*6* 6	*2* 2	*345* 403	*9* 11	*16* 106	*73* 190	*27* 44		*8* 15	*5* 9	1	1	
Semicytherura elongata n. sp.																								1	1										
S.? miurensis (HANAI)				*5* 12	4	9	*26* 226						*6* 15	*3* 5	*3* 4	*4* 9	1	*75* 114	*6* 9					*1* 3	*2* 6	*61* 89	*11* 111	*4* 15				*2* 4	*1* 6	1	
S. skippa (HANAI)																								*2* 2											
Loxoconcha (L.) japonica ISHIZAKI																							1												
L. (L.) optima ISHIZAKI																							2												
L. (L.) pulchra ISHIZAKI																							19	*1* 4	3		2	76						3	
L. (L.) uranouchiensis ISHIZAKI																											*6* 8	3	1			1	1	5	
Cytheromorpha acupunctata (BRADY)						1	*5* 16						*2* 10		1	*1* 2			*2* 2							*4* 4	*4* 21	*8* 16	*20* 30		1	*25* 33	*34* 50	*17* 29	*4* 9
Xestoleberis hanaii ISHIZAKI								3		*1* 2	2	*1* 3											*1* 1				1								
X. setouchiensis OKUBO					7			8			4											*1* 2	*3* 3	*4* 7	*4* 7		3				3				
Platymicrocythere tokiokai SCHORNIKOV																								*13* 13											
P. sp.																								*116* 117											
Paradoxostoma japonicum SCHORNIKOV					2			*1* 1							*4* 11									1	*1* 3	*1* 1									
P. ovulare KAJIYAMA																								1									1		
Cytherois zosterae SCHORNIKOV			1	*2* 3	2		*1* 9	*8* 20							*2* 3		1	*1* 2							*4* 5		*5* 12	*3* 4				2			
Sclerochilus mukaishimensis OKUBO					2																			2	*3* 7		4								*1* 1
Neopellucistoma inflatum n. gen. n. sp.																								1											
Cytheroma? sp.																								*87* 89	*1* 1		*1* 1								
Total living specimens	21	9	-	9	-	3	46	12	439	2	-	8	10	8	104	40	2	78	12	1	-	1	40	233	440	96	57	103	319	2	20	46	49	86	70
Total specimens	27	61	1	24	26	28	409	134	443	3	8	128	53	37	347	60	53	126	17	1	1	3	95	272	529	202	416	383	632	2	55	69	74	143	125
Volume of sediments (cc)	60	20	1	20	20	5	5	4	25	2	15	50	5	15	10	30	5	40	25	40	20	0.5	5	140	60	40	15	15	20	10	70	15	10	2	2

Table 4

Number of individual specimens of each species collected from stations of the open sea coast area. (Cardinal numerals show the total number of specimens; italics indicate the number of living specimens.)

4-1

Species \ Station number	51	52	53	54	55	56	57	58	59	60	61
Cytherelloidea munechikai ISHIZAKI	6	1	5	2	1	7	1		4		
Neonesidea oligodentata (KAJIYAMA)	*3* 3		*1* 1				*1* 1		*1* 1	1	*3* 3
Propontocypris (P.) attenuata (BRADY)	1									1	2
Cardonocypris assimilis SARS	3				1						
Potamocypris producta (SARS)	1										
Clithrocytheridea? japonica (ISHIZAKI)	4		1	2							2
Portocythere sekiguchii n. sp.	*1* 2	2	6	*3* 3		*5* 6					
P. japonica (HANAI)	*1* 5		1	2		2				4	1
P. minuta n. sp.	*1* 1		2	*1* 1					2		
P. miurensis (HANAI)	11	*2* 3	7	*3* 7			*2* 3		*1* 8	*1* 2	1
P. subjaponica (HANAI)	*11* 47	*1* 11	*12* 41	*2* 7	3	*18* 43	*2* 5		*16* 39	*2* 8	3
Neocytherideis punctata n. sp.	1	2	2	2		*1* 1					
Murseyella japonica (HANAI)			1								
Callistocythere alata HANAI	*1* 1			1		*3* 4					
C. japonica HANAI	*52* 57	*3* 5	*95* 103	*8* 9		*17* 22			*84* 87	*1* 2	
C. reticulata HANAI											*1* 1
C. rugosa HANAI						*1* 1					
C. subjaponica HANAI		*1* 1									
C. sp.	3		3	1					2		
Cythere lutea omotenipponica HANAI						*1* 1					
Schizocythere kishinouyei (KAJIYAMA)						1					
Haraiborchella triangularis (HANAI)	1			1		1				1	
Aurila hataii ISHIZAKI	2								1		1
A. uranouchiensis ISHIZAKI	4										
Mutilus assimilis (KAJIYAMA)	8	2	3	3	1	3	1	1	2	1	3
Cornucoquimba rugosa n. sp.	2	*1* 1				*1* 2	*1* 1		*1* 2		
Proteoconcha tomokoae (ISHIZAKI)	7				1	1	1				
Trachyleberis scabrocuneata (BRADY)		1	1						1		
T.? tosaensis ISHIZAKI	1		3			1			1		
T. sp.	1										

4-2

Species \ Station number	51	52	53	54	55	56	57	58	59	60	61
Actinocythereis sp.	2			1							
Echinocythereis? bradyformis ISHIZAKI					1						
Neocytheretta sp.			2			1					
Bythocythere maisakensis n. sp.						1	1				
Bythoceratina elongata n. sp.				1							
B. hanaii ISHIZAKI				1							
Hemicytherura cuneata HANAI		*1* 1									
H. tricarinata HANAI	4	2	1	1		1			1		
Semicytherura? miurensis (HANAI)	*5* 12	*1* 1	*10* 17	*3* 4		*2* 4			*5* 8	2	
S. sp.			2								1
Kobayashiina hyalinosa HANAI	1								1		
Paracytheridea bosoensis YAJIMA	2	1		1							
Loxoconcha (L.) optima ISHIZAKI		2	1	2	2	3					
L. (L.) pulchra ISHIZAKI	16	1	9	1	2	7			2		2
L. (L.) uranouchiensis ISHIZAKI			2	1		1			2		
L. (L.) viva ISHIZAKI	2										
Phlyctocythere hamanensis n. sp.	2									1	
Xestoleberis cf. *dentata* SCHORNIKOV	*1* 1										
X. hanaii ISHIZAKI	1	1	2	1	2	2			2	2	*1* 1
X. setouchiensis OKUBO	1		2	1							1
X. sagamiensis KAJIYAMA	1										
Platymicrocythere tokiokai SCHORNIKOV										*24* 24	*43* 45
P. sp.				*1* 2						*2* 2	*82* 82
Paradoxostoma japonicum SCHORNIKOV	*2* 3		*1* 1								
P. ovulare KAJIYAMA	2								1		
Cytherois zosterae SCHORNIKOV	*1* 3		*1* 1						*1* 2		
Sclerochilus mukaishimensis OKUBO	1						1				*1* 1
Neopellucistoma inflatum n. gen. n. sp.	8	1	3	5	*3* 10				*1* 1	2	
Cytheroma? sp.	1	*1* 1								*5* 5	*108* 109
Total living specimens	79	11	120	21	3	49	6	-	110	35	239
Total specimens	235	40	223	63	24	116	15	1	170	58	259
Volume of sediments (cc)	34	320	40	360	3.5	360	360	50	42	520	400

Table 5

Number of individual specimens of each species collected from stations of the inner bay coast area. (Cardinal numerals show the total number of specimens; italics indicate the number of living specimens.)

Species \ Station number	63	64	66	67	68	69	70	71	72	73	74	75	76	79	80	82	84	85	86	87	88	89	90
Ilyocypris angulata SARS			1								*8* 11	*1* 3		*102* 192	*7* 13	*2* 2							3
Cyprididae n. sp.				*2* 12	*2* 5	2	*231* 239	*2* 3	*16* 51	*1* 2					*19* 135	*2* 15	*2* 10	*3* 19	*1* 1	*20* 112	*82* 128	*2* 7	3
Physocypria sp.		*3* 7												1	*53* 195								*3* 9
Candonocypris assimilis SARS	*2* 8	*97* 276	*1* 5			11		1			2	1	*47* 84	*7* 14	64	3							3
Stenocypris major (BAIRD)											*4* 4	*1* 1		*30* 50	*3* 4								
Cypridopsis vidua (O. F. MÜLLER)		*2* 3						1			1	1		*15* 47	*2* 37	1							4
Potamocypris producta (SARS)		*18* 23	*1* 2					*1* 8							*1* 3								1
Clithrocytheridea? japonica (ISHIZAKI)											6												
Pontocythere miurensis (HANAI)											1												
Spinileberis quadriaculeata (BRADY)											*1* 8												
Mutilus assimilis (KAJIYAMA)			*3* 3								1												
Cornucoquimba rugosa n. sp.											1												
Proteoconcha tomokoae (ISHIZAKI)											1												
Hemicytherura tricarinata HANAI			*1* 1								*1* 1												
Semicytherura? miurensis (HANAI)			2								*1* 4												
Loxoconcha (L.) pulchra ISHIZAKI											1												
Cytheromorpha acupunctata (BRADY)											*3* 19												
Xestoleberis cf. *dentata* SCHORNIKOV				1																			
X. hanaii ISHIZAKI			*22* 144							1													
X. setouchiensis OKUBO			*7* 20								1												
Paradoxostoma japonicum SCHORNIKOV			*1* 1								1												
P. ovulare KAJIYAMA											1												
Cytherois zosterae SCHORNIKOV			*1* 1	*1* 3							1					5	1						
Total living specimens	2	120	37	3	2	-	231	3	16	1	18	2	47	154	80	4	2	3	1	20	82	2	3
Total specimens	8	309	180	16	5	13	239	13	51	3	65	6	84	304	451	26	11	19	1	112	128	7	23
Volume of sediments (cc)	111	2	53	50	180	150	. 25	405	109	25	53	79	51	1	13	23	13	165	70	55	8	200	210

temperature and salinity. This summer stratification is a phenomenon that exerts a strong influence on the deeper life of the bay. The dissolved oxygen content of the bottom water dwindles to zero in depths greater than 8 m, where hydrogen sulphide is produced in summer.

Recently, due to the development of pisciculture, the water environment of the bay is rapidly changing by an increase in the relative amount of sea water. According to the calculations of Nonaka et al. (1973), the amount of water interchanged every year between the bay and the sea is increasing up to about 23.3% of the bay water, and the amount of fresh water received by the bay is 6% in a year.

Nutritive salts are carried by terrestrial water and accumulate in general in the bottom water layer. They tend to decrease from the mouth of the streams toward the mouth of the bay and are especially abundant in the northern part of the east bay.

The concentration of organic calcium carbonate, represented mainly by shell fragments, is low in the well-sorted sand of the southern bay and high in the silt of the northern bay, especially along the bay coast.

Zostera nana and *Z. marina*, the characteristic species among aquatic plants of this area, offer a refuge for aquatic animals, including ostracods. The former species is distributed in several patches along the coast of the bay, while the distribution of the latter is confined to the border area between the northern and the southern parts of the bay.

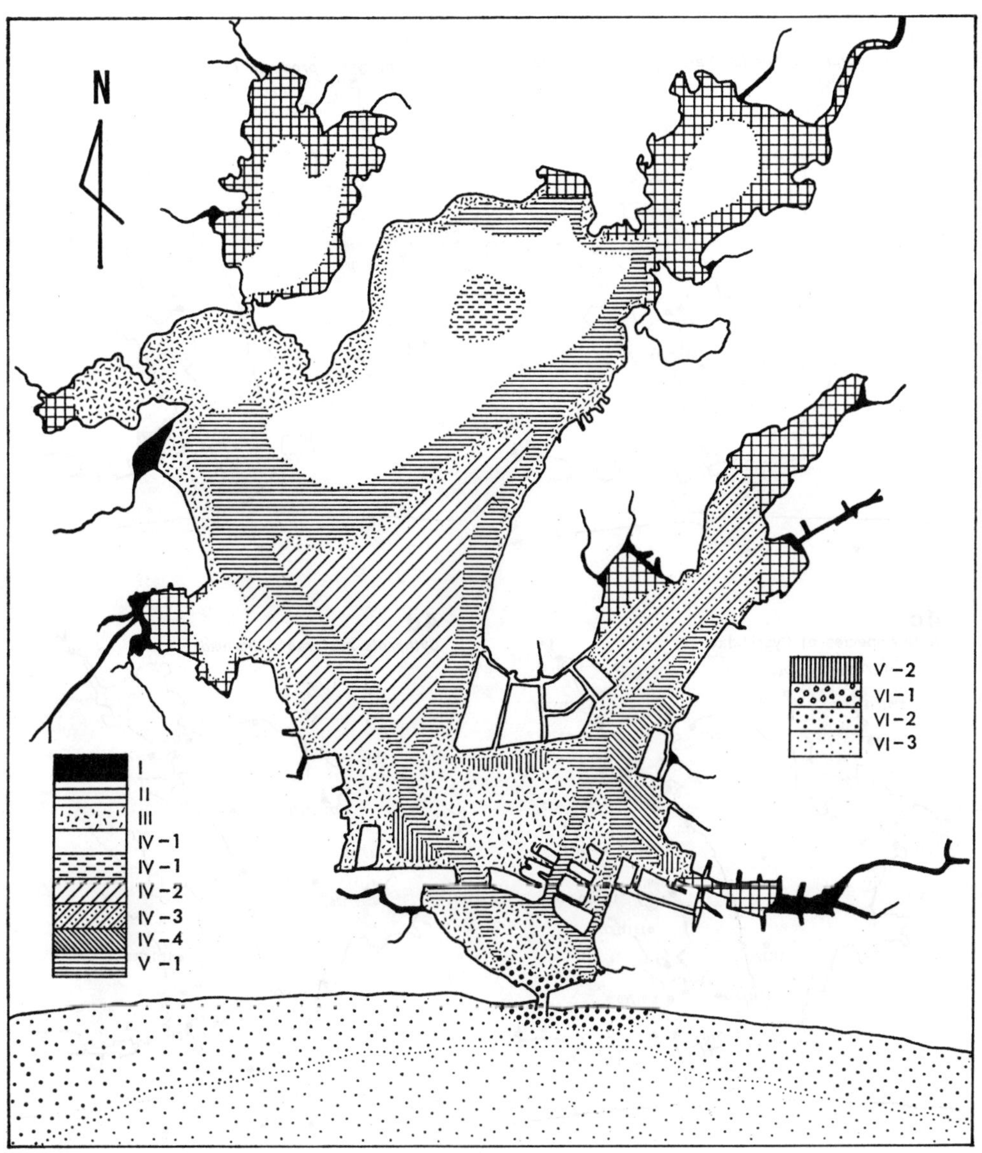

Text-fig. 3

Distribution of the various ostracod biofacies and sub-biofacies in the Hamana-ko region.

I: Fresh water pond and river biofacies; II: Brackish water inlet biofacies; III: Bay coast and sand bank biofacies, IV-1: Dy sub-facies; IV-2: Sand bottom sub-biofacies, IV-3: Silty bottom sub-biofacies; IV-4: Silty sand bottom sub-biofacies; V-1: Main channel sub-biofacies; V-2: Extension channel sub-biofacies; VI-1: Bay mouth sub-biofacies; VI-2: Open coast sub-biofacies; and VI-3: Offshore sub-biofacies.

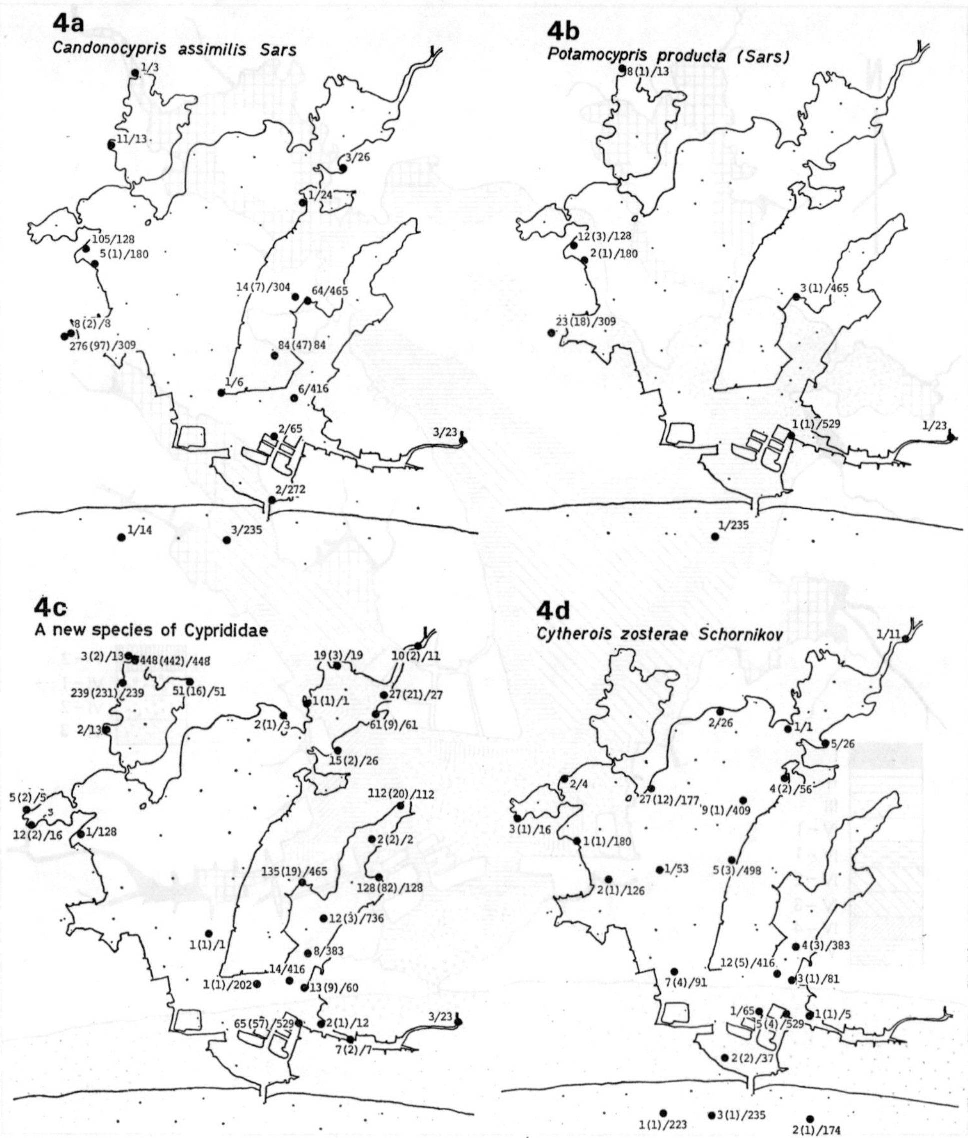

Text-fig. 4

Distribution of the four species characteristic in fresh to brackish water environments.

4a: *Candonocypris assimilis* Sars; 4b: *Potamocypris producta* (Sars); 4c: A new species of Cyprididae; and 4d: *Cytherois zosterae* Schornikov.

Denominators are the total number of individual specimens of all the species obtained from one station. Numerators indicate the total number of individual specimens of the characteristic species, with the number of live individuals of the characteristic species in parentheses.

Distribution of ostracods

Out of a total of 93 sampling stations in the Hamana-ko area, 70 stations yielded ostracods. The samples from the remaining 23 stations contained no ostracods. A total of 45 genera and 70 species were found. Living and dead specimens were distinguished, and a number of species from each station are listed in tables 2 through 5.

Distribution of ostracod species in the Hamana-ko area is dependent on the characteristics of the water (O_2 content, Eh, pH, salinity, etc.) and the substrate and bottom topography of the bay. Based on the characteristic species, 6 major biofacies and 7 sub-biofacies were recognized in the Hamana-ko area:

- I. Fresh water pond and river biofacies
- II. Brackish water inlet biofacies
- III. Bay coast and sand bank biofacies
- IV. Inner bay biofacies
  - (1) Dy sub-biofacies
  - (2) Sandy bottom sub-viofacies
  - (3) Silty bottom sub-biofacies
  - (4) Silty sand bottom sub-biofacies
- V. Channel and extension channel biofacies
  - (1) Main channel sub-biofacies
  - (2) Extension channel sub-biofacies
- VI. Bay mouth and open sea biofacies
  - (1) Bay mouth sub-biofaces
  - (2) Open coast sub-biofacies
  - (3) Offshore sub-biofacies

Distribution of these ostracod biofacies and sub-biofacies is shown in text-figure 3.

I. Fresh water pond and river biofacies: Many small ponds for breeding eels are found around the Hamana-ko. The number of fresh water ostracods in the ponds varies seasonally. Bottom mud of the ponds was obtained at stations 63, 64, and 76. *Candonocypris assimilis*, the dominant species of this biofacies, makes up 90–100% of the total number of specimens. Other subordinate members of this biofacies are such fresh water inhabitants as *Potamocypris producta* and *Cypridopsis vidua* and some slightly brackish water species of *Physocypria* sp.

River mouths are in general occupied dominantly by one or more of the above-mentioned four fresh water species derived from eel-breeding ponds. In addition to these species, fresh water *Ilyocypris angulata* and *Stenocypris major* (which are characteristic in running water), a subordinate new species of Cyprididae from the brackish inlet biofacies, and *Mutilus assimilis*, *Xestoleberis hanaii* from the bay coast biofacies are present. At stations 71, 80, and 90, the assemblage, consisting of three fresh water species—*Candonocypris assimilis*, *Potamocypris producta* and *Cypridopsis vidua*—characteristic of eel-breeding ponds, and the new species of Cyprididae, dominant in brackish water environments, is different from the assemblage in eel-breeding ponds in the domination of *Potamocypris producta* and *Physocypria* sp. over *Candonocypris assimilis* and the presence of brackish water species. The assemblage in station 22 is also characterized by fresh water species from the eel-breeding ponds with some bay coast species, *Mutilus assimilis* and *Xestoleberis hanaii*. Although stations 74 and 75 are located close to the mouth of the bay, such fresh water species as *Ilyocypris angulata*, *Candonocypris assimilis*, *Stenocypris major*, and *Cypridopsis vidua* bear a very high ratio to the marine water species: a ratio ranging from 28 to 100%. Inference of the direct supply of abundant fresh water species from an adjacent eel-breeding pond to the nearby rigorous marine environment may be supported by the paucity of live specimens and the abundance of the dead carapaces. In conclusion, the minor environmental condition in this fresh water biofacies seems to be strongly controlled by the condition of the water, i.e., salinity, agitation, etc., and the effect of the substratum; grain size distribution of sediment is likely to be quite insignificant.

II. Brackish water inlet biofacies: The coastline of Hamana-ko, characterized by drowned valley topography, has several small inlets with very narrow mouths. In the entire area of the inlets with narrow mouths and the inner half of the inlets with wide mouths, a brackish water environment is produced by the limited circulation of marine water. These areas are occupied exclusively by the new species of Cyprididae. Examples can be seen in stations 2, 3, 19, 45, 68, 70, 72, 85, 86, 87, 88, and 89. In stations 5 and 84 where partial growth of *Zostera* can be observed, *Cytherois zosterae* appears subordinately to the dominant new cypridid species. In Station 82, where the effect of the fresh water is significant, a small amount of dead carapaces of *Ilyocypris angulata*, *Candonocypris assimilis* and *Cypridopsis vidua* are found. On the other hand, a certain amount of *Xestoleberis* cf. *dentata* and *X. hanaii* derived from bay coast biofacies are also found at stations 67 and 73. In conclusion, brackish water inlet biofacies are found on clay or silt substrata partially overgrown with *Zostera* and are characterized by monotonous ostracod assemblage dominated by the new cypridid species.

III. Bay coast and sandbank biofacies: This biofacies is found in three areas: a narrow sand zone along the rocky coast of the main body of the bay excluding the inlets, a narrow sand zone along the northern edge of the sand flat, and the wide sand bank area just inside the mouth of the bay. In these areas, circulation of the fresh oceanic water with high salinity, entering through the narrow mouth of the Hamana-ko and running along the channel leading deeply into the inner half of the bay, has a strong influence on the nature of the substratum as represented by coarse sand and gravel. The bottom is commonly covered by various kinds of algae. Characteristic species of this biofacies are *Mutilus assimilis*, *Aurila hataii*, *Xestoleberis* spp. and *Loxoconcha* spp. *Mutilus assimilis* dominates in stations 15, 25, 26 and 28, all of which are located in the area where the channel biofacies terminates. *Hemicytherura tricarinata* and *Semicytherura*? *miurensis* derived from channel biofacies constitute the subordinate members in these stations. Typical examples can be found at stations 25 and 28 on the northern edge of the wide sand flat of the central part of the bay. In stations 15 and 26 of the narrow zones along the rocky coast of the main body of the Hamana-ko, where algal growth is noticeable, *Xestoleberis* spp., *Cytherois zosterae*, and *Paradoxostoma japonicum* can be added to the members of subordinate species. At stations 35 and 38 in the wide area just inside the mouth of the bay, *Aurila hataii*, *Xestoleberis setouchiensis* Okubo, *Mutilus assimilis*, and *Loxoconcha* (*L.*) *pulchra* are found in nearly equal numbers. Besides the four dominant species, *Pontocythere* spp., *Cythere lutea*

Text-fig. 5

Distribution of the four species characteristic in the bay environment (5a, b) and in channel environment (5c, d).

5a: *Spinileberis quadriaculeata* (Brady); 5b: *Cytheromorpha acupunctata* (Brady); 5c: *Hemicytherura tricarinata* Hanai; and 5d: *Semicytherura*? *miurensis* (Hanai).

Denominators are the total number of individual specimens of all the species obtained from one station. Numerators indicate the total number of individual specimens of the characteristic species, with the number of live individuals of the characteristic species in parentheses.

omotenipponica, *Callistocythere* spp., and *Neonesidea oligodentata*, derived from the open sea biofacies, occur as subordinate species. The distribution of *Neonesidea oligodentata* is limited to the area close to the mouth of the bay. Stations 10, 20, 21, and 66 are characterized by the dominance of *Xestoleberis* spp. Subordinate members consist of *Hemicytherura tricarinata* and *Semicytherura*? *miurensis* from channel biofacies. A small amount of *Paradoxostoma japonicum* and *Cytherois zosterae* from the algal habitat and some fresh water species are also found. These stations are located at the terminal area of the channel biofacies. In conclusion, bay coast and sand bank biofacies of Hamana-ko are an extension of the ostracod biofacies characteristic of the tide pool, rocky shore, and nearby coarse sand found along the open sea coast of the Pacific.

IV. Inner bay biofacies: Biofacies of the main part of the bay and inlets can be subdivided into four sub-biofacies, which correlate significantly with the nature of substratum.

(1) Dy sub-biofacies: As explained in the preceding chapter, in the area of black silt and clay with abundant humus, the deficiency of dissolved oxygen and production of toxic hydrogen sulphide in summer does not permit inhabitation by animals, including ostracods. An exceptional case was found only in station 12 where a few live specimens of *Hemicytherura tricarinata* occur in this sub-biofacies. Ostracods found in this biofacies consist of dead and more-or-less corroded carapaces of such species as *Hemicytherura tricarinata*, *Semicytherura*? *miurensis*, *Cytheromopha acupunctata*, *Mutilus assimilis*, and *Spinileberis quadriaculeata* from the adjacent biofacies. No ostracods are found in the dy from stations 1, 4, 6, 7, 9, 11, 14, 16, 17, 18, 24, 30, 31, and 81.

(2) Sandy bottom sub-biofacies: The central part of the main bay is occupied by a wide flat area with very well-sorted sand. Rapid sedimentation of sand carried from the open sea by a tidal current characterizes this area (Ikeya and Handa, 1972). Ostracods are extremely poor in the number of individuals as well as in the number of species. A small number of *Spinileberis quadriaculeata*, *Semicytherura*? *miurensis* and *Cytheromorpha acupunctata* are found at stations 33 and 34. Other species, a new cypridid species and *Physocypria* sp. seems to be derived from the adjacent biofacies.

(3) Silty bottom sub-biofacies: This sub-biofacies occupies the main part of the largest inlet of the Hamana-ko bay called Inohana-ko. The flat silt bottom is in water approximately 2 m deep; the water has a relatively low salinity (ca. 23‰) and is inhabited by dominant species of *Spinileberis quadriaculeata*, as exemplified by stations 44, 49, and 50. A subordinate species of this sub-biofacies is *Cytheromorpha acupunctata*; other species occur only rarely.

(4) Silty sand bottom sub-biofacies: In the southern part of the Inohana-ko inlet, a channel extending in the northeast direction separates the area into two: an area of northwestern silty bottom sub-biofacies, as explained above, and an area of the southeastern silty sand bottom sub-biofacies. The latter sub-biofacies is found on the bottom, which is coarser in grain size and higher in salinity (ca. 26‰) than the silty bottom of the former sub-biofacies. In this sub-biofacies, dominant and subordinate species relative to the silty bottom sub-biofacies become reversed and *Cytheromorpha acupunctata* becomes the dominant species, and *Spinileberis quadriaculeata* becomes

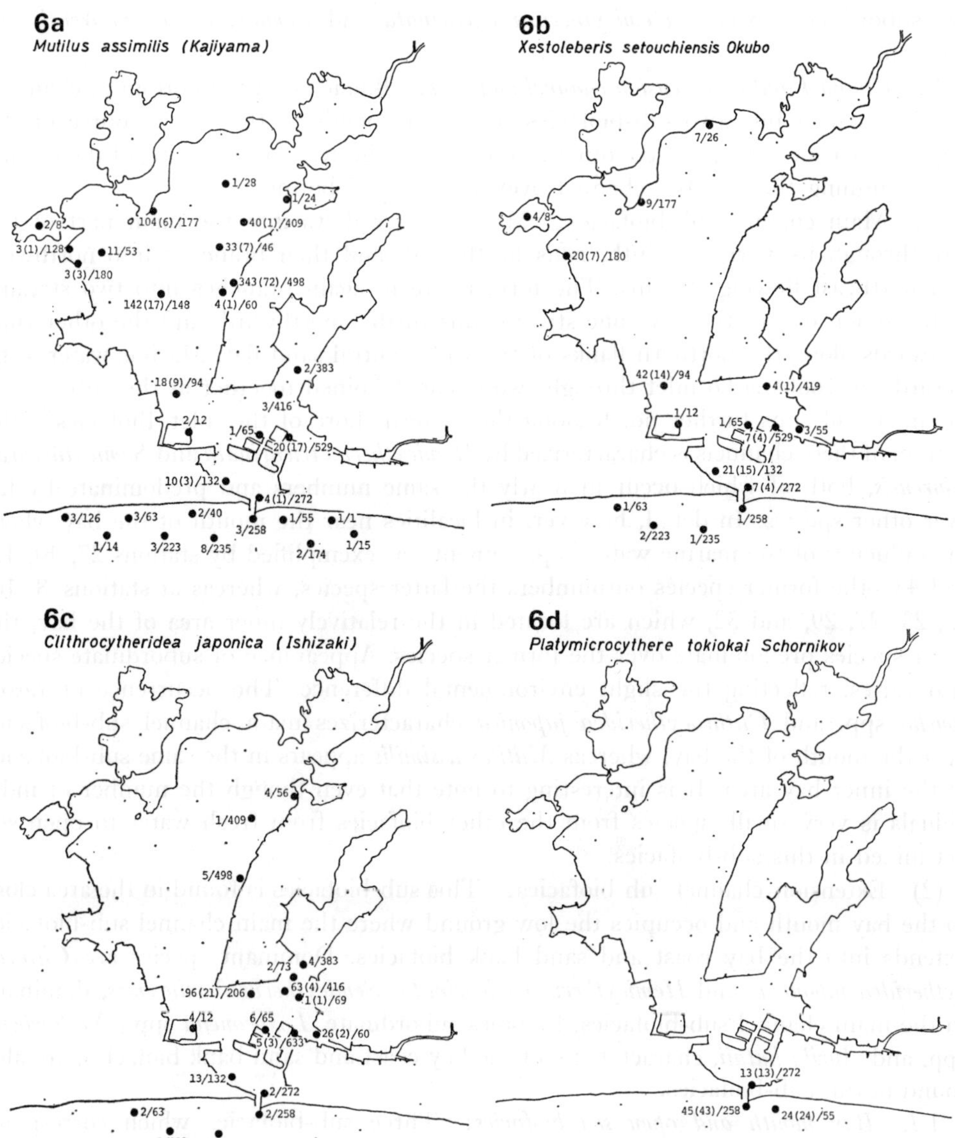

Text-fig. 6

Distribution of the four species characteristic in the coastal water environment.

6a: *Multilus assimilis* (Kajiyama); 6: *Xestoleberis setouchiensis* Okubo; 6c: *Clithrocytheridea japonica* (Ishizaki); 6d: *Platymicrocythere tokiokai* Schornikov.

Denominators are the total number of individual specimens of all the species obtained from one station. Numerators indicate the total number of individual specimens of the characteristic species, with the number of live individuals of the characteristic species in parentheses.

the subordinate species. *Hemicytherura tricarinata* and *Semicytherura? miurensis* are rarely found.

V. Channel and its extension channel biofacies: Biofacies on the bottom of channels can be divided into two sub-biofacies: one corresponds to the water course of the inflow of marine water, and the other can be seen in the areas of spreading of the marine water running through the channel over a flat sandy bottom.

(1) Main channel sub-biofacies: Marine water flows into the main part of the bay through its narrow mouth, runs northward, and then branches into northwest and northeast flowing streams. The former stream again branches into two streams in the inner area of the bay: one stream runs further northward, and the other runs eastwards along the northern flanks of the wide central sand flat. Marine water runs toward the Inohana-ko inlet through two channels joined together at the entrance of the inlet and runs further north along the eastern shore of the inlet. Biofacies of the bottom of these channels is characterized by *Hemicytherua tricarinata* and *Semicytherura? miurensis*, both of which occur in nearly the same numbers and predominate by far over other species. In detail, however, in localities near the mouth of the bay where the influence of the marine water is prominent—as exemplified by stations 27, 40, 42, and 43—the former species outnumbers the latter species, whereas at stations 8, 10, 13, 23, 27, 29, and 32, which are located in the relatively inner area of the bay, the latter species predominate over the former species. Appearance of subordinate species also varies, reflecting the slight environmental difference. The occurrence of *Loxoconcha* spp. and *Clithrocytheridea japonica* characterizes main channel sub-biofacies near the mouth of the bay, whereas *Mutilus assimilis* appears in the same sub-biofacies of the inner bay area. It is interesting to note that even though the number of individuals is very small, species from the other biofacies from fresh water to open sea get mixed in this sub-biofacies.

(2) Extension channel sub-biofacies: This sub-biofacies is found in the area close to the bay mouth and occupies the low ground where the main channel sub-biofacies extends into the bay coast and sand bank biofacies. Dominant species are *Clithrocytheridea japonica* and *Hemicytherura tricarinata*. *Semicytherura? miurensis*, dominant in the main channel sub-biofacies, becomes subordinate. *Loxoconcha* spp., *Xestoleberis* spp. and *Aurila hataii*, characteristic of the bay coast and sand bank biofacies, are also found in this sub-biofacies.

VI. Bay mouth and open sea biofacies: Three sub-biofacies which correspond well to the differences in substratum are distinguishable in the area from the mouth of the bay toward the open sea.

(1) Bay mouth sub-biofacies: This sub-biofacies is distributed from just inside to just outside the mouth of the bay called Ima-gire, where the swift stream of tidal current makes the bottom exclusively of the gravelly substratum. Three species, *Platymicrocythere tokiokai*, *P.* sp. and *Cytheroma?* sp., dominate in this sub-biofacies. In the area inside the mouth exemplified by station 39, the order of dominance is in descending order: *Platymicrocythere* sp., *Cytheroma?* sp., and *Platymicrocythere tokiokai*; in the area just outside the mouth exemplified by station 61, the order is *Cytheroma?* sp., *Platymicrocythere* sp., and *P. tokiokai*; and in the area farther outside the mouth exemplified by station 60, *Platymicrocythere tokiokai* predominates over *Cytheroma?*

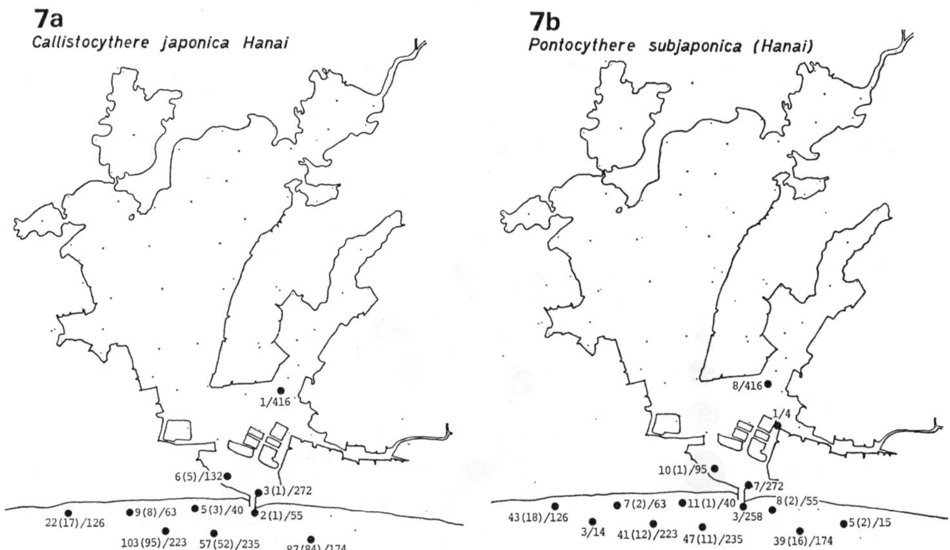

Text-fig. 7

Distribution of the two species characteristic in the shallow open sea environment.

7a: *Callistocythere japonica* Hanai; 7b: *Pontocythere subjaponica* (Hanai).

Denominators are the total number of individual specimens of all the species obtained from one station. Numerators indicate the total number of individual specimens of the characteristic species, with the number of live individuals of the characteristic species in parentheses.

sp. These three extremely small species are found alive and perhaps live in the interstitial water between gravels. The large number of subordinate species makes the species diversity of this sub-biofaices the highest among all the biofacies.

(2) Open coast sub-biofacies: Runoff material from the Tenryu River is transported westwards along the coast of Enshu-nada Sea by the coastal current; its medium sand fraction is deposited near the shore and fine sand fraction offshore. Medium sand found along the open sea coast of 0–10 m in water depth is occupied mainly by *Pontocythere* spp. *Callistocythere* spp., which are characteristic among many subordinate species. Examples can be seen in stations 52, 54, 55, 56, 57, and 58. Considerably high species diversity is also found in this sub-biofacies.

(3) Offshore sub-biofacies: Offshore sub-biofacies occupies an extensive area of fine sand, 10–20 m in water depth, outside the area of open coast sub-biofacies. In this sub-biofacies, the order of dominance in the open coast sub-biofacies becomes reversed and *Callistocythere* spp. predominate over *Pontocythere* spp. Species subordinate in number of individuals but characteristic of this sub-biofacies are *Cytherelloidea munechikai*, *Neonesidea oligodentata*, *Neocytherideis punctata* n. sp. and *Munseyella japonica*. All of these species are representatives of the shallow warm water inhabitants of the open sea of the Kuroshio region. Altogether, these species bring species diversity up to a very high level. The areal distribution of fourteen characteristic

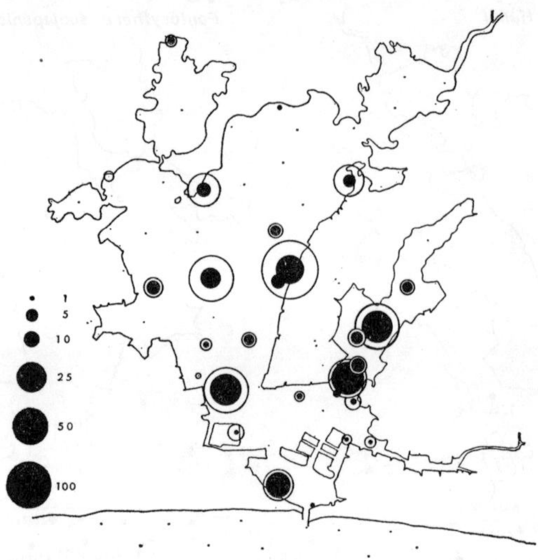

Text-fig. 8

The standing crop size of ostracods, based on the samples collected by Phleger-type core sampler and shown by the number of individual specimens of all the species per unit sample (approx. 10 cc) of the surface sediment.

The circle diameters are proportional to the numbers shown in the left-hand column of the legend. White circle: Total number of individual specimens. Black circle: Total number of live individuals.

species is shown in text-figures 4 through 7.

It is interesting to note that some species which have their counterpart species in tropical and subtropical area (e.g., *Trachyleberis* sp., *Actinocythereis* sp., and *Neocytheretta* sp.) are represented only by young molts in the Hamana-ko area; these species are not rare and are found in the upper Pleistocene to Recent warm water sediments in Japan (e.g., Aburatsubo Cove and upper Pleistocene deposits in Chiba Prefecture). Adult forms are, however, extremely rare; in fact, adults of some species have not yet been found in Japan. The explanation may be the difference of temperatures necessary for survival and for reproduction of these species, as suggested by Hazel (1970) in the discussion of temperature control of ostracod distribution. Environment suitable for reproduction of these species may be restricted to small areas, and the mortality rate of young molts may be comparatively high in the Hamana-ko area.

Ecological analysis of ostracod fauna

Standing-crop size of ostracods: Among 50 samples collected by a Phleger-type core sampler for quantitative analysis, 27 samples contained ostracod specimens. From these samples, the number of individuals per unit sample (ca. 10 cc) of the surface sediment is calculated to give the density distribution of ostracods. The total number of individuals per unit sample ranges from 1 to 151, and the maximum number of

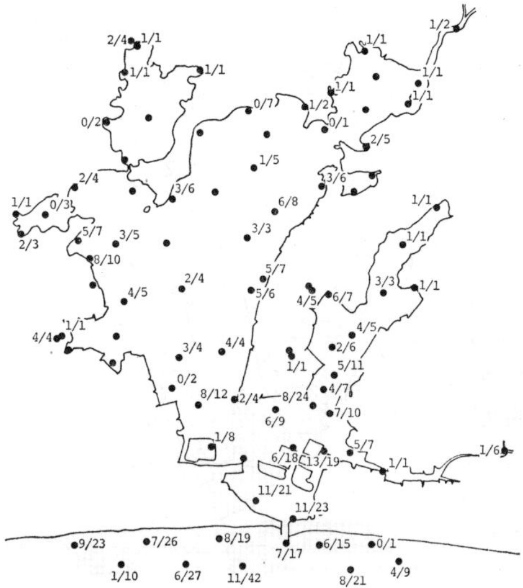

Text-fig. 9

Distribution of number of species.

Numerators are the number of species collected alive and denominators are the total number of species.

living specimens per unit sample is 46. Thus, it can be said that at maximum 4 to 5 living ostracods inhabit 1 cm^2. The densities of the total number of specimens as well as the ratio between living and dead specimens vary from station to station. A relatively high density of the total number of ostracods is found in the patches of moderately coarse sediments near the channels, where the living specimens are also found in high density making up 16 to 63% of the total number of specimens, and suggest a high productivity of ostracods in these areas. The ostracod density is low in the northern brackish water area and high in the southern marine water area. Low density is also observed in the sand bank of the southern area.

It is interesting to note that these results are contrary to those of foraminifers, i.e., the density of benthonic foraminifers is high in the northern half of the bay, because of the abundant occurrence of those with agglutinated tests (Ikeya, 1977). The standing crop size of ostracods is shown in text-figure 8.

Species diversity of ostracods: Among the seventy species of ostracods belonging to 45 genera which were identified in this paper, 44 species belonging to 31 genera were found alive. The number of total and living individuals from each station are tabulated for each taxon in tables 2 through 5 with the data of the volume of sediments used.

The total number of species per station ranges from 42 to 1, and the number of living species per station ranges from 13 to 0. The total number of species is largest in the area adjacent to and on both sides of the mouth of the bay, where the interaction of coastal and bay waters is detectable from hydrographic data. The number of total and living species is moderately large at the stations along the shore of Enshu-nada, where the coastal current is strong, and along the channel of the bay, where the in-

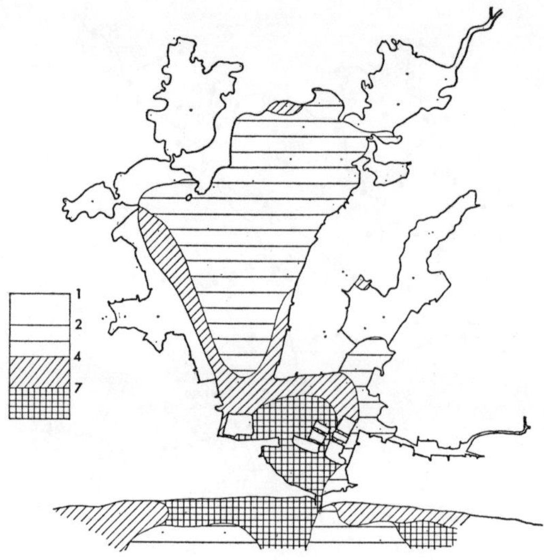

Text-fig. 10

Area distribution of diversity index values calculated by Yule-Simpson's formula, based on the total number of specimens.

fluence of marine water is reduced. The smallest number of species is found in the brackish water area, especially in the inner most part of the bay, where the ostracod fauna is simple and composed of only a few species.

Numerical distribution of the living species per total species is given in text-figure 9. Among the species whose empty carapaces are found commonly in the area along the coast of Enshu-nada, the following have never been found alive, suggesting that they are allochthonous being transferred from adjacent areas by coastal current: *Cytherelloidea munechikai*, *Callistocythere* sp., *Paijenborchella triangularis*, *Proteoconcha tomokoae*, *Trachyleberis*? *tosaensis*, and *Loxoconcha* (*L.*) *optima*.

Yule-Simpson's diversity index based on the number of species and the number of individuals in each species (Yule, 1944; Simpson, 1949) is adopted to analyze the correlation between ostracod biomass and the environments. Calculations of the numerical diversity values are based on the total number of specimens, because living specimens are too few to obtain values within the limits of confidence reasonable for discussion. For the same reason, when the total number of specimens is less than 10 individuals, the values are eliminated from the distribution map. For further explanation of this index, see Ikeya (1971, p. 186).

Area distribution of the diversity indices is shown in text-figure 10. The distribution pattern of the diversity indices agrees on the whole with the distribution pattern of biofacies. In the inner bay area under brackish water conditions, the diversity values are less than 2, and the value increases toward the mouth of the bay. Patchy distribution of high indices in the inner area of the bay is regarded as the result of the inflow of ostracods from the fresh water ponds. Fluctuations of the indices in the Enshu-nada area, together with the occurrence of allochthonous ostracods, may suggest the strong

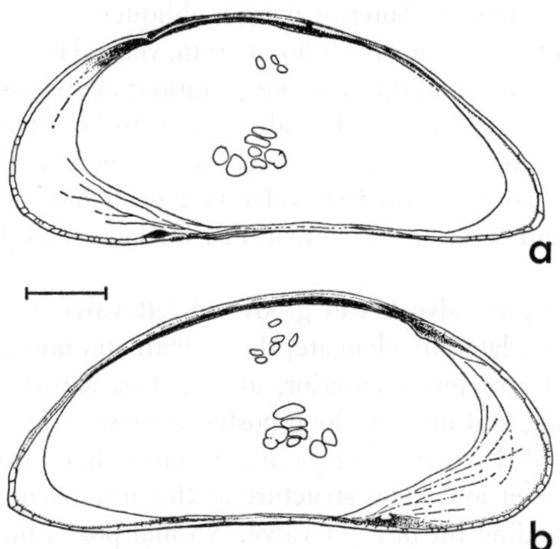

Text-fig. 11
Internal views of Cyprididae n. sp. (scale: 0.1 mm). A complete carapace (IGSU-O-74); a: right valve; b: left valve.

sieving effect of the wave base and the coastal current on the accumulation of dead ostracod carapaces. The highest index value of 18.4 is found at station 54 on the Enshunada Coast.

Systematic description

Subclass Ostracoda Latereille, 1806
Order Podocopida Sars, 1866
Superfamily Darwinulacea Brady and Norman, 1889
Family Darwinulidae Brady and Norman, 1889
Genus *Darwinula* Brady and Robertson in Jones, 1885
Darwinula sp.

Specimen.—A broken right valve, IGSU-O-7. Hamana-ko St. 10.

Remarks.—Only a broken right valve with typical muscle scars characteristic of darwinulids is obtained from St. 10. Poor preservation does not warrant exact specific identification. This is the first reported occurrence of this genus from Japan.

Superfamily Cypridacea Baird, 1845
Family Cyprididae Baird, 1845
Cyprididae, n. sp.
Pl. 1, figs. 1–6; text-fig. 11

Illustrated specimens.—A complete carapace, IGSU-O-8 (Pl. 1, figs. 1a, b, 2a, b, 3, 6); a left valve, IGSU-O-9 (Pl. 1, fig. 4); a right valve, IGSU-O-10 (Pl. 1, fig. 5), Hamana-ko St. 72; a complete carapace, IGSU-O-74 (text-figs. 11a, b), Hamana-ko St. 70.

Remarks.—Exact identification of this species is still pending until detailed anatomy of appendages will become clear. However, since this species is dominant and characteristic in the brackish water environment of the Hamana-ko area, it seems desirable to make clear the specific characters of this species. Thus, a brief description of the carapace is given below.

Carapace elongate-subtriangular. Anterior margin obliquely rounded, dorsal margin slightly arched, inclined gradually toward dorsoposterior margin, ventral margin nearly straight sinuated slightly at middle. Posteroventral margin without serration; acutely rounded. Maximum height slightly anterior to middle. Viewed dorsally, sides appear to be spindle shaped with somewhat pointed anterior and posterior ends. Greatest thickness at middle. Viewed anteriorly, carapace nearly round and slightly compressed. Left valve larger, slightly overlapping the right valve along anterior, ventral, and dorsoposterior margins. Surface smooth with very small pits of normal pore openings.

Hinge adont; edge of right valve fits in groove of left valve along hinge margin. Muscle scars consisting of one large laterally elongate, three small and one large fused adductor scars, with two mandibular scars in lower anterior, arranged as shown in text-fig. 11. Marginal infold broad along anterior, and narrow along posterior margin. Vestibule very deep. Zone of concrescence not developed. Lists running parallel to inner margin distinct near anteroventral area of duplicature. Snap pit and knob structure at the anteroventral corner, tongue-shaped knob in left and corresponding pit in right valve. Normal pore canals simple with broad lip. Sensory hair slender and simple without branching.

Measurements of samples from St. 72 are as follows:

S		N	$\overline{X}$	Sd	V	OR
C	L	13	0.652	0.013	1.99	0.63–0.68
	H	13	0.275	0.007	2.55	0.26–0.28
	W	13	0.216	0.009	4.17	0.20–0.23

The species is similar to a certain species of Pontocyprididae in its attenuate outline. However, two mandibular scars are distinct in this species. Adductor muscle scars are arranged in a group of four, in which a surmounted cap scar is elongate horizontally and large. These characters of muscle scar suggest close affinity of this species to Cyprididae.

The species occurs abundantly at St. 2, 19, 40, 70, 72, 80, 87, and 88. All of these stations are located at the inner most extremities of the inlets where a brackish water environment dominates. For further information on occurrence of this species, see text-figure 4c.

Family Candonidae Kaufmann, 1900

Subfamily Cyclocypridinae Kaufmann, 1900

Genus *Physocypria* Vávra, 1898

Physocypria sp.

Pl. 1, figs. 7–10; Pl. 7, fig. 7.

Illustrated specimen.—A complete carapace, IGSU-O-11 (Pl. 1, figs. 7–10; Pl. 7, fig. 7), Hamana-ko St. 80.

Remarks.—Four large and two small adductor muscle scars arranged as in the species of genus *Cypria*. However, in detail, markedly higher right valve with finely serrated free margin safely places this species into genus *Physocypria*. This is the first report of occurrence of this genus from Japan. The species is especially abundant in St. 80 of slightly brackish water. Other localities include St. 64 and 90. At St. 34 and 79 dead carapaces are found rarely.

Family Cypridopsidae Kaufmann, 1900
Subfamily Cypridopsinae Kaufmann, 1900
Genus *Cypridopsis* Brady, 1868
Cypridopsis vidua (O. F. Müller, 1776)
Pl. 1, figs. 11–13; Pl. 3, fig. 12.

Illustrated specimen.—A complete carapace, IGSU-O-12 (Pl. 1, figs. 11–13; Pl. 3, fig. 12), Hamana-ko St. 80.

Remarks.—Outer surface has small shallow punctations. Normal pores are distributed between punctations and are of two types: a simple open type without lip and with hair, and an open type surrounded by fairly broad lip without hair. General outline, nature of overlap of the left valve over the right valve along ventral margin, shape of vestibules, narrower anterior infold in right valve, pattern of muscle scar distribution; even the slight downward projection along anterior part of ventral margin in inner view, all exactly coincide with those of *Cypridopsis vidua*. This is the first report of this species from Japan. Abundant at St. 79 in fresh water pond for eel culture. Other localities include St. 22, 64, and 80. Dead carapaces are found abundantly at St. 80 and rarely at St. 71, 74, 75, 82, and 90.

Superfamily Cytheracea Baird, 1850
Family Cytherideidae Sars, 1925
Subfamily Pontocytherinae Mandelstam, 1960
Genus *Pontocythere* Dubowsky, 1939
Pontocythere sekiguchii Ikeya and Hanai, n. sp.
Pl. 2, figs. 1–5; Pl. 7, fig. 1; text-fig. 12.

Type.—Holotype, a complete carapace, IGSU-O-13 (Pl. 2, figs. 1a, b, 2a, b, 3, 4; Pl. 7, fig. 1. L, 0.44; H, 0.17; W, 0.17), Hamana-ko St. 56.

Illustrated specimens.—A right valve, IGSU-O-14 (Pl. 2, fig. 5), Hamana-ko St. 54; a complete carapace, IGSU-O-69 (text-figs. 12a, b), Hamana-ko St. 52.

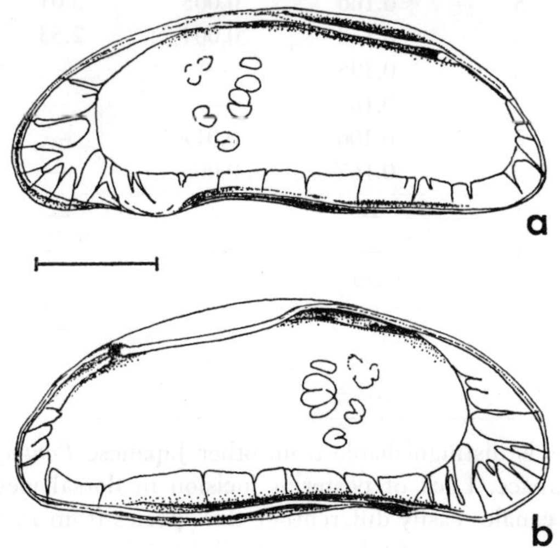

Text-fig. 12
Internal views of *Pontocythere sekiguchii* n. sp. (scale: 0.1 mm). A complete carapace (IGSU-O-69); a: right valve; b: left valve.

Diagnosis.—Pontocythere with relatively small carapace. Surface smooth with faint marginal wrinkles and approximately 65 scattered large punctations for normal pore openings. No posterior projections along posteroventral margin.

Description.—Carapace small, elongate and tumid. Anterior margin obliquely rounded, dorsal margin very gently arched, ventral margin sinuated at a point slightly anterior to its middle, posterodorsal margin gently arched sloping to the acutely rounded posterodorsal margin. Viewed dorsally, sides parallel, tapers more acutely to the anterior than the posterior. Viewed posteriorly, carapace nearly rounded. No posterior projections along posteroventral margin. Left valve larger, overlapping slightly the right along ventral, posterior and dorsoanterior margin. Surface smooth with approximately 65 scattered large punctations for normal pore openings. Marginal wrinkles parallel the anterior margin, distinct in anterior and ventral areas, disappear in central, dorsal, and posterior areas.

Hinge desmodout of *Cushmanidea* type. Central muscle scars consisting of slightly oblique row of four adductor muscle scars, a clover-shaped frontal scar and two relatively large mandibular scars. Of four adductor muscle scars, dorsomedian one triangular. Marginal infold broad along anterior margin, narrow along ventral and posterior margins. Vestibule moderately deep. Radial pore canals straight, moderate in number, approximately 12 along anterior margin. Normal pore canals large, sunken sieve type.

Dimensions.—Measurements of the pooled specimens from St. 51, 52, 53, 54, and 56 are as follows:

S			N	$\overline{X}$	Sd	V	OR
Adult	C	L	3	0.443	0.006	—	0.44–0.45
		H	3	0.177	0.006	—	0.17–0.18
		W	3	0.167	0.006	—	0.16–0.17
	RV	L	2	—	—	—	0.43–0.46
		H	2	—	—	—	0.17–0.18
Adult-1	C	L	5	0.384	0.011	2.86	0.37–0.40
		H	5	0.166	0.005	3.01	0.16–0.17
		W	5	0.158	0.004	2.53	0.15–0.16
	LV	L	2	0.395	—	—	0.37–0.42
		H	2	0.165	—	—	0.16–0.17
	RV	L	3	0.400	0.017	—	0.38–0.41
		H	3	0.167	0.006	—	0.16–0.17
Adult-2	C	L	1	—	—	—	0.31
		H	1	—	—	—	0.13
		W	1	—	—	—	0.13
Adult-3	C	L	2	—	—	—	0.26
		H	2	—	—	—	0.12
		W	2	—	—	—	0.12

Remarks—The species is distinguishable from other Japanese *Pontocythere* in having relatively small size of carapace. Lack of posterior incision in dorsal view and relatively small number of normal pore canales easily differentiate this species from *P. minuta* n. sp.: the only Japanese species whose carapace is of relatively small size.

Occurrence.—The species is characteristic of the well-sorted fine to medium-grained sand bottom of the shallow open sea at St. 51, 54, and 56. Dead carapaces are also found at St. 52 and 53.

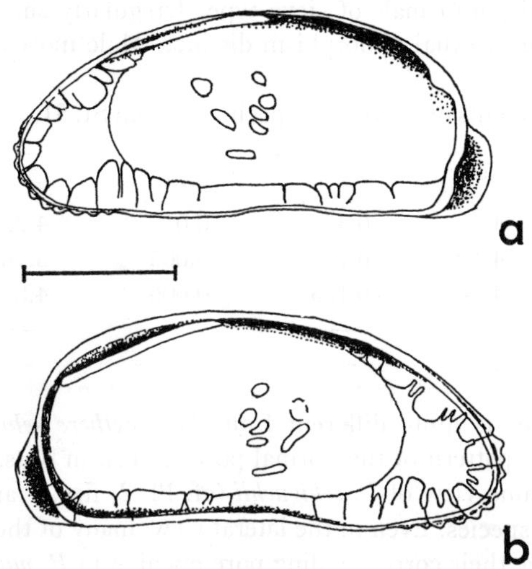

Text-fig. 13
Internal views of *Pontocythere minuta* n. sp. (scale: 0.1 mm). A complete carapace (IGSU-O-70); a: right valve; b: left valve.

Pontocythere minuta Ikeya and Hanai, n. sp.
Pl. 2, figs. 6–12; Pl. 3, fig. 11; text-fig. 13.

Type.—Holotype, a complete carapace, IGSU-O-15 (Pl. 2, figs. 6a, b, 7a, b, 8, 10, 11, 12. L. 0.36; H, 0.13; W, 0.12), Hamana-ko St. 53.

Illustrated specimens.—A left valve, IGSU-O-16 (Pl. 2, fig. 9; Pl. 3, fig. 11), Hamana-ko St. 53; a complete carapace, IGSU-O-70 (text-figs. 13a, b), Hamana-ko St. 54.

Diagnosis.—*Pontocythere* with small elongate carapace. Surface with marginal plications and more than 100 punctations of three kinds of origin. Punctae align in wrinkles in anterior and ventral area; lacking in dorsal area of carapace. Posterior projection along posteroventral margin large and distinct.

Description.—Carapace small, elongate and tumid, highest slightly posterior to middle. Anterior margin obliquely and narrowly rounded and serrate; slightly arched dorsal and slightly sinuated ventral margins running nearly parallel each other. Posterior cardinal angle distinct. Posterior contact margin obscured in its ventral half by distinct large caudal projection. Viewed dorsally, sides nearly parallel tapering acutely toward anterior. Posterior end with characteristic incision owing to conspicuous posterior projections. Viewed anteriorly, carapace nearly rounded. Left valve larger and overlapping right valve along dorsoanterior, dorso-posterior, and posteroventral margins. Surface with minute punctations except for dorsal area. Wrinkles confined to anterior and ventral areas, and running parallel with anterior and slightly oblique to ventral margins. Punctae tending to align in wrinkles of anteroventral area. Punctae consisting of normal pore openings, irregularly shaped sieves without normal pore openings, or simple pits.

Hinge desmodont of *Cushmanidea* type. Adductor muscle scars consisting of vertical row of four scars in which dorsal scar large, dorsomedian scar located somewhat posteriorly and ventromedian scar triangular. Frontal scar large and clovershape. Mandibular scar consisting of two elongate scars. Marginal infold widest anteriorly, narrow along ventral and posterior margin. Vestibule deep. Radial pore canal straight and moderate in number, at least five along anterior margin. Normal pore canals of sieve type. Irregularly shaped sieve without normal pore opening also present. Sexual dimorphism distinct. Male more elongate and slender than female.

Dimensions.—Measurements of the pooled specimens from St. 51, 53, 54, and 59 are as follows:

S			N	$\overline{X}$	Sd	V	OR
Adult	C	L	4	0.353	0.015	4.25	0.34–0.37
		H	4	0.133	0.005	3.76	0.13–0.14
		W	4	0.125	0.006	4.80	0.12–0.13
	LV	L	1	—	—	—	0.36
		H	1	—	—	—	0.13

Remarks.—This species is quite different from *Pontocythere sekiguchii* n. sp. in carapace outline. The distribution pattern of the normal pore as seen in dorsal view is quite similar to that observable in the same view of *P. sekiguchii* (cf. Pl. 2, figs. 1 and 6), suggesting a close affinity between the two species. Even in the lateral view, many of the normal pore openings of *P. sekiguchii* seem to have their corresponding pore opening in *P. minuta*.

Occurrence.—This species is characteristic of the well-sorted fine to medium-grained sand bottom of the shallow open sea at St. 51 and 54. Dead carapaces are also found at St. 53 and 59.

Subfamily Neocytherideidinae Puri, 1957

Genus *Neocytherideis* Puri, 1957

Neocytherideis punctata Ikeya and Hanai, n. sp.

Pl. 2, figs. 13–17; Pl. 6, fig. 11; Pl. 7, fig. 2; text-fig. 14.

Type.—Holotype, a complete carapace, IGSU-O-18 (Pl. 2, figs. 16, 17. L, 0.44; H, 0.16; W, 0.14), Hamana-ko St. 52.

Illustrated specimens.—A complete carapace, IGSU-O-17 (Pl. 2, figs. 13a, b, 14a, b, 15; Pl. 6, fig. 11; Pl. 7, fig. 2), Hamana-ko St. 54; a complete carapace, IGSU-O-71 (text-figs. 14a, b), Hamana-ko St. 51.

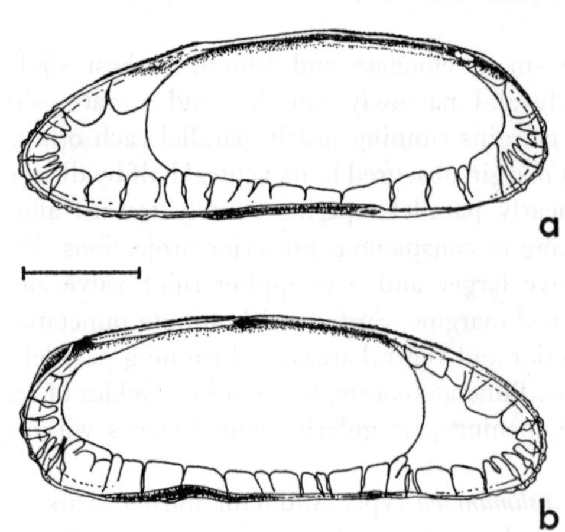

Text-fig. 14

Internal views of *Neocytherideis punctata* n. sp. (scale: 0.1 mm). A complete carapace (IGSU-O-71); a: right valve; b: left valve.

Diagnosis.—Neocytherideis with dense and large ornamental punctations, vertical ridges, and dorsomedian sulci.

Description.—Carapace elongate tumid. Anterior margin narrowly and obliquely rounded with long and slightly arched dorsoanterior margin. Short dorsal and long ventral margins nearly straight and parallel to each other. Posterior margin broadly rounded. Viewed dorsally, sides parallel, gradually tapering anteriorly and abruptly tapering posteriorly. Viewed anteriorly, carapace slightly compressed laterally. Left valve larger, overlapping right valve along dorso-anterior, dorsoposterior and ventral margins. Surface pitted with large ornamental punctations. Diameter of punctae larger in anterocentral area of carapace. General surface of shell ornamented by fine ridges and finer second-order pittings between ridges (Pl. 6, fig. 11). Fine ridges tending to be parallel, anterior margin slightly oblique to ventral margin and forming latticelike reticulum interlocking with vertical ridges in ventral area of carapace. Of two dorsomedian sulci, anterior one faint, and posterior one distinct with corresponding ridge on interior valve surface. Hinge desmodont of *Cushmanidea* type. Adductor muscle scars difficult to observe. Marginal infold widest anteriorly, moderately wide along ventral and posterior margins. Vestibule very deep developing anteriorly and posteriorly. Snap knob and pit structure on posterior part of ventral margin consisting of tonguelike projection on right valve and corresponding incision on left valve. Radial pore canals straight, approximately 10 along anterior margin, tending to tie up into a few bundles along posterior margin. Normal pore large, sieve type, opening not in punctae but on general surface.

Dimensions.—Measurements of the pooled specimens from St. 51, 52, 53, 54, and 56 are as follows:

S			N	$\overline{X}$	Sd	V	OR
Adult	C	L	5	0.448	0.008	1.79	0.44–0.46
		H	5	0.166	0.005	3.01	0.16–0.17
		W	5	0.144	0.005	3.47	0.14–0.15
Adult-1	RV	L	1	—	—	—	0.41
		H	1	—	—	—	0.13

Remarks.—Most *Neocytherideis* have carapace with smooth to pitted surface. Thus, the strong ornamentation explained in diagnosis easily differentiates this species from the other *Neocytherideis*.

Occurrence.—The species occurs at St. 56 in the medium coarse with well-sorted sand of the shallow open sea. Dead carapaces are also found at St. 51, 52, 53, and 54.

Family Leptocytheridae Hanai, 1957
Genus *Callistocythere* Ruggieri, 1953
Callistocythere sp.
Pl. 5, figs. 7a, b, 8, 9.

Illustrated specimen.—A right valve, IGSU-O-32 (Pl. 5, figs. 7a, b, 8, 9), Hamana-ko St. 53.

Remarks.—Specimens have *Callistocythere*-like outline with anterior and posterior marginal ridges in lateral view. Central muscle scars consist of a vertical row of four adductor muscle scars and one large heart-shaped frontal scar. Of four adductor muscle scars, ventral and dorsal ones are round and median two are elongate laterally. A crescent fulcral point located just above the central muscle scars. Normal pores are small and open on the ridges of the reticulum. Sieve-like structure of small pits is observed near the normal pore openings. Sensory bristles may have ring structure in stem and tassel in tip. Snap knob and pit structure present on the ventral sinuated margin, knob on right valve and pit on left valve. All these characters suggest that the specimens belong most likely to genus *Callistocythere* or its allied genera. It is highly probable from fine punctation and fine ridges of the carapace surface that the specimens represent a new species. However, since the extremely narrow marginal infold and noncrenulated entomodont hingement suggest young stage of the specimen, full description of this species is pending until adult specimens become available for study. Dead carapaces are found at St. 39, 41, 42, 51, 53, 54, and 59, suggesting bay mouth and open sea habitat of this species.

Family Hemicytheridae Puri, 1953

Subfamily Thaerocytherinae Hazel, 1967

Genus *Cornucoquimba* Ohmert, 1968

Cornucoquimba rugosa Ikeya and Hanai, n. sp.

Pl. 3, figs. 1–5; Pl. 7, fig. 3; text-fig. 15.

Type.—Holotype, a complete carapace, IGSU-O-19 (Pl. 3, figs. 1–5; Pl. 7, fig. 3. L, 0.61; H, 0.28; W, 0.34), Hamana-ko St. 56.

Illustrated specimen.—A complete carapace, IGSU-O-72 (text-figs. 15a, b), Hamana-ko St. 57.

Diagnosis.—*Cornucoquimba* with four large turret-like tubercles, distributed quadrilaterally in area behind muscle scar node.

Description.—Carapace subquadrate with a large round muscle scar node and four large turretlike tubercles. Highest at anterior cardinal angle. Anterior margin broadly and obliquely rounded with smooth margin. Dorsal contact margin straight, inclined toward posterior and obscured by dorsal ridge in its posterior half. Ventral margin nearly straight, slightly sinuate at middle. Posterior cardinal angle distinct. Posterior contact margin obliquely and narrowly rounded, obscured by posteriorly projected ridge in lateral view. Viewed dorsally, anterior and

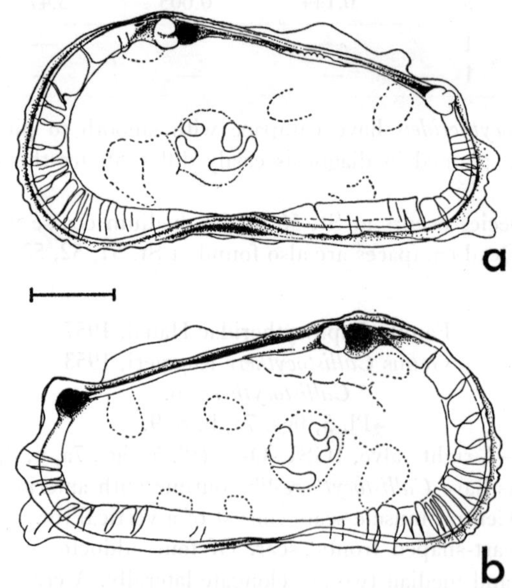

Text-fig. 15

Internal views of *Cornucoquimba rugosa* n. sp. (scale: 0.1 mm). A complete carapace (IGSU-O-72); a: right valve; b: left valve.

posterior sides compressed; inflated in appearance owing to central node and other large turret-like tubercles, incised in posterior end. Left valve slightly larger than right, especially at posterior cardinal angle. Surface characterized by large ornamental structures but no minor ornamentation. Eye tubercle round and prominent, located close to anterior cardinal angle. Anterior marginal ridge with sharp edge, starting from just above eye tubercle, running parallel and close to the anterior margin, separating into a row of fine elongate tubercles and terminating near anteroventral area. Strong ridge starting from anterodorsal area, running dorsoposteriorly, rising high enough to obscure dorsal contact margin and terminating near posterior cardinal angle. Another ridge appearing at just above muscle scar node, running anteroventrally, making an acute arch anteriorly and terminating in the ventral area. The ridge broad and high in anteroventral area. Posterior ridge projecting posteriorly at posterior cardinal angle, posterior end and posteroventral corner to obscure posterior contact margin in lateral view. Muscle scar node anterocentral, large, high, and cone shaped. Four large turret-like tubercles rising abruptly from carapace and castellated, distributed quadrilaterally in area behind muscle scar node.

Hinge holamphidont. Inside view of carapace characterized by one large subcentral cavity corresponding to adductor muscle node, one small cavity corresponding to eye tubercle, one anteroventral cavity corresponding to a broad ridge and four cavities corresponding to four quadrilaterally placed turret-like tubercles. Central muscle scars located inside muscle scar cavity. Adductor muscle scars visible in inside view of carapace consisting of one very long obliquely elongate scar and one round scar imediately above. Frontal scar consisting of one round dorsal and one heart-shaped ventral scars. Marginal infold moderately broad anteriorly and narrow posteroventrally. No vestibule. Radial pore canals with median swelling, numerous around lower half of anterior margin, less numerous along upper half of anterior margin and posterior margin. Normal pore canals sunken sieve type with overgrown orifice opening on ridges, and on flank of adductor muscle node and turret-like tubercles. Turret-like tubercle may be perforate.

Dimensions.—Measurements of the pooled specimens from St. 38, 39, 40, 51, 52, 56, 57, 59, and 74 are as follows:

S			N	OR
Adult	C	L	2	0.61–0.65
		H	2	0.28–0.32
		W	2	0.34–0.36
Adult-1	C	L	2	0.52–0.58
		H	2	0.28
		W	2	0.26–0.30
	LV	L	1	0.53
		H	1	0.29
	RV	L	2	0.53–0.57
		H	2	0.27–0.29
Adult-2	LV	L	2	0.39–0.43
		H	2	0.20–0.23
	RV	L	2	0.39–0.46
		H	2	0.21–0.23

Remarks.—*Puriana* spp. described by Hu and Yang (1975, p. 110, 111), Hu and Cheng (1977, p. 106, 107) and Hu (1976, p. 32–34; 1977, p. 94, 95) have certain feautures similar to this species, but diagnostic characters mentioned in the preceding lines will easily differentiate this species from Formosan *Puriana*. Gulf Coast species of *Puriana* is quite different from those of Formosa in surface ornamentation as well as in the characteristic lateral outline of posterior margin. South American *Cornucoquimba* has faint reticulation, strong crest-like projection of anterodorsal margin in side view, and characteristic oblique and linear arrangement of large posterodorsal, small posterocentral, and large posteroventral nodes. All of these characters are quite different from those of this species. The Japanese species, which does not fit quite well into any known genus, is tentatively assigned here to genus *Cornucoquimba*, but may belong to a new genus.

Occurrence.—The species occurs at St. 52, 56, 57, and 59 on fine to medium coarse sand bottom of the shallow open sea. Dead carapaces are found St. 38, 39, 40, 51, and 71 in the area around the mouth of the Hamana-ko Bay.

Family Trachyleberididae Sylvester-Bradley, 1948
Subfamily Trachyleberidinae Sylvester-Bradley, 1948
Genus *Trachyleberis* Brady, 1880
Trachyleberis sp.
Pl. 4, figs. 8a, b.

Trachyleberis sp.: Yajima, 1978, p. 398, 399, Pl. 49, figs. 1a, b.

Illustrated specimens.—A right valve, IGSU-O-24 (Pl. 4, figs. 8a, b), Hamana-ko St. 43.

Remarks.—Specimens are closely similar to the immature form of *Trachyleberis* sp. described and illustrated in detail by Yajima (1978). It may be interesting to note that, in this finely reticulate young form, spines are at the junction of the partition walls of the reticulum. Normal pore openings located on the walls or linked to the walls. Exact identification of this species is pending until adult specimens become available for study. Dead carapaces are found at St. 42, 43, and 51 in the channel sand and in shallow open sea sand, suggesting the derivation of this species from the nearby environment. In fact, in the Aburatsubo cove of Miura Peninsula, this young form occurs in the silty bottom of the cove.

Genus *Actinocythereis* Puri, 1953
Actinocythereis sp.
Pl. 4, figs. 7a, b.

Illustrated specimen.—A left valve, IGSU-O-23 (Pl. 4, figs. 7a, b), Hamana-ko St. 54.

Remarks.—Spines are enlarged more or less laterally, parallel to the margin of the carapace. Normal pores open on the general surface of the carapace, some open on the conules and sometimes at the flank of the spines. Three horizontal rows of spines, similar to that of "*Archicythereis*" assign this young molt to genus *Actinocythereis*. Because the morphological difference between adult and the last molting larvae is drastic in the so-called *Trachyleberis-Actinocythereis* group, pending the accumulation of enough specimens, identification of this young molt will remain tentative. *Trachyleberis* sp. from Uranouchi Bay (Ishizaki, 1968, p. 39, Pl. 9, fig. 9 later called *Actinocythereis* sp. by Hanai, et al., 1977, p. 50, 51) and *Actinocythereis kisarazuensis* (Yajima, 1978, p. 399, 400, Pl. 49, fig. 3, text-fig. 9–1) seem to have close affinity to this young molt. Dead carapaces are found at St. 51 and 54 in the well-sorted sand of shallow open sea.

Family Cytherettidae Triebel, 1952
Genus *Neocytheretta* van Morkhoven, 1963
Neocytheretta sp.
Pl. 4, figs. 9a, b.

Illustrated specimen.—A left valve, IGSU-O-25 (Pl. 4, figs. 9a, b), Hamana-ko St. 56.

Remarks.—Evenly and broadly rounded anterior margin, obliquely truncate posteroventral margin, slightly sinuate ventral margin, and long and straight hinge margin form a characteristic outline, similar to that observed in a certain species of Cytherettidae. Broad velate-like ridge runs along free margin overhanging to obscure ventral contact margin in lateral view. Both velate-like ridge and strong lateral ridge have lateral outgrowths which strut the ridges. Subcentral tubercle and posterodorsal irregularly shaped projection are prominent. Surface faintly reticulates and has several pore conules. Eye tubercle prominent, with corresponding grove inside of the carapace. A vertical row of four adductor muscle scars is observable. *Neocytheretta snellii* (Kingma, 1948) redescribed in detail by van Morkhoven (1963, p. 236–239) seems to have close affinity to this species in general outline and overall character of ornamentations. However, because extremely narrow marginal infold and lophodont-like hingement suggest the young molt of the specimens at hand, and because structure of the marginal infold, diagnostic character of genus *Neocytheretta* is not developed in young forms, thus exact specific identification is pending until adult specimens become available for study. Dead carapaces are found rarely at St. 40, 53, and 56, all of these stations are located in the area where the influence of warm water of the open sea is very strong.

Family Bythocytheridae Sars, 1926

Genus *Bythocythere* Sars, 1866

Bythocythere maisakensis Ikeya and Hanai, n. sp.

Pl. 3, figs. 6–10; text-fig. 16.

Type.—Holotype, a complete carapace, IGSU-O-20 (Pl. 3, figs. 6–10. L, 0.52; H, 0.28; W, 0.26), Hamana-ko St. 56.

Illustrated specimen.—A right valve, IGSU-O-58 (text-fig. 16), Hamana-ko St. 57.

Diagnosis.—*Bythocythere* with fine punctations and two groups of ridges on carapace surface, a group of four parallel ridges in anterodorsal area and a group of six ridges in ventral area of which lower three running on flank of ventral inflation.

Description.—Carapace subrhomboidal in lateral outline. Anterior margin obliquely rounded. Posterior caudal process obtuse, locating above middle of posterior margin. Dorsal and ventral margins straight and nearly parallel each other. Viewed dorsally, valves inflated fusiform, with maximum width slightly behind the middle. Sulcus broad and distinct. Strong anterior marginal ridge starting at the anterior cardinal angle, running parallel to anterior margin and terminating in anteroventral area. Flat area just behind marginal ridge narrow but distinct. Surface ornamented with two groups of thin but distinct costae-like ridges. A group of four parallel ridges rising in midanterior area running posterodorsally; three terminate in area of sulcus, whereas

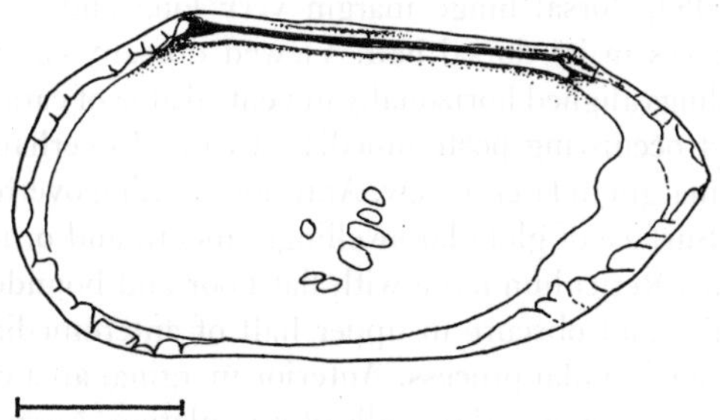

Text-fig. 16

Internal view of *Bythocythere maisakensis* n. sp. (scale: 0.1 mm). A right valve (IGSU-O-58).

one continues to run posteriorly forming an edge of dorsal flattened area and terminates in posterodorsal area. Another group of six ridges starting in midanterior area, running posteriorly making arch ventral; lower three long and strong, running on flank of ventral inflation of carapace, and terminating in posteroventral area; upper three short and faint. Eye spot obscure. Left valve slightly larger than right valve.

In inside view, carapace with anterior marginal swelling which corresponds to anterior flattend area of exterior surface, central swelling which corresponds to exterior sulcus postero-ventral depression which corresponds to posteroventral part of ventral swelling of exterior surface. Hinge lophodont. Adductor muscle scars on anterior flank of central ridge with a small round frontal and two distinct mandibular scars in front. Marginal infold relatively narrow. Vestibule developed anteriorly and posteroventrally. Radial pore canals not numerous. Normal pores numerous tending to align between ridges in general area and few in area of sulcus.

Dimensions.—Measurements of the pooled specimens from St. 56 and 57 are as follows:

	S		N	OR
Adult	C	L	1	0.52
		H	1	0.28
		W	1	0.26
Adult-1	RV	L	1	0.42
		H	1	0.22

Remarks.—*Bythoceratina* sp. A, described from Uranouchi Bay by Ishizaki (1968, p. 17, Pl. 3, fig. 11) resembles this species in its general outline. However, ventral swelling of this species is more feeble than that of the species from Uranouchi Bay.

Occurrence.—Dead carapaces are found rarely at St. 56 and 57 from the fine to medium sands of shallow open sea.

Genus *Bythoceratina* Hornibrook, 1952

Bythoceratina elongata Ikeya and Hanai, n. sp.

Pl. 4, figs. 5, 6; Pl. 7, fig. 4.

Type.—Holotype, a left valve, IGSU-O-22 (Pl. 4, figs. 5, 6: Pl. 7, fig. 4. L, 0.32; H, 0.15), Hamana-ko St. 54.

Diagnosis.—*Bythoceratina* with elongate lateral outline. Surface with three large globular swellings aligned horizontally in the ventral area of carapace. Anterior one overhanging anteriorly and posterior one overhanging ventrally to obscure ventral margin in lateral view.

Description.—Carapace thin, elongate oblong. Anterior margin evenly rounded, dorsal and ventral margins subparallel, dorsal hinge margin very long and straight, posterior end with large, obtuse caudal process in its dorsal half. Viewed dorsally carapace strongly inflated by three large globular swellings aligned horizontally in ventral area of carapace; middle one smallest and posterior one largest occupying posteromedian area and overhanging ventrally to obscure posterior half of ventral margin in lateral view. Anterior swelling overhanging anteriorly. Sulcus deepest in central area. Surface of globular swellings smooth and ornamented with pentagonal or hexagonal reticulations. Reticulum large with flat floor and bounded by low and thin walls. Reticulation becomes faint and obscure in upper half of anteromedian area, in central sulcus area and in posterior area of caudal process. Anterior marginal area compressed with approximately ten pore-conules growing over thin walls of reticulation. Normal pore canals large, sieve type forming gentle dome of sieve plate and growing over walls of reticulum obscuring walls. Setal perforation eccentric, running not always centrally on wall.

Hinge lophodont. Marginal infold broad, anteriorly and posteroventrally. Vestibule deep in upper half, and no vestibule in lower half of anterior margin. Dimensions: L, 0.32; H, 0.15.

Remarks.—Single specimen does not warrant exact description of marginal pore canals and muscle scars, yet inflated globular swelling, straight hinge and pentagonal to hexagonal reticulations prominent in the area of swelling suggest close affinity of this species with *Bythoceratina.* Very elongate lateral outline and lack of ventrolateral spine may differentiate this species from the rest of *Bythoceratina.*

Occurrence.—Only a dead specimen was found at St. 54 in well-sorted fine to medium sand of shallow open sea.

Genus *Semicytherura* Wagner, 1957

Semicytherura elongata Ikeya and Hanai, n. sp.

Pl. 5, figs. 4, 5; Pl. 7, fig. 5.

Type.—Holotype, a right valve, IGSU-O-29 (Pl. 5, figs. 4, 5; Pl. 7, fig. 5. L, 0.40; H, 0.22), Hamana-ko St. 39.

Diagnosis.—*Semicythrura* with caudal process being situated subventrally. Longitudinal costae parallel and well developed in lower surface, becoming subparallel and faint in medial and upper surface. Vertical inflation giving geniculate appearance ventrolaterally in anterior view. Intercostal punctations large and strong in anteroventral area.

Description.—Carapace subtrapezoidal and slender in side view, anterior margin broadly and obliquely rounded, dorsal and ventral margins subparallel, posterior caudal process being situated subventrally. Posteroventral area of carapace overhanging to obscure posteroventral margin in side view. Surface with longitudinal costae and intercostal punctations. Transverse costae faint. Among many longitudinal costae, one starting at middle of anterior margin, running posterolaterally on edge of ventral inflation of carapace and terminating at lower edge of caudal process. Longitudinal costae strong in ventral area becoming faint in dorsal area. Intercostal punctations large and strong in anteroventral area. Ventral inflation giving geniculate appearance ventrolaterally in anterior view.

Hinge lophodont of *Cytherura* type. Characters of marginal infold and marginal pore canals common with other *Semicytherura.* Normal pore opening open type with thickened rim, opening usually in intercostal area. Muscle scar field large, details of muscle scars could not be observed.

Dimensions.—Measurements of two valves from St. 39 and 40 are as follows: Adult, RV, L, 0.40; H, 0.20. Adult, LV, L, 0.33; H, 0.16.

Remarks.—The species differs from *Semicytherura polygonoreticulata*, described by Ishizaki and Kato (1976, p. 131, 132, Pl. 1, figs. 9, 10; Pl. 2, fig. 1; text-fig. 6.) from Furuya Mud, in having slender outline, different pattern of longitudinal ridges and intercostal punctations.

Occurrence.—Separate valves are found at St. 39 and 40 which are located at the mouth of Hamana-ko Bay.

Semicytherura sp.

Pl. 5, figs. 6a, b.

Illustrated specimen.—A left valve, IGSU-O-30 (Pl. 5, figs. 6a, b), Hamana-ko St. 53.

Remarks.—This young molt of *Semicytherura* sp. is closely similar in essential pattern of ornamental structure to that of *Semicytherura polygonoreticulata* Ishizaki and Kato from Furuya Mud and of *Semicytherura elongata* n. sp. from the Hamana-ko Bay. However, this young molt is closer to the former species than the latter, having a relatively larger size and higher position of the caudal process. Intercostal punctations are so large as to form reticulations between costae. Floor of each reticulum is so deep especially in posteroventral area that it is reflected on the inner surface of the carapace as a round swelling of the reverse side of the floor. Development of tegumen forms lateral growth of approximately 20 spines around the mouth of each reticulum. Normal pores simple open type, opening on the muri of the intercostal reticulations. Separate valves are found at St. 53 and 61, on the fine to medium coarse sand bottom of the shallow open sea.

Family Loxoconchidae Sars, 1925

Subfamily Loxoconchinae Sars, 1925

Genus *Phlyctocythere* Keij, 1958

Phlyctocythere hamanensis Ikeya and Hanai, n. sp.

Pl. 4, figs. 1–4; Pl. 6, figs. 12, 13; text-fig. 17.

Type.—Holotype, a complete carapace, IGSU-O-21 (Pl. 4, figs. 1–4; Pl. 6, figs. 12, 13. L, 0.41; H, 0.24; W, 0.19), Hamana-ko St. 51.

Illustrated specimen.—A complete carapace, IGSU-O-68 (text-figs. 17a, b), Hamana-ko St. 51.

Diagnosis.—Carapace inflated with gently arched hinge margin. Very narrow compressed area along anterior margin. Area of dorsal sulcus not sulcating but being represented by area of coarse punctations. Peripheral part of marginal pore canals does not reach margin in posteroventral flattened area.

Description.—Carapace small, thin and translucent, elliptical in lateral outline. Anterior margin evenly rounded, dorsal hinge margin gently arched, ventral margin slightly sinuated, dorsal and ventral margin subparallel, posteroventral margin evenly arched, and posterior margin with slightly upturned caudal process subdorsally. Surface smooth with very narrow

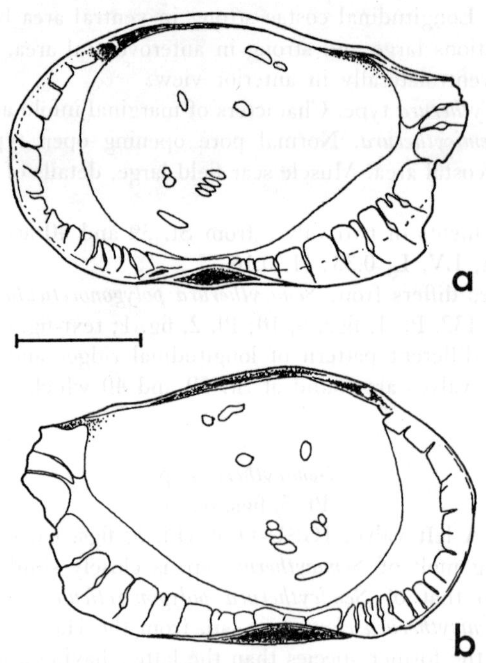

Text-fig. 17

Internal views of *Phlyctocythere hamanensis* n. sp. (scale: 0.1 mm). A complete carapace (IGSU-O-68); a: right valve; b: left valve.

compressed area along anterior margin, trapezoidal compressed area of caudal process, and with wide crescent-shaped, flatly compressed area along posteroventral margin. Carapace fusiform in dorsal view with a small anterior projection of narrow compressed marginal area and a posterior projection of compressed caudal process. Viewed dorsally, hinge margin straight; anterior free margin with anterodorsal slit opening; posterior free margin with posterodorsal overlapping of left valve on right and with opening of slightly upturned caudal process. Carapace oblong in anterior view. Normal pore openings presumably sieve type and not numerous. Five pore openings on posteroventral compressed area being aligned along margin. Area of dorsal sulcus not sulcating but being represented by area of coarse punctations.

Edge of left valve along hinge margin, fitting into groove of right valve. Reduced anterior element of gongylodont hinge observable. Marginal infold broad anteriorly and ventroposteriorly. Vestibule crescent shape and deep anteriorly, and shallow posteroventrally. Marginal pore canals straight, moderate in number, approximately 17 along anterior margin, eight along posteroventral margin, and two along margin of caudal process. Posteroventral marginal pores do not reach margin. Adductor muscle field relatively small. Muscle scars consisting of oblique row of four adductor scars with one large and one small frontal scar in front and two mandibular scars adjoined laterally to form one elongate scar. Eyes large and distinct through translucent carapace just below anterocardinal angle.

Dimensions.—Measurements of pooled specimens from St. 35, 51 and 60 are as follows:

S			N	$\overline{X}$	OR
Adult	C	L	3	0.397	0.39–0.41
		H	3	0.227	0.22–0.24
		W	3	0.180	0.17–0.19
	LV	L	1	—	0.37
		H	1	—	0.21
Adult-1	C	L	1	—	0.28
		H	1	—	0.16
		W	1	—	0.12
	RV	L	1	—	0.37
		H	1	—	0.17

Remarks.—Arched hinge margin and inflated carapace of this species suggest close relationship of this species to genus *Phlyctocythere* rather than to genus *Pseudocythere*. Present species resembles *Phyctocythere pellucida* (G. W. Müller, 1894) in general appearance, but differs from the European species in having less arched hinge margin in lateral view. *Loxocauda* Schornikov, 1969, may be synonymous with *Phlyctocythere*.

Occurrence.—The species occurs at St. 35 on the well-sorted channel sand strongly influenced by tidal current. Valves are also found at St. 51 and 60 on the well-sorted fine to medium sand of shallow open sea under the influence of warm current.

Family Xestoleberididate Sars, 1928
Genus *Xestoleberis* Sars, 1866
Xestoleberis setouchiensis Okubo, 1979

Xestoleberis setouchiensis Okubo, 1979a, p. 10–14, figs. 2a-f, 3a-r.

Remarks.—The species is conspecific with *Xestoleberis* sp. Hanai and Ikeya, in Hanai et al. (1977, p. 67).

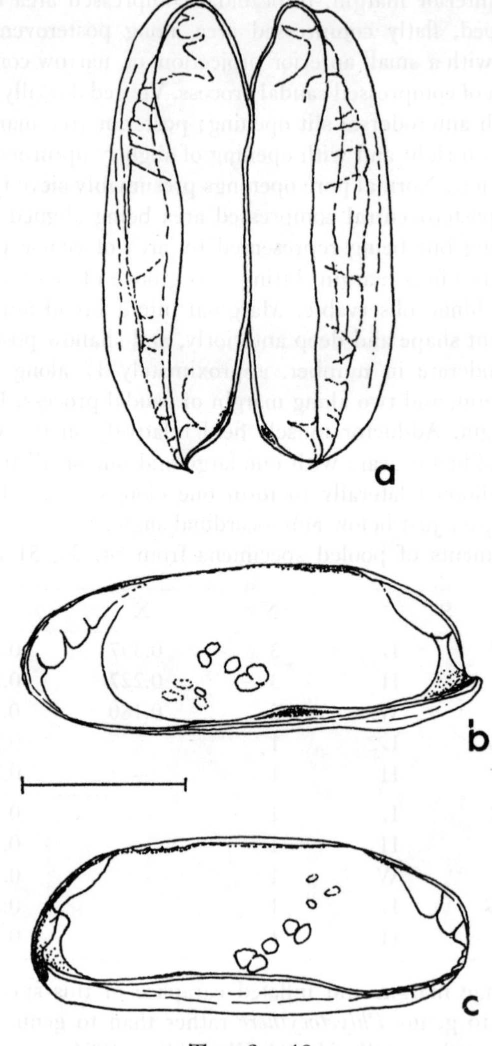

Text-fig. 18

Ventral and internal views of *Platymicrocythere* sp. (scale: 0.1 mm). A complete carapace (IGSU-O-75); a: complete carapace; b: right valve; c: left valve.

Family Microcytheridae Klie, 1938
Subfamily Microcytherinae Klie, 1938
Genus *Platymicrocythere* Schornikov, 1975
Platymicrocythere sp.
Pl. 5, figs. 1–3; Pl. 7, fig. 8; text-fig. 18.

Illustrated specimens.—A complete carapace of female, IGSU-O-28 (Pl. 5, figs. 2a, b); a complete carapace, IGSU-O-49 (Pl. 5, figs. 1a, b, 3; Pl. 7, fig. 8); a complete carapace, IGSU-O-75 (text-figs. 18a–c), Hamana-ko St. 39.

Remarks.—Because the basic concept of classification of this group has been based entirely on appendage structures, pending clarification of appendage structure of this species, exact specific identification can not be made. However, since the species has a characteristic outline which probably forms a part of the diagnosis of this possible new species, a brief description of the carapace outline will given below:

Carapace elongate trapezoidal in lateral view, with flat and wide ventral surface. Width larger than height. Anterior margin narrowly rounded, dorsal and ventral margins long and parallel, posterior margin evenly arched posterodorsally, meeting with ventral margin nearly at a right angle, leaving small posterior projection in lateral view. Viewed dorsally, carapace elliptical, sides parallel and broadly rounded anteriorly and posteriorly, contact edges along anterior margin appearing as a small projection at anterior end of each valve. Posteroventral corner of posterior contact margin with a hook-like projection in each valve to interlock both valves, and leaving a small projection at posteroventral corner in lateral view. Hinge margin straight; right valve overlapping left valve anterodorsally, and left valve overlapping right valve posterodorsally. Surface smooth, with open-type normal pore canal openings with moderately broad rim. Very faint fine ridges subparallel to anterior margin in anterior area of carapace. Marginal infold wide along anterior and ventral margins. Vestibule deep anteroventrally and posteroventrally. Marginal pore canals straight, not numerous along anterior margin, approximately 10 along ventral margin and a few along posterior margin. Sexual dimorphism distinct, female more slender than male.

The characters described above do not always agree well with the characters diagnosed by Schornikov (1975, p. 13) for genus *Platymicrocythere*. A deep excavation in the anteroventral region of carapace margin is absent. The present species occurs always associated with *P. tokiokai* in the Hamana-ko area. *P. tokiokai*, the type species of genus *Platymicrocythere*, is based only on the female specimens; the male form has not yet been described. In *P. labiata* from the Pacific coast of El Salvador, however, the male form is found very rarely and has been described as having the same carapace as female. (Hartmann, 1959, p. 219).

Measurements of male and female carapaces from St. 39 are given below:

S			N	$\bar{X}$	Sd	V	OR
Adult Female	C	L	9	0.278	0.008	2.88	0.26–0.29
		H	9	0.106	0.005	4.72	0.10–0.11
		W	9	0.120	0.000	0.00	0.12
Male	C	L	14	0.269	0.011	4.09	0.25–0.29
		H	14	0.094	0.005	5.32	0.09–0.10
		W	14	0.116	0.005	4.31	0.11–0.12

The species is abundant near the mouth of the Bay at St. 39 and 61. Dead carapaces are also found near St. 54 and 60 of Enshu-nada.

Family Paradoxostomatidae Brady and Norman, 1889

Subfamily Paradoxostomatinae Brady and Norman, 1889

Genus *Neopelluctstoma* Ikeya and Hanai, n. gen.

Type-species.—Neopellucistoma inflatum Ikeya and Hanai, n. sp.

Diagnosis.—Kidney-shaped carapace with smooth surface. Hinge modified lophodont with median element consisting actually of valve edge with a tooth at anterior and posterior terminals in left valve and corresponding median groove with anterior and posterior terminal sockets in right valve. Anterior hinge element consisting of a large elongate tooth of antislip nature in right valve and posterior hinge element consisting of fused two or three small teeth in right valve and corresponding sockets in left valve. Muscle scars consisting of oblique row of four adductor scars, two frontal scars adjoined and two mandibular scars adjoined laterally to form one elongate scar. Inner margin of infold projects upwardly in posteroventral area to form very deep vestibule. Vestibule also deep anteriorly and fused zone narrow along entire free margin. Normal pore canals sieve type, not numerous, and opening large punctations on interior surface. Marginal pore canals moderately numerous.

Remarks.—Pellucistoma Coryell and Fields, 1937, resembles *Neopellucistoma* in the distribution pattern of muscle scars. Hinge structure of both genera is essentially the same, but the location of antislip tooth is just below the anterior terminal of median hinge element in *Pellucistoma*, and just at anterior hinge element in *Neopellucistoma*. *Javanella* Kingma, 1948, is close to *Neopellucistoma*. In fact, dorsal edge of left valve fitting in the groove of right valve in *Javanella* is essentially the same structure as that of *Neopellucistoma*. However, both *Pellucistoma* and *Javanella* differ from *Neopellucistoma* by having a wide zone of concrescence and a caudal projection.

Neopellucistoma inflatum Ikeya and Hanai, n. sp.

Pl. 6, figs. 5–10; Pl. 7, fig. 6; text-fig. 19.

Type.—Holotype, a complete carapace of female, IGSU-O-35 (Pl. 6, figs. 5a, b, 6a, b, 7, 8, 10; Pl. 7, fig. 6. L, 0.54; H, 0.23; W, 0.21), Hamana-ko St. 53.

Illustrated specimen.—A complete carapace, IGSU-O-36 (Pl. 6, fig. 9); a complete carapace, IGSU-O-67 (text-figs. 19a, b), Hamana-ko St. 54.

Diagnosis.—Neopellucistoma with subtriangular lateral outline as illustrated in Pl. 6, fig. 6. Inner margin of infold strongly projects upwardly. Zone of concrescence narrow, thus vestibule large anteriorly and posteroventrally. Pattern of distribution of normal pores shown in Pl. 6, fig. 10.

Description.—Carapace subtriangular. Viewed laterally, anterior margin narrowly and obliquely rounded, dorsal margin strongly arched, ventral margin sinuated anteriorly and downwardly curved posteriorly, posterior margin slightly arched in its upper half and acutely

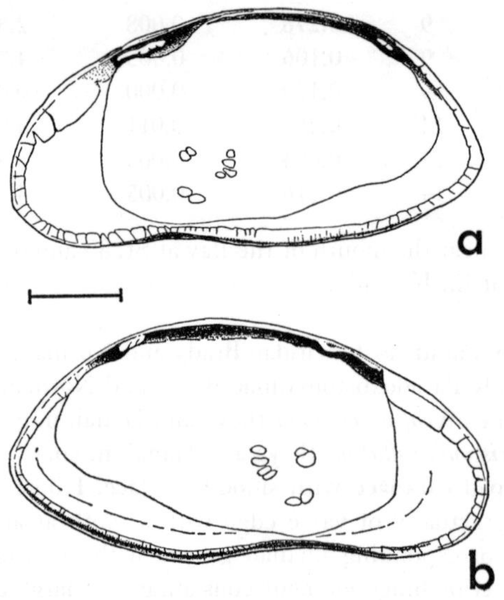

Text-fig. 19

Internal views of *Neopellucistoma inflatum* n. sp. (scale: 0.1 mm). A complete carapace (IGSU-O-67); a: right valve; b: left valve.

bending in its lower half: highest at a little posterior to the middle. Viewed dorsally, sides parallel, tapering gently toward anterior and somewhat quickly toward posterior. Right valve slightly larger than left valve, overlapping left along hinge margin. Surface smooth with few fine, sieve-type normal pore openings.

Hinge modified lophodont, consisting of elongate anterior tooth; long and arched median groove, and small posterior tooth in right valve; corresponding anterior socket, median ridge (actually edge of valve), and posterior socket in left valve. Median element with a socketlike anterior terminal in right valve and corresponding tooth in left valve, with an obscure socket at posterior terminal in right valve receiving fused two or three small teeth of left valve. Anterior hinge element large, elongate tooth of antislip nature in right valve receiving edge of left valve between tooth and edge of right valve. Muscle scar field moderately large located in anteroventral area of carapace, consisting of oblique row of four adductor scars with two frontal scars in front. Frontal scars often adjoined to form one large scar. Mandibular scar located below frontal scar and consisting of two scars, being adjoined laterally to form one elongate scar. Marginal infold broad anteriorly and posteroventrally. In posteroventral area, inner margin of infold strongly projects upwardly. Vestibule deep along anterior and posteroventral margins, thus fused zone narrow along entire free margin. Marginal pore canals moderately numerous, approximately 21 along anterior margin. Large and deep punctations characteristic of interior surface of valves not numerous, corresponding to sieve type normal pores of exterior surface. Sexual dimorphism distinct. Male slenderer than female and having lower carapace.

Dimensions.—Measurements of pooled specimens from St. 39, 51, 52, 53, 54, 55, 59, and 60 are as follows:

S			N	$\overline{X}$	Sd	V	OR
Adult Female	C	L	12	0.542	0.027	4.98	0.50–0.59
		H	12	0.264	0.012	4.55	0.24–0.28
		W	12	0.215	0.014	6.51	0.19–0.24
Female	RV	L	1	—	—	—	0.54
		H	1	—	—	—	0.27
Male	C	L	4	0.533	0.019	3.56	0.52–0.56
		H	4	0.238	0.013	5.46	0.22–0.25
		W	4	0.205	0.006	2.93	0.20–0.21
Male	C	L	4	0.550	0.012	2.18	0.54–0.56
		H	4	0.248	0.010	4.03	0.24–0.26
Adult-1	C	L	1	—	—	—	0.45
		H	1	—	—	—	0.22
		W	1	—	—	—	0.19
	RV	L	2	—	—	—	0.40–0.42
		H	2	—	—	—	0.20
Adult-2	LV	L	1	—	—	—	0.35
		H	1	—	—	—	0.16

Remarks.—Since *Neopellucistoma* is so far a monospecific genus, it is difficult to give a diagnostic character of the species. However, in paradoxostomatids, the lateral outline of the carapace, the features observable in the marginal infold, and the shape of male copulatory organ are usually characteristic of the species. In *Neopellucistoma*, the pattern of distribution of normal pores easily observable, especially in the inside view of the carapace, will also give one of the species specific characteristics.

Occurrence.—This species seems to be restricted to the area of well-sorted fine sand of Enshu-nada area (St. 55, 59). Dead carapaces are found in a wider area extending from shallow open sea of Enshu-nada to the entrance of the Hamana-ko Bay.

Family Cytheromatidae Elofson, 1939

Genus *Cytheroma* G. W. Müller, 1894

Cytheroma? sp.

Pl. 6, figs. 1-4; text-fig. 20.

Illustrated specimens.—A complete carapace, IGSU-O-33 (Pl. 6, figs. 1a, b, 4); a complete carapace, IGSU-O-51 (Pl. 6, figs. 2a, b, 3); a complete carapace, IGSU-O-76 (text-figs. 20a, b), Hamana-ko St. 61.

Remarks.—None of the existing genera seems to be able to accomodate this minute species. Pattern of muscle scars and nature of marginal infold suggest affinity with cytheraceans. However, the species has neither a flat venter, nor relatively well-developed hingement which is characteristic of microcytherids. Though we placed the species in the cytheromatids, the anteroventral hood and its extension along anteroventral contact margin is a unique feature among cytheromatids. Carapace of this species is not so transparent as that of paradoxostomatids, yet it is translucent. Surface is smooth under stereo-microscope, but under SEM, surface is minutely punctated with numerous large, and shallow fossae. Brief description of the species will be given below:

Carapace elongate bean-shaped, compressed anteroventrally to form area of hood. Viewed laterally, anterior margin broadly rounded, dorsal margin evenly arched, ventral margin sinuate at middle, posterior margin obliquely and narrowly rounded: highest at middle. Viewed dorsally, sides parallel, tapering gently toward anterior and posterior. Surface ultra-minutely punctate with numerous shallow, large and round fossae. Marginal infold moderately wide along anterior and posterior margins. Vestibule wide and deep anteriorly and narrow and deep posteriorly. Marginal pore canals straight, eight along anterior margin and seven along posterior margin. Muscle scars consisting of oblique row of four adductor scars, one round frontal scar, and one round mandibular scar. Normal pores open type and moderately numer-

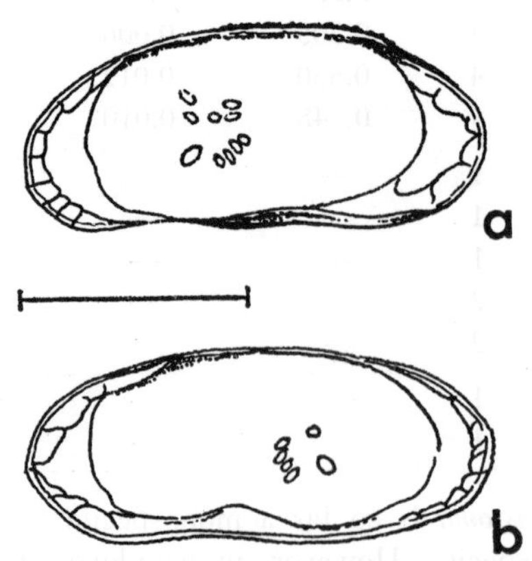

Text-fig. 20

Internal views of *Cytheroma*? sp. (scale: 0.1 mm). A complete carapace (IGSU-O-76); a: right valve; b: left valve.

ous.

Measurements of samples from St. 61 are given below:

	S	N	$\overline{X}$	Sd	V	OR
Adult	L	13	0.213	0.009	4.23	0.20–0.23
	H	13	0.093	0.006	6.45	0.08–0.10
	W	13	0.061	0.003	4.92	0.06–0.07

The species is abundant at St. 39 and 61, in the gravel bottom at the mouth of the bay; the species also occurs at St. 51, 52 and 60 in the well-sorted open sea and near the mouth.

Acknowledgments

The authors wish to express their appreciation to Drs. J. E. Hazel and T. M. Cronin of the U.S. Geological Survey, Reston, Virginia, U.S.A., Drs. P. M. Frydl and Michiko Yajima of the University of Tokyo for reading the original manuscript and giving us invaluable advice. We also wish to thank Mr. Yoshiyuki Sekiguchi of Kodama High School, Saitama Prefecture, for his assistance in collecting freshwater ostracods.

The facilities for sampling were offered by the Hamana-ko Branch, Shizuoka Fisheries Experimental Station. Financial support for this study was in part from a Grant-in-aid for Scientific Research in 1977 (project no. 234051), 1979 (project no. 434042) and 1979 (project no. 448028) from the Ministry of Education, the Government of Japan.

HOLOCENE OSTRACODS IN THE SOUTHERN BOSO PENINSULA

Paul M. Frydl
Mobil Oil Canada Ltd., Calgary, Canada

Abstract: Marine terraces along the coast of the tectonically uplifted southern Boso Peninsula (Honshu, Japan) are formed by sediments deposited during the Jomon (Flandrian) transgression from 9,000 to 5,000 yr B. P. Sixty samples of these sediments were taken in the southern Boso, and 20 samples of recent sediments were collected in the adjacent Tateyama Bay. Of the 115 ostracod species recorded, six species are described as new.

Six fossil ostracod biofacies and six recent ostracod biofacies were delineated using cluster analysis and principal component analysis. Comparison of the distribution of fossil and recent ostracods was used to infer the environments of deposition during the Jomon transgression. In most cases, deposition took place in drowned valley-type, long and narrow bays. The vertical change of fossil ostracod biofacies reflects the widening of the bays due to a rising sea level at the initial stages of the transgression and a gradual shallowing due to sediment infilling prior to the tectonic uplift of the area around 5,500 yr B. P.

The differences between the life, death, and fossil assemblages in samples on hand are examined, and the processes responsible for the changes and ways of detecting altered fossil assemblages are discussed.

Introduction

The working unit in most paleontological studies is one million years or, exceptionally, 100,000 years. Most neontological studies, however, last only a few years and very seldom extend over several decades. Changes taking place over millions of years and those taking place over just a few years are therefore well known. There is however a striking lack of information on changes taking place in the time span between those two extremes—that is, of the order of thousands and tens of thousands of years.

It is unfortunate that as a result of the postglacial marine transgression most of the marine sediments deposited in the interval are presently below sea level and are accessible to study only through material obtained by borings.

The southern part of the Boso Peninsula has been uplifted at least 20 to 30 m over the past 10,000 years and offers an excellent opportunity to study changes occurring over the time interval of several thousand years. Early Holocene sediments in the area were deposited during the Jomon (Flandrian) transgression: a period characterized by rapid sea level change, sedimentary facies shifts, and climatic variations. Changes in animal communities during this period of environmental fluctuation are of particular interest because of the availability of the time framework provided by C^{14} dating.

Because of their abundance in Holocene sediments from the southern part of Boso Peninsula, and the relative ease with which samples large enough for statistical analysis can be collected, ostracods were selected for this study.

Two aspects of Holocene geology of the area have already attracted attention for some time. One is the strikingly large diversity of hermatypic corals found in the uplifted sediments in the vicinity of Tateyama City (77 species of fossil corals in contrast with only 10 species found living in the present Tateyama Bay). It has been suggested that the climate optimum around 6,000 yr B.P. was responsible for the high diversity of the corals, and their decline has been related to the worsening climatic conditions around 4,000 yr B.P. The presumable effect of the climatic change on the associated molluscan fauna is much less pronounced (Nomura, 1932) and no substantial temperature changes were indicated by the foraminiferal faunas studied (Asano, 1936). The depth at which the corals lived and the reason of their irregular distribution throughout the area are still imperfectly understood. The study of the ostracod fauna within the larger time and space framework of the Holocene deposits in the entire southern Boso Peninsula might provide answers to some of these questions.

The second aspect of the Holocene geology of the area extensively studied in the past is the distribution and time of formation of the marine terraces. Because they are directly related to the relative sea level changes in the area, which were partly caused by tectonic movement, marine terraces offer important clues to the recent tectonic history of the area and activity of the Asian and Pacific crustal plates.

The elevation of the terraces is related to, but cannot be equated with, the position of sea level. The near absence of clear sea level indicators, such as peat intercalated with marine sediments, makes it difficult to determine the true position of sea level in spite of the abundant C^{14} datings available for the area. The distribution of ostracod assemblages is controlled by several factors, many of which are influenced by sea level variations; thus, changes in the ostracod assemblages may provide an indication of the relative sea level movement.

The bias introduced into the fossil record by differential preservation of its components is especially important in the case of microfossils, which are, due to their small size, particularly affected by destructive processes. The comparison of composition of the live and dead assemblages in the present Tateyama Bay and the fossil assemblages from similar environments in the Holocene sediments offer many insights into the changes that take place as a live assemblage is changed into a fossil assemblage.

Previous work

Corals: The unusual occurrence of hermatypic corals in the sediments in the southern part of Boso Peninsula has attracted the attention of investigators for over 100 years. Naumann (1877) was the first to investigate the coral-rich sediments near Numa, which he believed to be of Tertiary age. Yabe and Sugiyama (1932, 1935) examined the deposits as a part of their study of corals in Japan. They listed 46 species of corals from sediments near Numa and Koyatsu villages and suggested that they are of Holocene age. They also reported occurrence of 7 species of reef-building corals in the present Tateyama Bay. Subsequent C^{14} datings of corals from the Numa bed (Hoshino, 1967; Konishi, 1967) confirmed the Holocene age of the sediments.

Hamada (1963) found additional 29 species of coral in Koyatsu, and Eguchi and Mori (1973) discovered two more coral species in Koyatsu, bringing the total to 77 species. Hamada (1963), on the basis of comparison with similar coral fauna in Tanabe

Bay (Kii Peninsula), argued that the corals in Koyatsu lived in depths of about 10 m.

Molluscs: Yokoyama (1911) reported the occurrence of several tropical species of molluscs and later (1924), described 124 species of molluscs from the coral-bearing deposits and estimated their age as Pleistocene. Nomura's study (1932) of molluscs from the Boso and Miura Peninsulas included also the coral-bearing deposits of the southern Boso, which he regarded as of Holocene age. He suggested that the molluscan fauna indicates a slightly warmer climate than that of the present Boso Peninsula, being perhaps as warm but not warmer than the present sea off Kii Peninsula in southwest Japan. Matsushima (1979), in a study of southern Kanto postglacial molluscan assemblages (which included the southern Boso Peninsula), pointed out the appearance of tropical species *Ostrea pauluciae* and *Tellinimactra edentula* during the peak of the postglacial transgression about 6,500 to 6,000 yr B.P. He explained their disappearance around 4,000 yr B.P. by a change of sedimentary environment rather than a lowering of temperature.

Foraminifers: The microfauna of the Numa coral bed was investigated by Asano (1936), who failed to find species characteristic of coral reefs in tropical regions, such as *Amphistegina radiata*, *Operculina venosa*, etc., and concluded that the climate at the time of deposition was not markedly different from the present one.

Marine terraces: Naruse and Sugimura (1953) briefly described the Holocene sediments exposed along the Tomoe River. Yoshikawa and Saito (1954) investigated the Holocene terraces and the bathymetry of the shallow sea near Chikura on the west coast. Sugimura and Naruse (1954, 1955) studied the distribution of marine terraces in the southern Kanto region and pointed out the correlation between the elevation of terraces and the amount of uplift during historical earthquakes. Yonekura (1975) further studied the relationship between uplift (as indicated by terrace distribution) and tectonic movements. Yokota (1978) investigated the distribution of terraces and described the sediments forming them along the southeast coast of the southern Boso Peninsula. Nakata et al. (1980) reexamined in detail the distribution of marine terraces in the Boso Peninsula south of the 35° N parallel and emphasized crustal movements as the main cause of relative sea level changes in the area.

General geology

Boso Peninsula projects from the northern third of Honshu for about 80 km into the Pacific Ocean, forming the northern flank of the Tokyo Bay. It is divided into the Northern, Central, and Southern Geological Provinces.

The area under investigation in this study occupies the tip of the peninsula south of 35°15′ N and nearly coincides with the Southern Geological Province (text-fig. 1). Siltstones of the Hota Group, which belong to the Central Geological Province, occur to a limited extent in the northern part of the area. The central and southern parts of the investigation area are occupied by low rounded hills (200 m) with flattened summits formed by rocks correlated with the Miura and Kazusa Groups. In the low lying flat coastal areas, the Neogene rocks are covered by a thin (0 to 40 m) veneer of unconsolidated Holocene sediments forming four marine terraces. The Neogene rocks are tightly folded, with fold axes trending southwest to northeast and south to east.

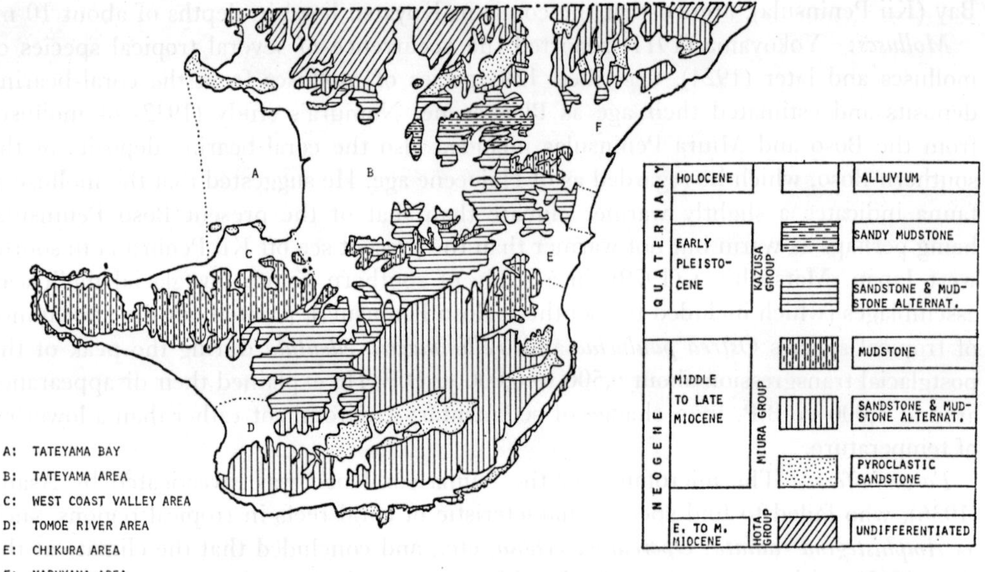

Text-fig. 1
Geology of the southern Boso Peninsula and location of the study area within the investigation area in southern Boso Peninsula.

Several active faults trending east to west are present in the northern part of the area. The center of the synclinorium is occupied by a siltstone-sandstone alternation of the Toyofusa Formation, which is correlated to the Kazusa Group and is separated from the Miura Group by an unconformity. Rocks of the Miura Group form the flanks of the synclinorium, with volcanic sandstone and mudstone-sandstone intercalation of the Chikura Formation in the southeast and soft siltstone of the Nishizaki Formation in the southwest.

The unconsolidated Holocene sediments consist of marine silt and clay, partly overlain by and interfingering landward with nonmarine silt, clay, sand, and conglomerate. The marine sediments are the subject of the present study.

Tectonic setting: The Boso Peninsula occupies a tectonically highly unstable area in the close proximity to the triple junction between the Philippines, Pacific, and Asian crustal plates. In historical times, it has been a site of several large earthquakes, two of which were of magnitude 8 on the present-day Richter scale. The two large earthquakes in the Genroku (1703) and Taisho (1923) eras were accompanied by considerable crustal deformation and destructive tsunamis.

The southern part of the Boso Peninsula lies 20 km northeast of the junction between the Philippine and Asian plates, and both earthquakes resulted in an uplift of several meters. During the Genroku earthquake, maximum uplift reached 6 m, and locally the coastline retreated up to 800 m (Imamura, 1925), producing a 3 to 6 m high terrace girdling the southern part of the peninsula. During the Taisho earthquake, the uplift was smaller, reaching a maximum of 2 m (Omura, 1926). No terrace was produced, owing to the brevity of the interval preceding the second uplift.

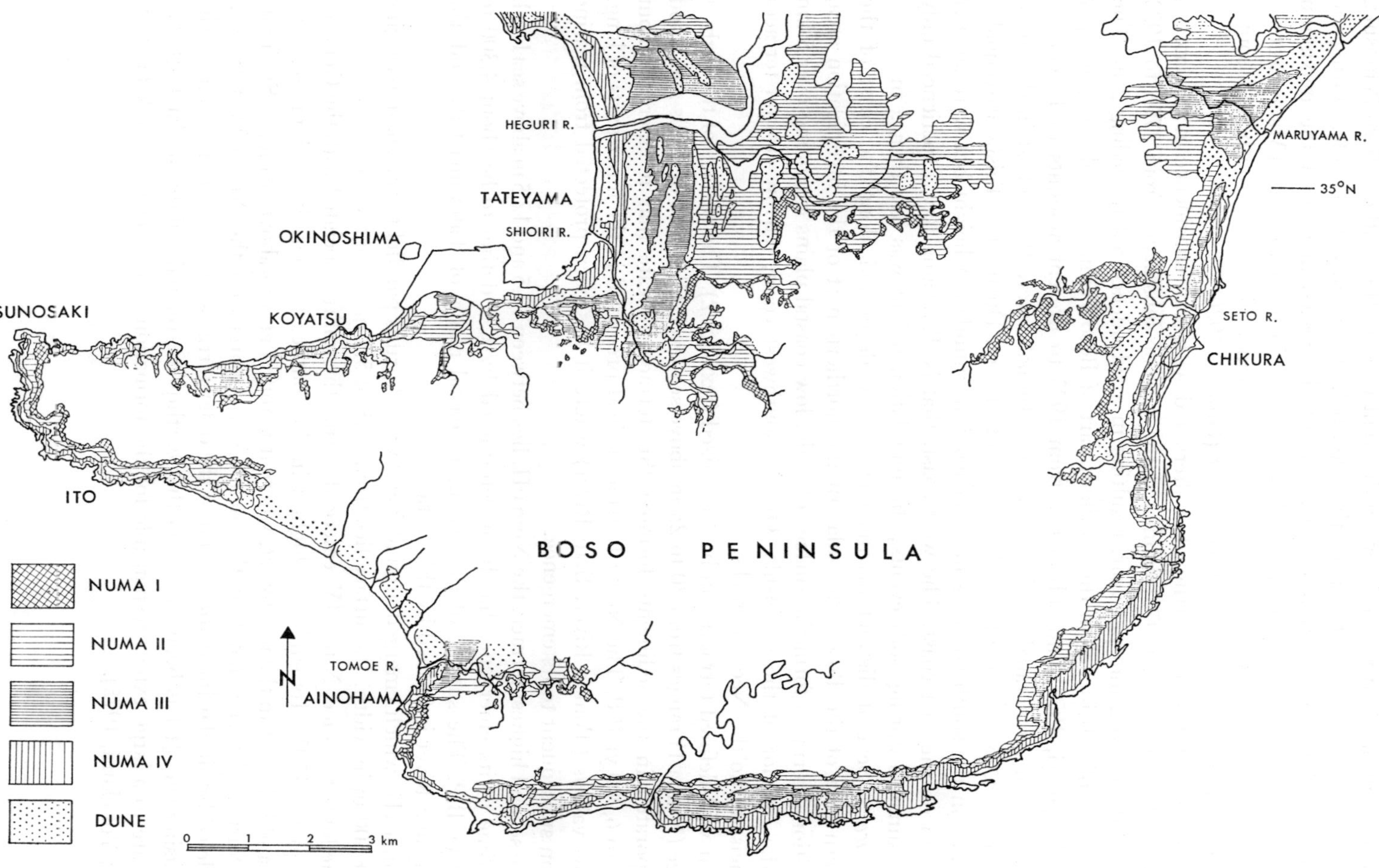

Text-fig. 2

Terrace distribution on the southern part of Boso Peninsula. [The part south of 35°N is after Nakata et al., 1980].

The amount of uplift was different in different parts of the southern Boso Peninsula, and areas of maximum uplift during the two earthquakes did not coincide. During the Genroku earthquake, the southeastern part of the area experienced the maximum uplift, while the area of minimum uplift was located in the norhthwest (Matsuda et al., 1978). The Taisho earthquake resulted in maximum uplift in the south and southwest and minimum uplift in the northeast (Omura, 1926).

Rapid uplift during the earthquakes alternated with slow subsidence in the interim periods. Matsuda et al. (1978) estimated the rate of subsidence from 1703 to 1923 to be about 0.2 to 2 mm per year, resulting in a subsidence not exceeding 2 m. The present rate of subsidence in the southern part of Boso Peninsula is about 2 mm per year (Kanamori, 1973) and subsidence from 1923 to present amounts to 15 cm.

The tsunamis generated by the Genroku earthquake were the highest in the south and southeast, exceeding an inundation height of 5 m. During the Taisho earthquake, tsunamis on the south coast again exceeded 5 m inundation height but were smaller than 2 m on the east coast. The west coast, that is Tateyama Bay, experienced only small tsunamis during both events; the inundation height was less than 5 m.

Terraces: Several distinct marine terraces are developed along the coast of the southern part of the Boso Peninsula. In the northern part of the investigation area, depositional terraces form the surface of wide, low coastal plains extending over 5 km into the interior of the peninsula. On the southwest coast, narrow erosion terraces are incised into the Neogene rocks.

Four well-defined terraces can be recognized (text-fig. 2). The surface of the highest terrace (Numa I) ranges from 20 to 25 m above sea level. C^{14} dates of shells and wood incorporated in the sediments forming this terrace indicate that it emerged about 5,000 to 6,000 yr B.P. The Numa I terrace is best preserved in the inner parts of long, narrow valleys (Tomoe River, Seto River) where it has been protected from marine erosion subsequent to its emergence.

The second highest terrace, the Numa II, lies between 17.5 and 12.5 m above sea level. The time of its emergence has been determined by C^{14} dating to be about 4,300 to 4,000 yr B.P. The surface of the extensive coastal plains of Tateyama City and the Maruyama River is formed by this terrace.

Numa III, which emerged around 2,800 yr B.P., runs parallel to the coast averaging about 500 m in width. Its surface lies from 12.5 to 5.5 m above sea level.

The lowest terrace, Numa IV, emerged during the uplift associated with the Genroku earthquake (1703). Its surface is highest in the southeast, where it lies 5 m above sea level and is lowest in the northwest, where it is found at 3 m above sea level. Sugimura and Naruse (1954) noted that the inclination apparent on the Numa IV terrace is paralleled by similar inclination trend of the older terraces, increasing in steepness with increasing age. It has been suggested that earthquake-related uplifts similar to that in 1703 are to a large extent responsible for the formation of the terraces (Nakata et al., 1980; Yonekura, 1975).

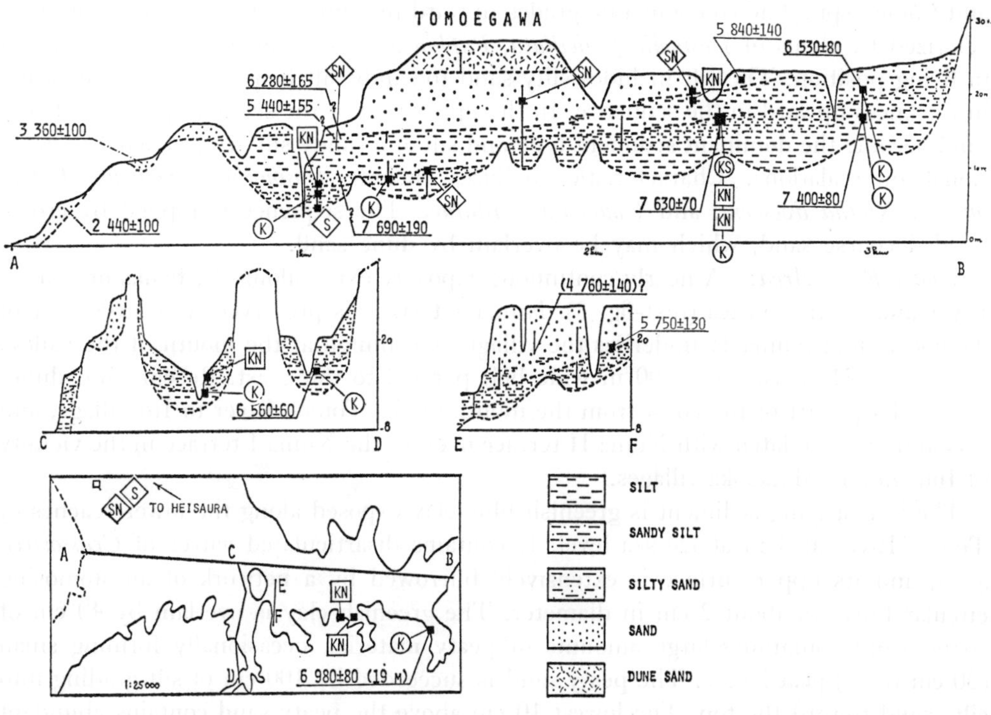

Text-fig. 3
Distribution of Jomon transgression sediments in the Tomoe River Area.
Location of samples and biofacies to which they belong are also indicated.

Holocene marine sediments

Approximate reconstruction of southern Boso Peninsula coastline at the time of higher relative sea levels reveals that at least three physiographic sedimentary environments were present. The northern part of the area was occupied by east- and west-facing, broad opened bays, similar to the present Tateyama Bay. The southeastern and western parts of the area were characterized by long, narrow drowned valley-type bays. Finally, stretches of straight coastline framed by beaches and sand dunes, similar to the present coast existed, particularly during periods prior to and after the relative sea level reached its maximum.

Five areas (text-fig. 1), where exposure of Holocene marine sediments is good and which are believed to represent the range of variation within the larger investigation area, were selected and studied in detail.

Sediments in all areas examined conform to, or are part of, the following coarsening upward sedimentary sequence. Abraded and extensively bored Neogene bedrock, often in the form of a wave-cut bench, is overlain by 10 to 50 cm of angular conglomerate derived from local bedrock. Macrofossils typical of this part of the sequence are *Barnea manilensis inornata* and *Saxidomus purpuratus* (boring in the bedrock and conglomerate) and attached forms *Spondylus* spp., *Pretostrea imbricata*, *Barbatia* spp., and *Chama* spp. The conglomerate grades upward into greyish blue massive silt characterized by shells of *Dosinella penicillata*, *Paphia undulata*, and *Ostrea denselamellosa* in living position. The silt includes silty sand to sandy silt beds, generally increasing in thickness upward in the sequence. In the upper part of the sequence, sand and silty sand become the dominant sediment type and silt and sandy silt are restricted to occasional intercalations. Characteristic molluscs include *Dosinorbis japonicus*, *Fulvia mutica*, *Niotha livescens*, and *Umbonium costatum*. The sequence is topped by cross-bedded coarse sand, which may be overlain by dune sand.

Tomoe River Area: A nearly continuous exposure exists along the banks of Tomoe River and its tributaries (text-fig. 3). Numa I terrace is preserved in a large part of Tomoe valley; Numa II underlies the village of Daijingu at the mouth of the valley; and Numa III exists as a 500 m wide belt parallel to the coast. Recent sand dunes cover a large part of the coast from the mouth of the Tomoe River to Ito village, and a sand dune correlated with Numa II terrace overlies the Numa I terrace in the vicinity of Inuishi and Matsuoka villages.

The lowest lying sediment is greenish blue clay exposed along the lower reaches of Tomoe River at 3 m above sea level. It contains disarticulated valves of *Crassostrea gigas*, and its upper surface is extensively burrowed by a network of anastomosing circular burrows about 2 cm in diameter. The greenish clay is overlain by 40 cm of coarse sand containing large amounts of peaty material, occasionally forming small (20 cm thick) peat lenses. The peaty sand is succeeded by 100 cm of silt grading into silty sand toward the top. The lowest 10 cm above the peaty sand contains abundant articulated *Ostrea denselamellosa*. The upper part of the bed contains *Batillaria zonalis*, *B. multiformis*, and articulated valves of *Paphia undulata*. At 4.4 m above sea level the grey sandy silt is overlain by a 15 to 20 cm thick layer of brown silty clay extending for at least 400 m laterally. The lower contact is abrupt and sharp, while the upper surface contains burrows, similar to those in the greenish clay, which penetrate to the depth of about 10 cm and are filled with medium grey sand of the overlying bed. Microscopic examination revealed a high percentage (about 70% by volume) of pyrite in the sand size residue. The brown clay layer is overlain by up to 15 m of poorly exposed silty sand and sandy silt alternation, containing *Dosinorbis japonicus*, *Fulvia mutica*, *Paphia undulata*, *Macoma incongrua*, and *Babylonia japonica*.

In the central part of the valley, in the vicinity of Nagaoka village, the lowest sediment is 2+ m thick, firm, massive blue-grey silt with *Dosinella penicillata* in living position. Five to 15 cm thick layers of subangular cobbles and molluscan shells (*Chama reflexa*, *C. lazarus*, *Serpulorbis imbricatus*, *Pretostrea imbricata*, *Spondylus butleri*) showing signs of transport occur at 30 to 60 cm intervals throughout the silt bed. A dip of the layers ranges from 0 to 15° away from the valley sides, and the layers can be occasionally traced to lenses of similar composition, up to 1 m thick. Fossil content and the character of the sediment below and above these layers does not show any change.

The silt gradually changes to sandy silt and then at about 13 m above sea level abruptly to sand with occasional interbeds of silty sand. The sand reaches thickness of 5 m and consists of alternating layers of fine black and coarse light brown sand with horizontal parallel laminations. Fossils occurring in the sand are *Ochetoclava kochi*, *Niotha livescens*, *Dosinella penicillata*, and *Dosinorbis japonicus*. The parallel laminated sand is over-

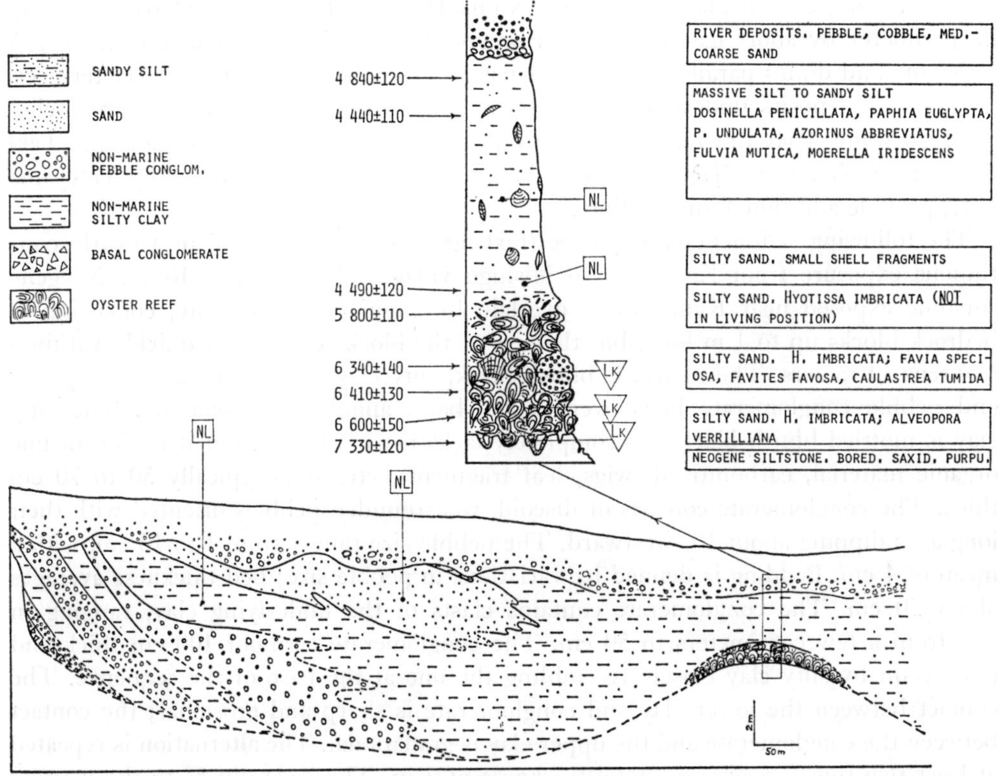

Text-fig. 4

Distribution of Jomon transgression sediments along the banks of Heguri near Nishigo, Tatayama Area. Location of samples and biofacies to which they belong are also indicated.

lain by well-sorted, grey coarse sand with low angle cross-laminations. The upper part of this sand is characterized by abundant fossils of *Umbonium costatum*. The *Umbonium*-rich bed is succeeded by over 1 m of poorly sorted, strongly bioturbated, medium brownish yellow sand, which is overlain by parallel laminated medium sand with occasional shallow straight vertical burrows. The parallel laminated sand is overlain by cross bedded coarse sand reaching up to 1 m in thickness. The cross bedded sand is covered by massive reddish-brown dune sand reaching up to 3 m in thickness.

Bivalve boreholes in the bedrock are found up to 25 m above sea level. A similar sequence is found in the upper part of the valley and tributary valleys, except that the lower silt unit persists up to a higher elevation and the upper sand unit is thinner or absent.

C^{14} dates indicate that most of the sediments forming the Numa I terrace were deposited between 7,600 yr B.P. and 5,800 yr B.P. The date of 7,690 yr B.P., obtained at 10 to 11 m above sea level, indicates that the lowest clay and peaty sand exposed nearby at 3 m above sea level are substantially older.

Tateyama City Area: The northeastern part of the investigation area is occupied by a low lying coastal plain formed by Numa II and III terraces and fringed along its perimeter by short narrow valleys commonly preserving the Numa I terrace. A series of sand dunes parallel to the coast partly covers the Numa II and III terraces.

Holocene deposits underlying the plain are best exposed along the lower reaches of Heguri and Taki Rivers, and several outcrops are found along the banks of Taki River up to Kokonoe village. Drill hole data, mostly in the area occupied by Tateyama City, provide additional information.

The following sedimentary sequence (text-fig. 4) can be observed in a nearly continuous exposure from Kamehara to Nishigo villages along Heguri River. Neogene siltstone exposed near Kamehara is overlain by angular conglomerate, consisting of bedrock blocks up to 1 m long, but the size of the blocks diminishes quickly within a few tens of meters. The matrix is bluish black, silty clay. An alternation of silty clay and pebble conglomerate beds overlies the basal angular conglomerate. The silty clay is mottled bluish black and completely lacks marine fossils, but it is rich in fine organic material, carbonitized twigs, leaf fragments, etc. It is typically 50 to 70 cm thick. The conglomerate consists of discoid, well-rounded pebbles oriented with their long axes dipping about 10° westward. The pebble size ranges from 1 to 10 cm, with a mean of 4 cm. Bedding is defined by variations in pebble size; bed thickness averages about 20 cm. The conglomerate contains clasts of the underlying clay, ranging in size from a few centimeters to 50 cm. The conglomerate is about 100 cm thick and is overlain by silty clay closely resembling the one at the base of the sequence. The contact between the lower clay and conglomerate is sharp and erosional; the contact between the conglomerate and the upper clay is gradational. The alternation is repeated at least two times, reaching a total thickness of 3 m. The beds dip 5° to the west.

The uppermost clay bed is overlain by a 20 cm thick bed of angular pebble conglomerate, which is overlain by blue-grey marine sediments ranging from silt to fine sand. The silty fine sand reaches a thickness of 3 m and contains abundant fossils of marine molluscs, many of which are in living position. The fossil assemblage is dominated by *Dosinella penicillata*, *Paphia undulata*, *Ostrea denselamellosa*, *Saxidomus purpuratus*, and *Macoma sectior*. The sediments appear massive and are probably intensely bioturbated.

The silty sand is abruptly overlain by poorly sorted medium to coarse sand with a silty matrix. The sand, consisting of well-rounded volcanic rock fragments and subangular shell fragments, contains silty clay interclasts and discontinuous clayey seams. It reaches 2 m in thickness and contains the following fossils: *Fulvia mutica*, *Mactra chinensis*, *Azorinus abbreviatus*, *Petricolirus aequistriatus*, *Pinna bicolor*, and *Novathaca schencki*. The coarse sand abruptly grades laterally into blue-grey silty sand near Nishigo village and is overlain by it. Sediment characteristics and molluscan fossils in this silty sand closely resemble the characteristics of the silty sand underlying the coarse sand unit.

Fine silty sand containing characteristic *Dosinella penicillata* shells in living position crops out along Taki River and also Heguri River downstream of Nishigo Bridge, where it overlies an oyster reef. Near Kokonoe village, in the eastern part of the plain, medium silty sand to sand is present, suggesting an eastward coarsening of the sediment. Near Kokonoe, where the contact between Holocene sediments and bedrock is exposed,

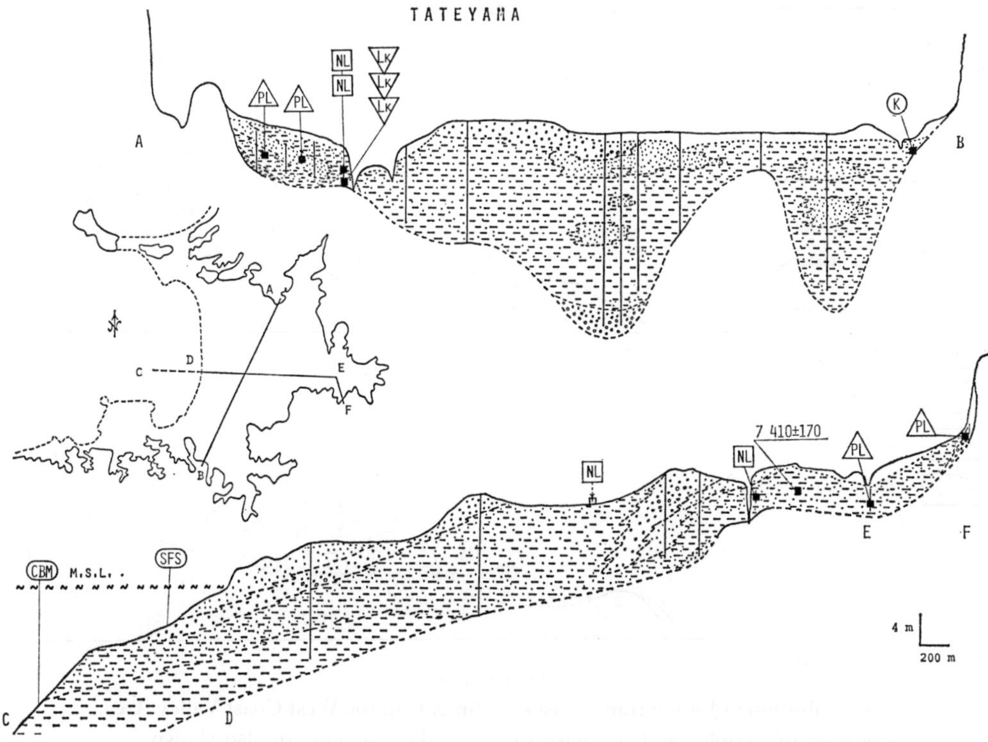

Text-fig. 5

Distribution of Jomon transgression sediments in the Tateyama Area. Location of samples and biofacies to which they belong are also shown.

the silty sand includes numerous blocks of bored bedrock and rock dwelling shells, such as *Chama reflexa* and *Pretostrea imbricata.*

Boring data are available only in the form of simplified well logs recording coarseness of sediment and the presence or absence of fossils. They indicate that the thickness of unconsolidated sediments in the area exceeds 25 m in some places but averages about 15 m.

The sediments penetrated by the borings consist of fossiliferous sandy silt and silt interrupted by and partly overlain by medium to coarse sand (text-fig. 5). The fossiliferous sandy silt and silt are probably similar to the *Dosinella penicillata* silt and silty sand exposed along the Heguri River.

The sand is partly dune sand occurring on the surface and partly marine sand occurring in the form of lenses and westward (seaward) dipping tongues in the marine sandy silt. Conglomerate and coarse sand found at the base of the two deepest boreholes are tentatively interpreted as fluvial deposits similar to those found at the base of the sequence along the Heguri River. Their nonmarine origin is indicated by the absence of fossils, presence of organic matter in the immediately overlying silt, and the inferred proximity of bedrock.

Oyster reef: A one meter high and at least 200 m long oyster reef is exposed along

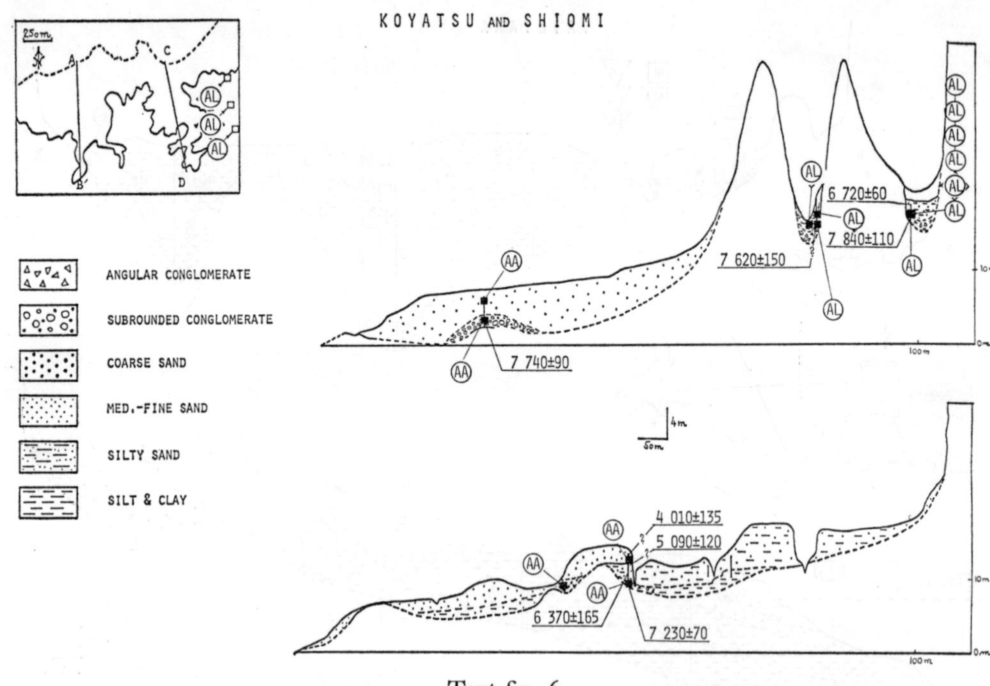

Text-fig. 6

Distribution of Jomon transgression sediments in the West Coast Valley Area. Location of samples and biofacies to which they belong are also shown.

both banks of Heguri River about 200 m upstream from the confluence of Heguri and Taki Rivers. The reef rests on a flattened bedrock whose top is presently 2.5 m above sea level and which slopes gently to the west (seaward). The bedrock is a soft, bluish grey siltstone. Its surface is extensively bored. Three types of boreholes can be found. The first type is two, three, and more branched circular holes (diameter 15 to 20 mm), which penetrate vertically to a depth of 30 cm, where they become horizontal and connect with other vertical holes. The bedrock is bored intensively down to 6 cm. No fossils were found in the borings, but based on the character of the boring, they are assumed to be of crustacean origin. The second is circular borings of the same diameter; but borings that are straight and seldom deeper than 4 cm are also common. In some cases the boring shell *Barnea manilensis inornata* is still found in the holes. The third is hemispherical borings with wide circular to elliptical openings, diameters ranging from 2 to 10 cm, depths from 10 to 15 cm. They commonly contain both valves of the bivalve *Saxidomus purpuratus*. Sediment trapped between the valves of these shells is silty medium coarse sand.

The oyster reef can be divided into three parts. In the lowest part (20 to 40 cm) oysters are predominantly in vertical living position. *Pretostrea imbricata* is by far the most common; other molluscs present include *Barbatia lima*, *Acra boucardi*, *Chlamys nobilis*, and *Serpulorbis imbricatus*. The coral *Alveopora verilliana* is commonly attached to the shells but other corals are absent. In the central part of the reef, *Alveopora* disappears, but at least 14 species of coral are present and coral becomes volumetrically as abundant as oyster shells. Several large—1 m in diameter—colonies of *Favia speciosa* and *Favites favosa* are present in living position. Fragments of *Dosinella penicillata* shells occur in the sediment surrounding the corals. The highest part of the reef is characterized by an absence of coral except for a few transported fragments. Oyster shells are mostly horizontal—that is, not in living position. Occasional shells of *Dosinella penicillata* with both valves present also occur in this part. The matrix in the lower parts of the reef is formed by sandy silt and it changes upwards into silty sand.

Directly above the oyster reef is a 15 cm thick layer, rich in shell fragments, which are mostly *Dosinella penicillata*. Shell fragments become fewer upwards, and the sediment changes into blue-grey silty sand with *Dosinella penicillata* in living position.

West Coast Valley Area: The narrow, relatively short valleys incised into the Neogene formations along the west coast and the eastern rim of the coastal plain are filled with marine sediments and often preserve the Numa I terrace. Natural exposures are typically small and few, and the following discussion is largely limited to two valleys: Koyatsu and Shiomi, where exposures are best and to an extent complementary (text-fig. 6).

At the lower part of Koyatsu valley a 1 m thick boulder conglomerate bed 1 m above sea level and is overlain by 3 m of medium silty sand. The transition between the conglomerate and sand is gradual. Both beds contain marine molluscan shells; in the conglomerate *Chama reflexa*, *Serpulorbis imbricatus*, *Barbatia bicolorata* and boring *Barnea manilensis inornata* are common. The sand bed contains *Macoma incongrua*, *Fulvia mutica*, and *Notochione jedoensis*.

In the upper part of the Koyatsu valley, two outcrops reveal a more than 3 m thick coral-rich bed unconformably overlying bored Neogene siltstone. Both outcrops are located adjacent to the side of the valley. The coral bed can be divided into two parts. The lower part contains numerous colonies of *Favia speciosa*, *Favites favosa*, and *Caulastrea tumida*, many of which are in living position. The upper part of the bed consists of blue-grey silty sand, in most places weathered to light brown, containing numerous fragments of platy corals, among which *Echinophyllia aspera* is predominant.

Bivalve boreholes found in the Neogene siltstone exposed nearby are abundant up to 25 m above sea level, and isolated borings are found up to 27.5 m above sea level.

In the neighboring Shiomi valley, outcrops occur in the central parts of the valley. The thickest bed (from 7.5 to 10 m above sea level) consists of massive silty sand with *Panopea japonica*, *Dosinella penicillata*, *Clementia papyracea*, and *Paphia undulata* in living position. Fragments of transported coral (*Lobophylia japonica*, *Echinophyllia aspera*) are present but rare. In the upper parts of the valley the silty sand bed rests on bluish grey silt with articulated *Crassostrea gigas*; in the parts of the valley closer to the sea, it is underlain by angular siltstone conglomerate containing the shells of *Chama reflexa*, *Pretostrea imbricata*, *Ostrea denselamellosa*, *Niothe jedoensis*, *Spondylus* sp. and *Serpulorbis imbricatus*. Some *Oulastrea crispata* is present, but other corals are absent. At 10 m above sea level, the silty sand bed is overlain by sand of the Numa III terrace; the sand bed and terrace are separated by a 20 cm bed of coarse sand and rounded pebbles.

C^{14} dates obtained from samples from the basal horizon range from 7,840 to 7,230

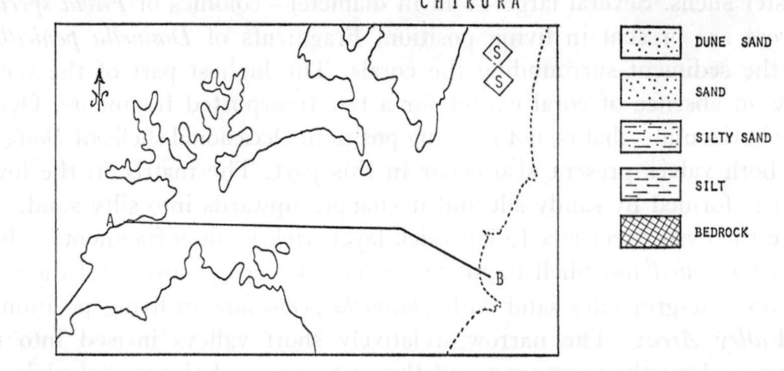

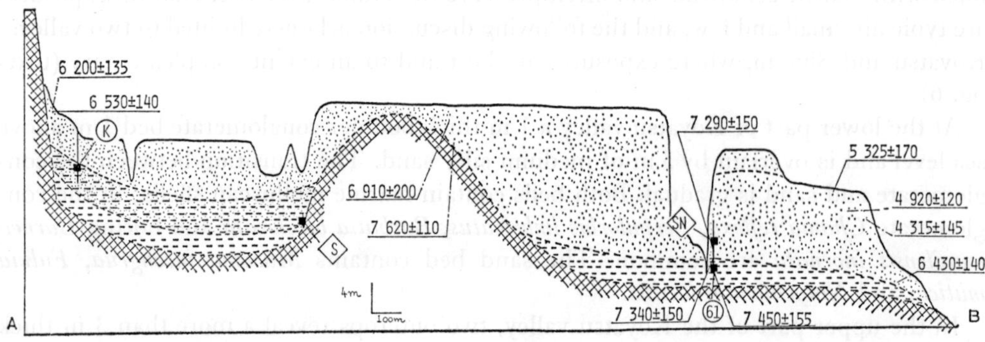

Text-fig. 7

Distribution of Jomon transgression sediments in the Chikura Area. Location of samples and biofacies to which they belong are also shown.

yr B.P. (TK-337, N-3088) and those from the upper horizon range from 6,720 to 4,010 yr B.P. (TK-329, TH-218, TH-272, TH-219) as listed in tables 3 and 4.

A small outcrop of coral-bearing sediments resembling the upper part of the Koyatsu coral bed is found along a stream north of Kamisanakura village. It contains mainly platy corals embedded in silty sand matrix. Molluscan shells from the top of this bed were dated at 5,500 yr B.P.

Several very small outcrops consisting mainly of transported coral fragments in silty sand matrix are found in the vicinity of Numa village. The outcrop of the type locality of Numa coral bed is presently covered by a pond and is mostly inaccessible.

Chikura Area: On the west coast, the Numa I terrace is particularly well preserved in the valley of Seto River near the town of Chikura. The river cuts through Numa I terrace in the upstream parts of the valley and then through Numa II, III, and IV terraces developed in a nearly 1 km wide coastal zone. Numa II terrace is partly obscured by a wide sand dune, which brings it to level with the Numa I terrace surface.

Dark greyish blue massive silt crops out at 4.9 m above sea level in the downstream part of the valley at Baba (text-fig. 7). Shells of *Scapharca inequivalvis*, *Pinctada margaritifera*, and *Dosinella pencillata* are preserved in living position. The silt reaches a thickness of at least 120 cm and is overlain at 6.1 m above sea level by coarse-to-medium, dark-to-light-grey cross-laminated sand. The 260 cm thick, cross-laminated sand bed is overlain by a 100 cm thick alternation of yellowish brown sand and silty sand. The silty sand beds are about 20 cm thick and have abrupt lower and uneven upper contact surfaces caused by extensive burrows filled with sand of overlying beds. The sandy beds are devoid of fossils, but the silty sand beds contain shells of *Solen strictus*.

At 9.7 m above sea level, the silty sand and sand alternation is succeeded by over 170 cm of coarse, well-sorted sand with inclined beds of rounded pebble conglomerate about 15 cm thick.

Several small outcrops in the upper and central parts of the valley reveal burrowed silt to sandy silt with poorly preserved molluscan shells resting on silty clay with abundant organic material. The fossiliferous silt is covered by nonfossiliferous medium to coarse sand, reaching up to 4 m in thickness.

C^{14} dates of samples from this area range from 7,450 to 5,570 yr B.P. (see TH-273, TH-274, TH-275, N-2089, TH-106, TH-277, TH-282, GaK-3067 in table 4).

Maruyama Area: Maruyama and Nuruishi Rivers in the northwestern part of the study area flow through a low coastal plain formed predominantly by the Numa II terrace.

Massive silty clay, similar to the lowest silty clay at Baba, Chikura Area, crops out from 1 to 4 m above sea level along the downstream part of the Maruyama River. Sparse fossils consisting mainly of *Dosinella penicillata* are found. Along the Nuruishi River, bluish grey massive silt containing carbonized wood fragments crops out at 7 m above sea level. Shells of *Cyclina sinensis* in living position are common. At 7.75 m above sea level the silt is overlain by a 14 cm thick bed consisting of alternating thin layers of peat and small, rounded gravel and clay. It is succeeded by an 80 cm thick bed of sandy silt coarsening upwards to silty sand. Clay draped ripples are present in the central part of the bed. Further upstream, mottled greyish blue silt with leached out shells is overlain at 14.9 m above sea level by fine sand with a subparallel, discontinuous, 1 to 2 cm thick laminae of clay. The fine sand bed is 60 cm thick and devoid of fossils.

At a distance of 2 km from the shoreline, both rivers cut through a more than 13 m thick alternation of nonmarine clay rich in organic matter and rounded pebble conglomerate beds.

Recent environment

The southern part of the Boso Peninsula is characterized by straight to gently curved coastline consisting of beaches and rocky shores. The east coast of Tateyama Bay is formed by a medium to coarse sandy beach bordered on the north by 50 m high cliffs of the Daibutsu Cape fringed by a narrow boulder beach. The southern border of the bay is formed by the island of Okinoshima connected by a low sandy bar to the Japan Self-Defense Forces base, constructed by landfill around a small island of Takenoshima. From Okonoshima to Sunozaki, the coast is formed by sandy beaches occasionally interrupted by rocky promontories. Neogene siltstones crop out along the shore at Sunozaki Cape, forming a more diverse coastline with shallow, short bays and scattered small rocky islands. From the village of Ito to Ainohama, the coast is formed by a long sandy beach. The southeast coast of the peninsula from Ainohama to Chikura is char-

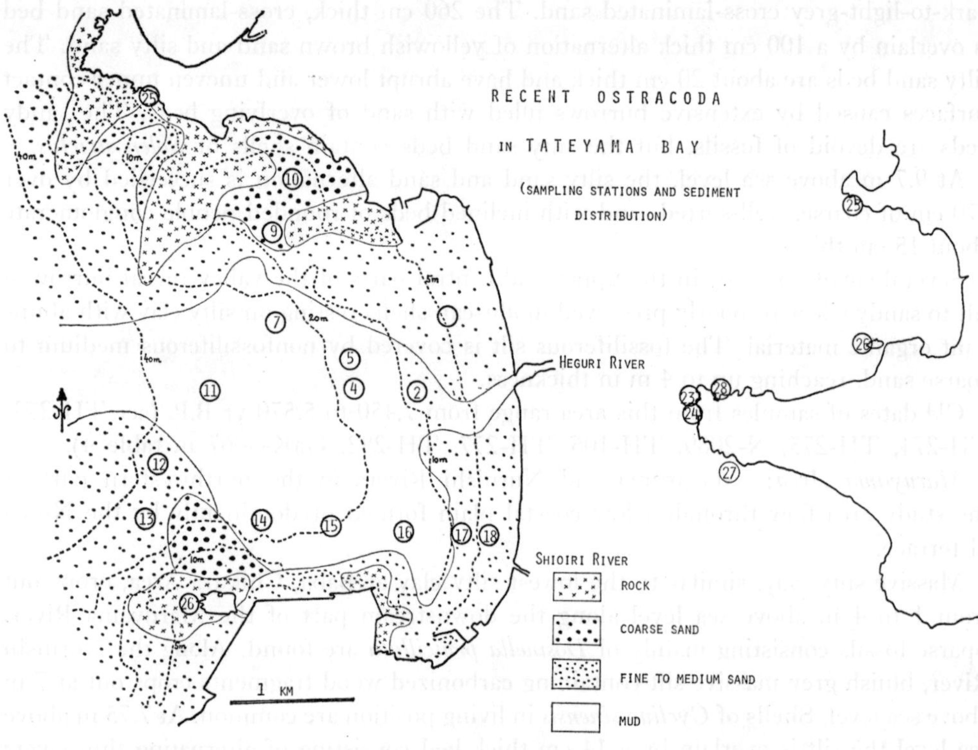

Text-fig. 8
Location of sampling stations, bathymetry and sediment distribution in Tateyama Bay.

acterized by a rocky shore bordered by a zone of low rocky reefs. The northeastern coast of the area is formed predominantly by sandy beaches.

The sea floor around the peninsula slopes down to about 20 to 30 m, forming a shallow platform 1 to 1.5 km wide. A second, deeper platform surrounds the first at a depth of 100 m. It is about 5 to 10 km wide but is most extensive near the Sunozaki Cape, from which it extends west for about 20 km. The platform is cut by several submarine valleys believed to have been dissected by rivers during a lower stand of sea level. At the edge of the platform the sea floor slopes steeply down to depths of 600 to 1,000 m.

Sediment distribution in Tateyama Bay is shown in text-fig. 8. The central part of Tateyama Bay is occupied by silt and sandy silt. The bottom along the northern fringe of the bay, to a depth of 10 to 15 m, is formed by rocky reefs surrounded by coarse sand. Along the eastern and southern shore of the bay, fine to medium sand extends from the beach to depths of about 10 to 15 m, where it merges with the central bay silt. An area of coarse sand lies northwest of Okinoshima at a depth of about 30 m.

The sea bottom from Okinoshima, around Sunozaki Cape to Ainohama town, consists of bedrock outcrops surrounded by fine to coarse sand. Information available for the east coast of the area, from Ainohama to Maruyama , is very scarce but suggests that the bottom sediments are predominantly sand and rock.

Text-fig. 9

Distribution of living and fossil hermatypic corals in southern Boso Peninsula. Stars indicate sites visited by the author; solid circles show localities given in literature but not found by the author. Hollow star gives the locality of live corals in Tateyama Bay.

Large rivers are concentrated in the northern part of the area. The Heguri and Shioiri Rivers empty into Tateyama Bay. The Nuruishi, Maruyama, and Seto Rivers flow into the sea along the northeastern coast. The Tomoe River is the only large river on the south coast.

Sea water temperature off the southern Boso Peninsula is influenced by the warm Kuroshio current flowing from the southwest. The main branch of the current flows past the southwest coast at a speed of 1 to 3 knots. A smaller branch, moving at about 0.5 knots, enters the Tateyama Bay from the north, and flows clockwise to join the main branch near Sunozaki Cape. The strength and direction of the flow of the Kuroshio current are variable and occasionally the cooler water masses of the northern Oyashio current extend to the southern part of the peninsula, thus adversely affecting the organisms living on the east and south coast of the area.

In Tateyama Bay, February and March are the coldest months (13°C), and August and September are the warmest (25°C) (Hamada, 1963). The predominant wind direction in the area is southwest. The maximum tidal range is 175 cm.

Corals

Corals in southern Boso Peninsula represent the northernmost occurrence of Holocene hermatypic corals in Japan. Jomon transgression corals occur partly in situ and partly as transported fragments. Living corals are found in the Tateyama Bay.

Fossil corals are found in marine sediments on the north and southwest coast of the southern Boso Peninsula. In situ hermatypic corals form an important constituent of the oyster reef in Tateyama area, which was described in a previous section. With the exception of the above locality, all in situ coral occurrences so far reported are located in the West Coast Valley area. At least 12 coral localities have been given in literature over the past 40 years, but because of temporary nature of many outcrops (construction sites, excavations, etc.) only some of them could be examined by the present author. Coral localities are shown on the map in text-fig. 9. It is apparent from the map, with the probable Jomon coastline drawn in, that most outcrops lie in the inner parts of small coastal bays. The distribution appears erratic, because corals are found in one valley but not in the neighboring one, which is apparently identical.

The type locality of the Numa coral bed is in the valley of Numa. Most of the outcrop is presently covered by a pond, but an exposure about 30 cm high and 10 m long at 18 m above sea level exists along the eastern edge of the pond. *Stylocoeniella hanzawai* and *Favia speciosa*, which appear to be in living position, are embedded in silty sand matrix. About ten large colonies, which were unearthed during the construction of the pond, are preserved in a local municipal monument. Included are *Caulastrea tumida*, *Podabacia elegans lobata*, and *Montastrea curta*; some of the colonies reach a diameter of 70 cm.

An outcrop about 1 m high and 30 m long is found at 17 m above sea level along the banks of a stream flowing into the Shioiri River at Kamisanakura. The outcrop lies about 500 m upstream from the confluence where the stream approaches the right side of the valley. Corals and shells are embedded in silty sand matrix. *Lima sowerbyi*, *Barbatia lima*, and *Pretostrea imbricata* are the most abundant molluscs. The coral assemblage is dominated by *Echinophyllia aspera*, most of which occurs as 5 to 10 cm fragments, even though a few nearly entire colonies which appear to be in living position are found. About 20 cm fragments of *Goniopora* sp. are not in living position, but are found in clusters, suggesting that they are nearly in situ, having probably broken off from the mother colony. Large massive colonies of *Favia speciosa* and *Favites favosa* are absent.

Two outcrops found in the Koyatsu valley are the largest found to date. The outcrops have been described in detail by Hamada (1963). The smaller of the outcrops, located in the upper part, left side of the valley, lies between 14 and 16 m above sea level, where Neogene siltstone is exposed in several places. The bedrock is overlain by angular conglomerate containing shells of *Crassostrea gigas* and *Serpulorbis imbricatus*. Occasional small *Oulastrea crispata* colonies are found attachgd to the rocks. Large colonies of *Favia speciosa*, *Favites favosa*, and radial fingerlike growths of *Caulastrea tumida*, all in living position, are attached directly to the bedrock or rest on the conglomerate and coral rubble. The corals are surrounded by greyish blue silty calcareous sand. The in situ corals are overlain by a 50 to 100 cm thick bed of silty sand containing abundant shells (especially *Barbatia bicolorata*, *Pretostrea imbricata*, and *Lima sowerbyi*) and fragments of predominantly platy corals: *Echinophyllia aspera*, *Physophyllia ayleni*, *Pectinia lactuca*, *Lobophyllia japonica*, and *Lobophyllia robusta*. Massive corals are uncommon, few *Favia* colonies (not in living position) occur in the upper part of the bed. *Caulastrea* is absent except for a few worn fragments. Most of the platy corals are present as 5 to 10 cm fragments, but several well-preserved colonies, reaching the diameter of up to 40 cm, and one 80 cm colony in living position, were found. All the corals, including the fragments, are extremely well preserved and do not show any signs of transportation.

The larger outcrop lies between 13 and 18 m above sea level and consists mainly of platy corals embedded in silty sand matrix, resembling the uppermost bed of the smaller outcrop.

The area of distribution of transported coral fragments partly overlaps but is wider than that of the in situ corals. Sand containing coral fragments and isolated coral rubble is found in fields in the valleys neighboring the Numa outcrop, suggesting that extensive coral-bearing sediments are present in the area. Worn fragments of *Echinophyllia aspera*, *Lobophyllia japonica*, and a large, well-preserved colony of *Montastrea curta* were found in silty sand in the center of Shiomi valley, about 500 m west of Koyatsu. Slightly worn, transported fragments of *Caulastrea tumida* and *Favia speciosa* were found in sediments in the Tomoe River valley. The sediments are silty sand located in the central part of the valley at 16 m above sea level, containing numerous *Ostra nipponica*, apparently in situ. Several small *Oulastrea crispata* colonies attached to pebbles, but apparently slightly transported, were also found.

Hermatypic corals presently living in Tateyama Bay have been found to the northwest of the island of Okinoshima in depths exceeding 5 m. The corals are attached to bedrock occurring as low (1 m), long ridges formed by steeply dipping resistant beds of Neogene volcanic sand and extending from the shore to depth. The present day corals differ from the Jomon transgression corals not only in their ecology (virtually open coast versus inner bay) but also in their growth form. Although coral colonies exceeding 4 m in diameter have been reported from Tateyama Bay (Yabe and Sugiyama, 1935), the colonies the present author was able to examine were smaller than 1 m and on large parts of the surface coral tissue was absent, and living tissue occurred in patches. Smaller specimens, which can be examined in detail, revealed a repeated overlapping of coral and calcareous algae, occurring in 5 to 10 mm thick layers. Apparently, under unfavorable conditions, portions of the coral die and are overgrown by calcareous algae. When conditions improve again, the faster growing coral expands from surviving patches and covers the algae. No similar overgrowth has been observed in the Jomon transgression specimens, suggesting much more favorable conditions.

Hamada (1963) has already pointed out the similarity between the special distribution of Numa corals and distribution of recent corals in Tanabe Bay, Kii Peninsula. A similar pattern where the most luxuriant coral growth is concentrated in patches along the perimeter of bays is found in several other localities—for instance, in Kamae Bay in northeast Kyushu (Utinomi, 1971).

It is probable that the availability of hard substrate, in the form of bedrock outcrop, is the main reason for this distribution, although poor circulation and low transparency

Table 1

Fossil ostracods taken from 60 samples collected throughout the investigation area.

Table 1–1

Biofacies	A	A	A	A	A	A	A	A	A	A	A	A	Lk	Lk	N	N	N
Sample number	1	3	5	7	8	9	11	15	23	24	26	36	40	41	43	44	51
1 Neonesidea oligodentata (KAJIYAMA)	9	6			1	1	1	1	1		1	2	2	1			1
2 Darwinula sp.																	
3 Argilloecia lunata n. sp.	6		2	10	5	4		1			4	3	10	15			
4 Propontocypris(Propontocypris) sp.																1	
5 Aglaiocypris nipponica OKUBO				1			6								1		
6 Paracypris sp.					1							1	2	1			
7 Pontocythere japonica (HANAI)																	
8 Pontocythere minuta IKEYA & HANAI																	
9 Pontocythere sekiguchii IKEYA & HANAI																	
10 Pontocythere subjaponica (HANAI)		1							4				2	10	11	17	30
11 Parakrithella pseudadonta (HANAI)	7	6	1	21	14	1	27	6	5	3		34	14	8	32	11	9
12 Neocytherideis reticulata IKEYA & HANAI																	
13 Munseyella japonica (HANAI)														13	2		
14 Callistocythere alata HANAI	1												1				
15 Callistocythere hayamensis HANAI	2		2	3	2	3		9	10	1	2	11	2	2	1	1	2
16 Callistocythere nipponica HANAI																	
17 Callistocythere numaensis n. sp.	8	5	2	18	1	6	7	1		4	2	1					
18 Callistocythere reticulata HANAI														1			
19 Callistocythere rugosa HANAI									2	2							
20 Callistocythere setouchiensis OKUBO									2			1					
21 Callistocythere tateyamaensis n. sp.	1			1				1			1			2			6
22 Callistocythere undata HANAI	3			1	1		15	2	8	5				2			1
23 Callistocythere undulatifacialis HANAI																	
24 Callistocythere japonica HANAI																	
25 Callistocythere subjaponica HANAI																	
26 Callistocythere minor HANAI																	
27 Tanella pacifica HANAI																	
28 Cythere lutea omotenipponica HANAI	1			1					11	10		1		1	14	9	9
29 Schizocythere kishinouyei (KAJIYAMA)																1	1
30 Hanaiborchella triangularis (HANAI)	1													2			3
31 Neomonoceratina sp.																	
32 Spinileberis quadriaculeata (BRADY)	11	15	5	38	12	6	5	2	25	20	2	1	16	3	7	23	6
33 Hemicythere sp.																	
34 Aurila hataii ISHIZAKI									1								
35 Aurila kiritsubo YAJIMA	4							1	1	1			1				
36 Aurila munechikai ISHIZAKI								1						4		1	
37 Aurila uranouchiensis ISHIZAKI		1				1		1					1		4		1
38 Aurila sp. A	41	42	24	102	51	43	21	53	13	36	32	25	1		5		
39 Aurila sp. B	7		8	19	17	8	10	11	30	22	6	8	14	1	2	3	6
40 Mutilus assimilis (KAJIYAMA)		3			2	8	20				3		7	3			2
41 Pseudoaurila japonica (ISHIZAKI)	2	13	5	20	18	6		10	2	4	4	19		1			1
42 Cornucoquimba tosaensis (ISHIZAKI)	4			2						2	4	1		1	1		2
43 Coquimba ishizakii YAJIMA															3	1	
44 Proteoconcha tomokoae (ISHIZAKI)																	
45 Trachyleberis scabrocuneata (BRADY)				1		1			6			1	1	2	2		1
46 Trachyleberis straba n. sp.																	
47 Trachyleberis sp.																	
48 Acanthocythereis? niitsumai (ISHIZAKI)																	3
49 Cletocythereis bradyi HOLDEN									2					1			1
50 Ambocythere japonica ISHIZAKI	9	10	6	22	21	7	5	8	1	6	5	6	6	15	2		
51 Echinocythereis? bradyformis ISHIZAKI	5			1	1								1	1			2
52 Echinocythereis? bradyi ISHIZAKI													2	1		8	2
53 Lixouria nipponica YAJIMA																	
54 Ruggieria (Keijella) bisanensis (OKUBO)				1			2					1	1				
55 Basslerites obai ISHIZAKI																	
56 Bythoceratina elongata IKEYA & HANAI																	
57 Bythoceratina hanaii ISHIZAKI																	1
58 Bythoceratina sp.																	
59 Pseudocythere frydli YAJIMA														2			
60 Hemicytherura cuneata HANAI													3	3	4		7
61 Hemicytherura kajiyamai HANAI									1					3		3	
62 Hemicytherura tricarinata HANAI		1							6	1		1					
63 Semicytherura henryhowei HANAI & IKEYA	1			4	3			4	1	4			1			1	1
64 Semicytherura? miurensis (HANAI)			1											4		4	1
65 Semicytherura mukaishimensis OKUBO									1					1	1		
66 Semicytherura polygonoreticulata ISHIZAKI & KATO	1							2	1								
67 Semicytherura skippa (HANAI)																	
68 Semicytherura tetragona (HANAI)																	2
69 Cytheropteron miurense HANAI														3	2		
70 Cytheropteron uchioi HANAI																	
71 Kangarina hayamii YAJIMA																	
72 Kobayashiina hyalinosa HANAI								1						1			
73 Paracytheridea bosoensis YAJIMA														4	1	2	2
74 Paracytheridea neolongicaudata ISHIZAKI					1												
75 Loxoconcha japonica ISHIZAKI	64	35	44	49	49	26	11	39	64	10	12	21		5			3
76 Loxoconcha kattoi ISHIZAKI										1			29	35	7		4
77 Loxoconcha laeta ISHIZAKI					2						1	3					
78 Loxoconcha optima ISHIZAKI																1	
79 Loxoconcha uranouchiensis ISHIZAKI	9	4	13	23	10	4	10	8	26	6	7	10	8	14	13	7	42
80 Loxoconcha viva ISHIZAKI													6	8	16	26	12

Table 1-2

	NN	PL	A	A	A	KS	SN		PL	N	Lk	KS	KS	A	A	A	KS	KS	KS	N	KS	N	N	N	SN	SN	KS	KS	KS	KS
species	52	53	55	56	57	59	60	61	63	67	71	76	83	84	87	88	96	99	103	104	106	107	110	111	112	114	118	120	122	123
1	3	1	6	12	12		9					1	1		8	7		1	2	8	9	4	8	7	11	1				
2																		1										1		
3			1					1			6				1					1	2									
4			1	3															1						1					
5				1									1																	
6				1							6	2									4			1					1	
7							8					1								3			4	1	7					
8							3						1							1				1		1				
9								15				1																		
10	23	57	9	6			13		88	28	5		2		1					2				1	6	1				
11	4		6	10	15		6			17	14	1			8	2			3	2	20	43	8	3				10	3	5
12								2																		1				
13							3	1		1	3	3										1		1		4				
14		2	1	11	4																									
15					6										3	5							1							
16																										1				
17	2																					4	4							
18											1													2						
19		5																						1						
20							1													1	4	3	4					1		
21				2	10				2																					
22	3	6	1	3	4		2	2	2		1	1	1		4	2	1							3	6					
23																							3			3				
24		6					5													1			1	2	3	1			1	
25																														
26	1																													
27																											1			
28	8	32		3	2		31		12	20	2	7	2		1	6				3			1	5	10	1			2	
29		4	1	1			12	7	2	2	1	25								2			5	14	25	18			1	
30			2	1	3					1	1																		1	
31								1																		1				
32	50	19			3	43		47	21	9	6	15	57	47	1	2	54	13	20	6	24	15	11	4	1	19	28	24	22	14
33																														
34		1	1				18	11				11								3				7	21	9				
35	3	1		1	5		1				1																			
36	3	5	7	20	2		8			4					18	14				3				4	6					
37				1			6				6	3								1				2	3	1				
38	10		7	15	3		7	10	2	5		5		112	3	10		1		5		1	6	6	9	6		1		
39	20	7	15	25	10		7		7	7	18	23	46	5	35	49	15	7	10	4	11	8	13	7	12	6	8	13	6	43
40	2	8	4		8	2	12	4			5			2	5	3				1	11	2	6		4	4		1	1	1
41														9											1	2		1		
42					2			2												1			1			2				
43	1						5			1														1	2					
44								1																						
45	4		4	7	4		12		3	1	6		3		2	4	3	7	19	10	7	15	17	6	2			19	4	1
46																					7	3	2		1	1				
47	3	1		2	2						1				2	3				1		2	13	5				4	10	
48															1								1						5	
49																														
50	3				1		1		1	1	7			1						1										
51	1	2	4	1	4				2				1					1	1	2	16	11	14	4	2			6	8	
52	8		5		3		3	1		4	2		3				15	13	10	18	54	21	9	11	1		3	43	19	8
53		2																												
54	5	1			1	1	1	23	6			29	66				118	123	48	26	34	47	5	19	4	1	33	56	37	74
55																														
56				1																										
57			1	1	3		4								4	8				2				1	2					
58							1																		1					
59																														
60	1	2		11	1		15			1	3	7	3			3							9	5	11	7				
61			6				2	14				8								8				1	7					
62		2									1	11							13								6	2	2	1
63	1		3	6	1		1		1	3	3	5			9	10		1		1	4	4	2	1	2					2
64	1		2	6			2					1			1	1				1			1		1	1				
65	1		2				3	3	1	2														2	1					
66															2															
67		1					1					1																		
68	1																													
69		1		1	1			2		1	1	2			3								2							
70								1																						
71							1		1																					
72			1							1			1				8	1		2	1				1			1		
73	2		1	1	1		3	1	1	1	2	1													1	3				
74																				1					1					
75	9		30	64	18								21	6	32	43				10	1		12		2	11	43		2	
76			2		1		4	8			40	4				3							1	4	12	12				
77	2		2	2								1			3	1				3			1	2	2	1				
78							3																		4					
79	31	19	12	20	18		7		15	11	11	9	10	1	26	21	19	14	42	12	18	15	60	9	5	27	55	13	42	48
80	10	2							2	19	12	1	5							9	1	28	1	8				20	4	

Table 1-3

Sample number	1	3	5	7	8	9	11	15	23	24	26	36	40	41	43	44	51
81 Loxoconcha sp.																	
82 Loxocorniculum mutsuense ISHIZAKI	17	9		19	30	22	4	8	1	4	23						
83 Phlyctocythere hamanensis IKEYA & HANAI					1							1					
84 Cytheromorpha acupunctata (BRADY)			3	14	1	5	1		7	5	1			2	2		12
85 Nipponocythere bicarinata ISHIZAKI			2	3	2						2		11	5	34	45	12
86 Nipponocythere hastata n. sp.		1	1	1	1	2	2	2			2	4	1	6			
87 Xestoleberis hanai ISHIZAKI	16	24	16	15	33	2	14	20	20	2	21	40	30	45	11	6	7
88 Xestoleberis opalescenta SCHORNIKOV													8	3	10	9	8
89 Xestoleberis sagamiensis KAJIYAMA	23	16	4	8	12			20			6			2	4	4	2
90 Xestoleberis setouchiensis OKUBO	1		3	4	7		2	6			7	1					
91 Xestoleberis suetsumuhana YAJIMA	1	1		1								3	9	1			
92 Cobanocythere? japonica SCHORNIKOV																	
93 Platymicrocythere? sp.																	
94 Paradoxostoma hartmanni OKUBO	1																
95 Paradoxostoma brunneum brunneatum SCHORNIKOV																	
96 Paradoxostoma coniforme KAJIYAMA			1				1									1	
97 Paradoxostoma inabai OKUBO																	
98 Paradoxostoma faccidum SCHORNIKOV	1			12	3												
99 Paradoxostoma honssuense SCHORNIKOV							5										
100 Paradoxostoma yatsui KAJIYAMA													1				
101 Paradoxostoma sp. 1				1	1	5											
102 Paradoxostoma sp. 2																	
103 Paradoxostoma sp. 3														3			
104 Neopellucistoma inflata IKEYA & HANAI			1												3		1
105 Sclerochilus mukaishimensis OKUBO																	1
106 Sclerochilus sp.		2					6							2			
107 Cytherois asamushiensis ISHIZAKI	1		3	3	1			1				8	1	4			
108 Cytherois violacea SCHORNIKOV																	
109 Cytherois zosterae SCHORNIKOV	5		17	18	6	12	26	5		2	4	7	1	4	1	1	1
110 Cytherois? sp.	1							5				1					
111 Paracytherois tosaensis ISHIZAKI	5	1	1		2		22			1	3	4	2	3			
Total number of individuals	270	196	166	438	311	173	220	230	254	153	155	220	195	255	196	186	210
Total number of species	33	20	22	32	32	21	23	28	27	23	24	29	32	47	28	24	38
Weight of sample picked (in grams)	5	5	1.25	3.4	2.36	1	1.4	2.5	·25	10	1.25	5	2.5	2.5	6.25	7.5	3.75

species	52	53	55	56	57	59	60	61	63	67	71	76	83	84	87	88	96	99	103	104	106	107	110	111	112	114	118	120	122	123
81		2						20				15								1							2		6	
82	4	2	3	.6			1					4		1	19						2	3	1	2	4	1		3		
83																					1									
84	10	11	1		2		3		25	3	1	25	31				12	13	15	12	14	7	8	20	5		1	23	29	25
85	34	1	3	2	2				18	28	10							10	4	20	21	42	3	24	1		4	10		16
86				1							2				1											1				
87	10	6	22	55	15		21	2	8	9	31	9	4	7	24	17				9	10	2	15	15	18	3		3	1	3
88	3		1	1	2				1	7	6																			
89	1		3	18	1				1	6				9		6								1	1					
90		1	8	6					1			1			1						1	1	7	1				2		
91				1	1						9									1			1		1	3				
92											1																			
93			2																											
94			1																											
95				1																										
96			1																						1					
97																1														
98																							1							
99																														
100				5			2	3								1									1					
101																1														
102							1	1																						
103				1																										
104				2											1								3						3	
105			1					6				2			1	1	1				1					2				
106																														
107											1				1	2					7	2	1					4	1	
108			3	4											1		7													
109			1	1		3	3	1		2		13			10	8											1			1
110			1								2																			
111	2			2			1	1			5					2	5	3	4		3	4						1	1	
	280	210	188	345	171	49	253	191	223	195	233	241	260	200	232	239	271	209	192	198	288	288	266	214	223	156	185	262	212	242
	37	30	42	45	35	4	43	28	24	28	37	34	19	11	31	29	12	15	14	38	27	25	40	40	45	33	12	24	25	14
	5	1.25	1.25	1.25	2.5	80	3.75	5	1.25	1.25	10	1.25	10	10	5	3.75	10	7.5	7.5	2.5	2.5	1.25	5	5	3.74	15	1.13	3.75	6.25	10

Table 1-4

	N	N	N	SN	KS	KS	SN	KS	N	N	KS	SN	N
species	125	126	127	128	132	134	142	144	145	151	157	162	175
1	6	4	19	26			9		9	6		19	
2													
3	1		1									3	
4				3					1				
5			1										
6	2	2	2							2		2	
7	7	5	4	8						5	1		
8			6						1				
9													
10	19	4	4	9			21		16		12	4	59
11	11	13	1				3		3	3		7	5
12													
13							1				1	.2	
14				3							1		
15													
16													
17	1												
18													
19	1		1										
20												2	
21				1						1			
22	1		3	5			5		1	7		2	2
23													
24	1								2				2
25									2				
26			1	1						2			
27													
28	7	4	8	7			14		10	1	4	2	30
29	3	1	9	46	1		19		6	5	5	1	7
30													
31													
32	5	2	12		58	11		51	9	2	115		35
33												1	
34	3		9	15			6		1	6	1	1	
35	1						3						
36	6	2	6	9					3	4	2	2	2
37	1			1			3						3
38	4	2	3	4	1		7	1	4	10		9	2
39	8	7	8	4	1	2	7	9		7	3	24	11
40	5	3	5	5			1		6	6		10	4
41	2		2	4					1	1		1	1
42	3		2				1				1		2
43				4			1		8				
44													
45	4	3	11						9	19		7	
46			1							2			
47		2					8		9	1			1
48													
49				1									
50	2	1		1					1				
51	1							6		11	1	3	
52	14	6	16					53	9	7		4	12
53													
54	41	33	31		11	66		54	12	15	10	3	4
55										1		1	
56													
57	1		6	3			4		3	5		7	
58				2					1				
59									1				
60	3	4		3			25	1	2	10	2	2	7
61			12			2			3				
62					1		1						1
63	1		1				1			4		6	
64	2						17						
65		1					4		2	1		2	1
66													
67													
68													
69	1										2	3	
70													
71			1				1						
72								2	2	3		1	
73				1			1			1			
74				2						1			
75	5	5	2	3	3	102	5		1	9		5	4
76	1	1	1	5			3			1		7	
77	7	1	3	5					1	9		20	1
78	1			3			3		3	1			
79	7	3	10	1		6	10	12	24	24	3	2	26
80	31	31	11	1			3	22		8		5	8

species	125	126	127	128	132	134	142	144	145	151	157	162	175
81			1		1	1		1			9		1
82	3	1					2		3	9			2
83	1	·1											
84	2	2	19		6	26	1	29	15	6		2	29
85	20	31	11				4			16	5	8	38
86													
87	16	8	15	10			19	5	12	23		19	6
88							1				1		
89				1							2		
90	1	1	1	1			1		1	2		1	
91	1	1		1			1			1		2	
92													
93													
94													
95													
96						1			1				
97							2						
98									2				
99			1										
100			1										
101													
102													
103												1	
104		1	1										
105		2					7					1	
106	1											2	
107	1	1										1	
108			1									4	
109	1	1				3	2	3		2·		3	
110			1										
111		4				5							
	267 48 375	194 36 25	247 44 188	199 35 10	83 9 80	225 11 5	227 39 25	249 14 5	200 39 .5	261 43 25	180 20 10	214 44 75	317 29 5

Table 2 Recent ostracods from 12 samples in Tateyama Bay. Figure to the left of a solidus (/) refers to the number of live specimens, figure after the solidus, or when the solidus is absent, gives the total number of specimens, including live and dead.

Table 2–1

Biofacies	SFS	CBM	CBM	CBM	SCS	SCS	DM
Sample number	2	4	5	7	9	10	11
1 Neonesidea oligodentata (KAJIYAMA)	3	1		6		3	
2 Propontocypris sp.							
3 Aglaiocypris nipponica OKUBO	1						
4 Pontocythere japonica (HANAI)					10/25	1	
5 Pontocythere minuta Ikeya & Hanai							
6 Pontocythere subjaponica (HANAI)	11/39	1/20	7	10			3
7 Krithe japonica ISHIZAKI			3/7	2/5			28/78
8 Parakrithella pseudadonta (HANAI)	4	1	4	4	1	1	1
9 Munseyella japonica (HANAI)							
10 Callistocythere alata HANAI							
11 Callistocythere hayamensis HANAI							
12 Callistocythere rugosa HANAI							
13* Callistocythere setouchiensis OKUBO							
14 Callistocythere undata HANAI			2	1		1	1
15 Callistocythere minor HANAI							
16 Cythere lutea omotenipponica HANAI	2						
17 Schizocythere kishinouyei (KAJIYAMA)						1	
18 Hanaiborchella triangularis (HANAI)	1					1	
19 Spinileberis quadriaculeata (BRADY)	5	7/24	5/25	3			3
20 Aurila hataii ISHIZAKI		1				1	
21 Aurila kiritsubo YAJIMA							
22 Aurila munechikai ISHIZAKI	1	3		5	1	1	
23 Aurila uranouchiensis ISHIZAKI	1	2	1	5	6	1	
24 Aurila sp. A	3	3			1		
25 Aurila sp. B	3		3	2	3	2	2
26 Mutilus assimilis (KAJIYAMA)		10				4	
27 Pseudoaurila japonica (ISHIZAKI)							
28 Cornucoquimba tosaensis (ISHIZAKI)							1
29 Coquimba ishizakii YAJIMA		1		1	2/3	1	
30 Trachyleberis scabrocuneata (BRADY)		1/2	2				
31 Trachyleberis sp.							
32 Cletocythereis bradyi HOLDEN							
33 Echinocythereis? bradyformis ISHIZAKI	4	1	3				
34Echinocythereis? bradyi ISHIZAKI	1	1	1/8	3			
35 Lixouria nipponica YAJIMA							2/2
36 Ruggieria (Keijella) bisanensis (OKUBO)		2/15	3/13	9			
37 Basslerites obai ISHIZAKI		3/4	3/15	9			12/45
38 Bythoceratina hanaii ISHIZAKI	1	2	1	7		5	
39 Hemicytherura cuneata HANAI							
40 Hemicytherura kajiyamai HANAI	8	8	5	8	4	1	
41 Hemicytherura tricarinata HANAI							
42 Semicytherura henryhowei HANAI & IKEYA		2		4		2	
43 Semicytherura? miurensis (HANAI)	19/26			3	59/70	1/4	
44 Semicytherura mukaishimensis OKUBO	4	5	2	5			
45 Semicytherura sabula n. sp.						35/35	
46 Semicytherura skippa (HANAI)		2					
47 Semicytherura tetragona (HANAI)	1			45/49	4/4		
48 Cytheropteron miurense HANAI	1						
49 Cytheropteron uchioi HANAI							
50 Kobayashiina hyalinosa HANAI			1				1
51 Paracytheridea neolongicaudata ISHIZAKI	1						
52 Loxoconcha japonica ISHIZAKI	1					1	
53 Loxoconcha kattoi ISHIZAKI							
54 Loxoconcha laeta ISHIZAKI	7	4	2	13		2	2
55 Loxoconcha optima ISHIZAKI							
56 Loxoconcha tosaensis ISHIZAKI	1						
57 Loxoconcha uranouchiensis ISHIZAKI	17	1	5				2
58 Loxoconcha viva ISHIZAKI	9	4/53	6/39	1/53		5	1/21
59 Loxoconcha sp.							
60 Loxocorniculum mutsuense ISHIZAKI							
61 Cytheromorpha acupunctata (BRADY)	3	6/14	3	5			
62 Nipponocythere bicarinata (BRADY)	9	26/64	6/22	4/9			10/23
63 Xestoleberis hanaii ISHIZAKI	11	12	6	17		13	
64 Xestoleberis opalescenta SCHORNIKOV	2	1	3			2	1
65 Xestoleberis sagamiensis KAJIYAMA							
66 Xestoleberis setouchiensis OKUBO							
67 Xestoleberis suetsumuhana YAJIMA							
68 Platymicrocythere sp.					25/25	49/53	
69 Paradoxostoma assimile OKUBO							
70 Paradoxostoma brunneum brunneatum SCHORNIKOV		1		1			
71 Paradoxostoma coniforme KAJIYAMA		1					
72 Paradoxostoma inabai OKUBO		1	1				
73 Paradoxostoma faccidum SCHORNIKOV	1						
74 Paradoxostoma honssuense SCHORNIKOV							
75 Paradoxostoma pedale HIRUTA							
76 Paradoxostoma yatsui KAJIYAMA	1			1			
77 Paradoxostoma sp. 1							
78 Sclerochilus mukaishimensis OKUBO		1					1
79 Cytherois asamushiensis ISHIZAKI							
80 Cytherois zosterae SCHORNIKOV	1	1/2	6		22/25	23/40	
81 Cytherois sp.				5			
82 Paracytherois tosaensis ISHIZAKI							
83* Callistocythere tateyamaensis n. sp.							
Total number of individuals	172	265	186	194	198	185	187
Total number of species	32	32	25	27	12	24	16
Volume of sample picked (in cc)	15	10	5	2.5	20	80	12.5

Table 2-2

	DS	DS	CBM	CBM	CBM	SFS	SFS	IR	IR	IR	IR	IR	Z
species	12	13	14	15	16	17	18	23	24	25	26	27	28
1		1	1	1	1	3	7	4	4	1	1/2	1/12	1/28
2		3	1	1									
3		2											
4						7	2/17					10	3
5								1				2	1
6	8	4	11	7	6	1/25	2/15						
7	1/3	1/1	2/29	8/36	29								
8	9	2	3/11	1		1/2	1					3/8	
9	3	7	2	1									
10	2	2	1										1
11	7	5	2	2	1								
12							1					1	
13									1		9/11		
14	1	1/5					2	1	1		2/3	1	1
15			1			1	1						
16		1	2		2	2	3						1
17		3		1							3	4	3
18		3	1	1									
19	1		4	4	3/14	8/19	1						
20							1	2				5	8
21	44	44	12		3								
22	1	1	4			2	11				3	1	4
23	21	5	5			1	4			1		2	2
24		1	1		3	4		2		1		7	7
25				1/5		8						3/7	9
26		3	4				7	8	6/30		7/21		6/17
27										1		1	
28		2											1
29	1	1		2		3	2	1				1	1
30	6	12	9			1/3							
31			4	3									
32													1
33						1/2	2						
34			2	3	3								
35	1		5										
36			1	1	21	2/15	1						
37	7		21	2/13	4/24	1/4							
38							4			1		1	
39	3	12	7	5	1	5							
40							12	2	7	1	4	9	6
41							5	4					
42	2	4	2	2			8	6			7	5	5
43	2/3	1/7	3	2	2	3/10	3/20						1
44	1	2	5	1	1		4				9		3
45													
46													
47							1					1	
48	13	1	2				1						
49	14	2											
50	1	2	16	6	4	2							
51			1									1	1
52	1	5	5	1		2	2		1		1		14
53												4	1
54		2	3	1		8	10		1		3	3	8
55												1	
56	1	1				4	5						
57	2	4	7	2		9	18						6
58	1/28	1/18	6/35	5/50	52	1/18	15						
59		1											
60							1	1/8	2/4		2/2		3
61			1		2	4/17							
62				2/11	3/18	5	8						
63	1	5	9	3	4	8	33	50/135	50/105	5/22	20/56	60/107	41
64	1		4	1		3	5						
65	3	8	3	1									
66			2	1		4	4	5	4		1/1	8	
67							2						
68		1/3				1							
69												3/3	
70			1				2	21/24		133/133	40/50	13/15	4
71			1										
72	1												
73								4	20/27	50/60	20/27	1/5	
74		1				1							
75								1	16/16		1		
76		1				2	2				1	4	4
77							1						
78													
79						2/2							
80	1/2	6	1	3	1/2						1		2/4
81													
82							2						1
83*		6		1									
	192	199	242	173	194	202	241	208	201	221	206	230	189
	31	41	42	32	20	33	39	16	12	9	19	28	31
	7.5	7.5	5	2.5	5	5	10	—	—	—	—	—	5

of water may contribute to keeping corals out of the central, deeper parts of the bays. The same factors were probably responsible for the distribution of corals in the southern Boso Peninsula during the Jomon transgression. At the beginning of the transgression, when the temperature became high enough, the sea level already reached at least 15 m. The centers of the drowned valley bays must have been covered with soft sediment; the only firm substrate for corals to settle and grow was available along the perimeter of the bays where bedrock was exposed. Low visibility may have been another cause preventing corals from settling in the deeper parts of the bay. As the sea level rose even further at a rate far exceeding the speed of coral growth, corals found new substrate ready on the outcrops of bedrock exposed by rising sea water along the bay edges. The deeper environment favored platy forms, such as *Echinophyllia aspera*, which are presently found at depths of 5 to 15 m.

Large platy corals, such as *Echinophyllia*, are attached by a very small area relative to the size of the entire colony. The attachment is also the oldest part of the colony and is most likely to be weakened by bioerosion (boring sponges, bivalves). Large colonies would therefore easily break off (during storms, earthquakes, etc.) and fall down the slope, breaking into smaller fragments. This accounts for the fragmented state of the fossil platy corals as well as for the lack of signs of extensive transport.

The edges of the bays represented not only a suitable habitat for the corals but also a site where corals growing over a large area accumulated in a relatively small area at the foot of the slope, producing thicker deposits with increased potential for preservation and exposure.

Ostracods

Methods: Methods explained in Ikeya and Hanai (1980, MS) were employed for preparation of 20 living and 60 fossil ostracod samples collected throughout the investigation area.

Sediment samples from Tateyama Bay were collected on March 4, 1980, using a modified Ockelmann-type bottom net sampler (Ikeya and Hanai 1980, MS) operated from a fishing boat. Samples of intertidal ostracods were collected from March 31 to May 6, 1980 by strongly agitating calcareous and soft algae from the intertidal zone in a large plastic bag filled with sea water and collecting the washed off material on a No. 200 (73 μm) mesh. In the case of recent ostracods, the presence of dried appendages was used as the criterion for distinguishing individuals alive at the time of collection.

Data on the occurrence and relative abundance of ostracods were analyzed using both R and Q mode arithmetic average cluster analysis (CA) and Q mode principal component analysis (PCA). The cluster analysis program used calculates correlation coefficient matrix and then forms clusters by comparing arithmetic averages of similarity coefficients between units. Variables were limited to 62 most commonly occurring species in the analysis of fossil ostracods, and to 25 most commonly occurring species in the case of recent ostracods. The program used in the principal component analysis is Hitachi CPCOMP program modified to accomodate up to 100 variables. The number of variables considered was limited to the 66 most commonly occurring species in the case of fossil samples and to the 46 most commonly occurring species in the case of

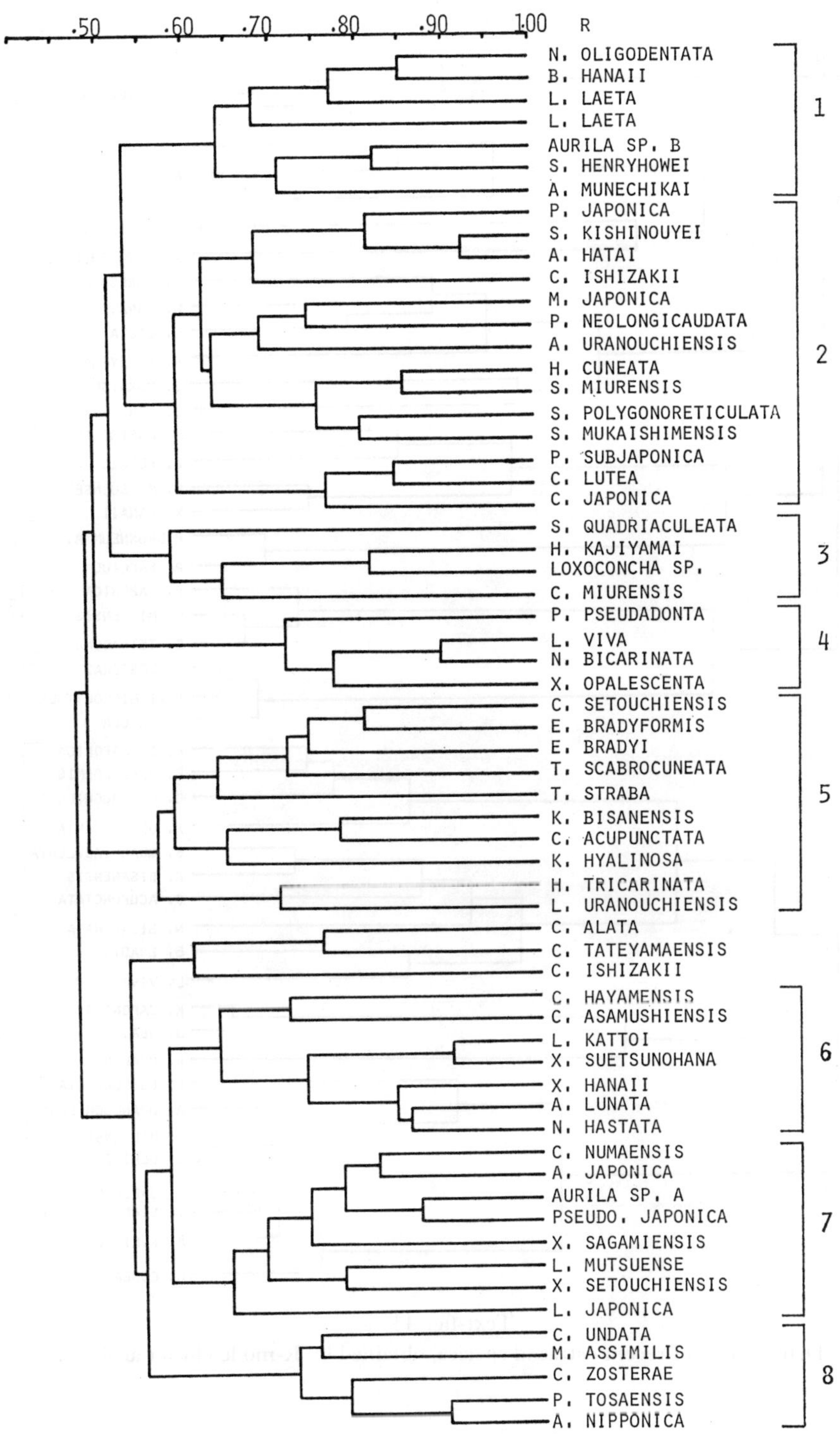

Text-fig. 10
Dendrogram of fossil ostracod species, obtained by R-mode cluster analysis.

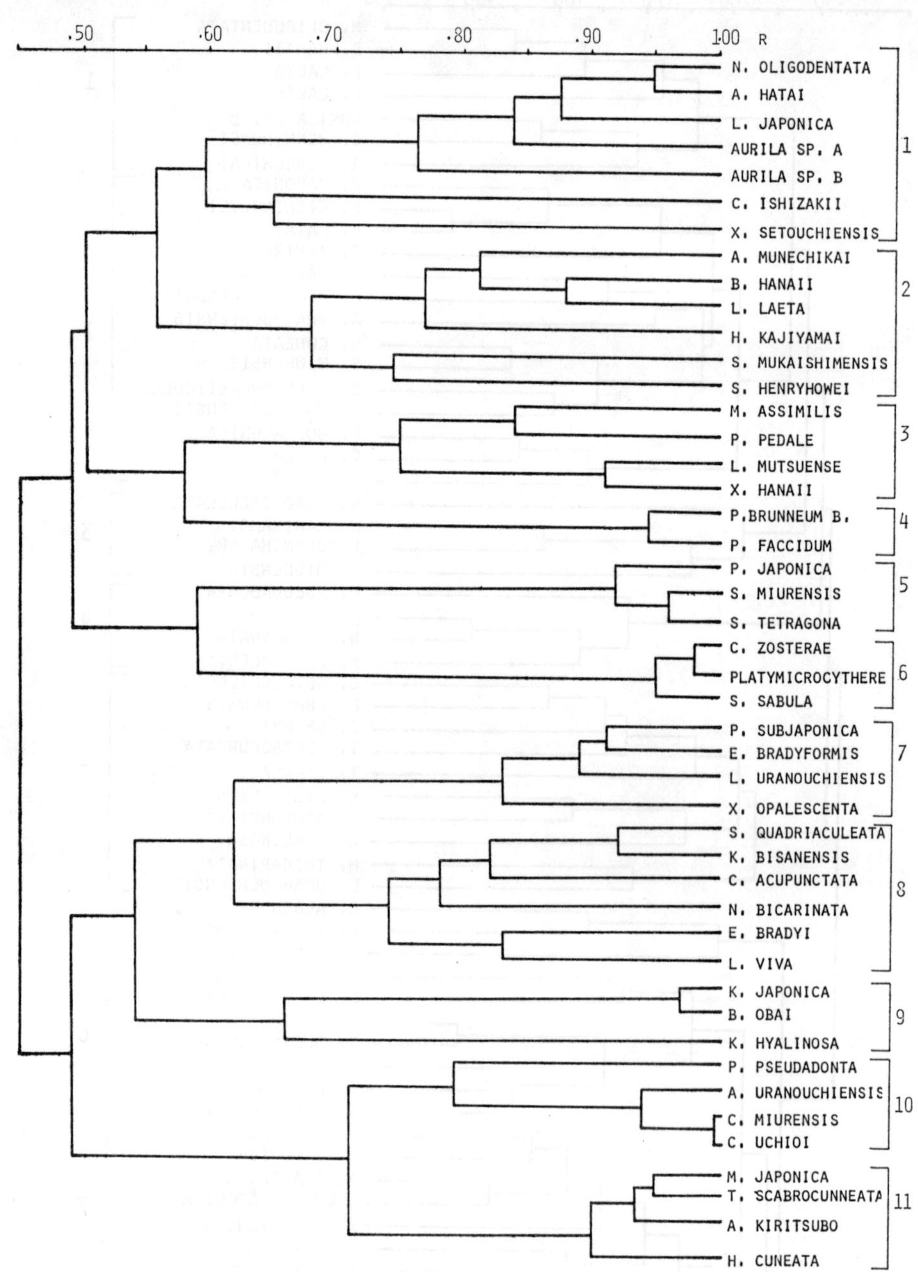

Text-fig. 11

Dendrogram of recent ostracod species, obtained by R-mode cluster analysis.

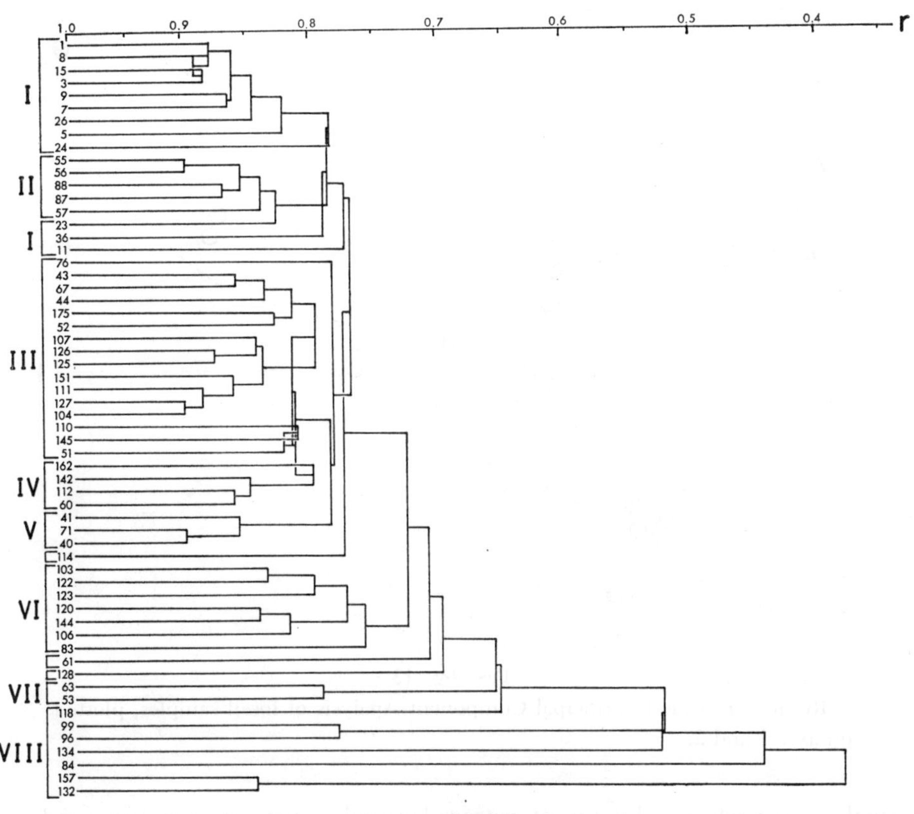

Text-fig. 12

Dendrogram of samples of fossil ostracods, obtained by Q-mode cluster analysis.

recent ostracods. The computer program computes the correlation coefficient *r*, based on the percentage of species in each sample. It then extracts a number of eigenvalues and eigenvectors from the matrix made up of correlation coefficients expressing similarity between the samples. It also calculates the coordinates of individual samples on the eigenvectors. Samples can be then plotted on various combinations of the eigenvectors (coordinate axes).

Species associations: Results of cluster analysis of fossil ostracods are shown in text-fig. 10. Eight clusters (referred to hereafter by arabic numerals from 1 to 8) can be distinguished. Comparison with text-fig. 11, showing the results of cluster analysis of recent ostracods from Tateyama Bay, will show several important differences. Individual species are connected at a higher similarity level, and clusters are connected at a lower similarity level in text-fig. 11 than in text-fig. 10. In other words, in the case of recent ostracods, similarity within clusters is higher and similarity among clusters is lower than in the case of fossil ostracods. Furthermore, it is apparent that even though clusters of similar composition are present in both cases, some clusters are markedly different.

These differences stem from a number of factors, including differences in sampling,

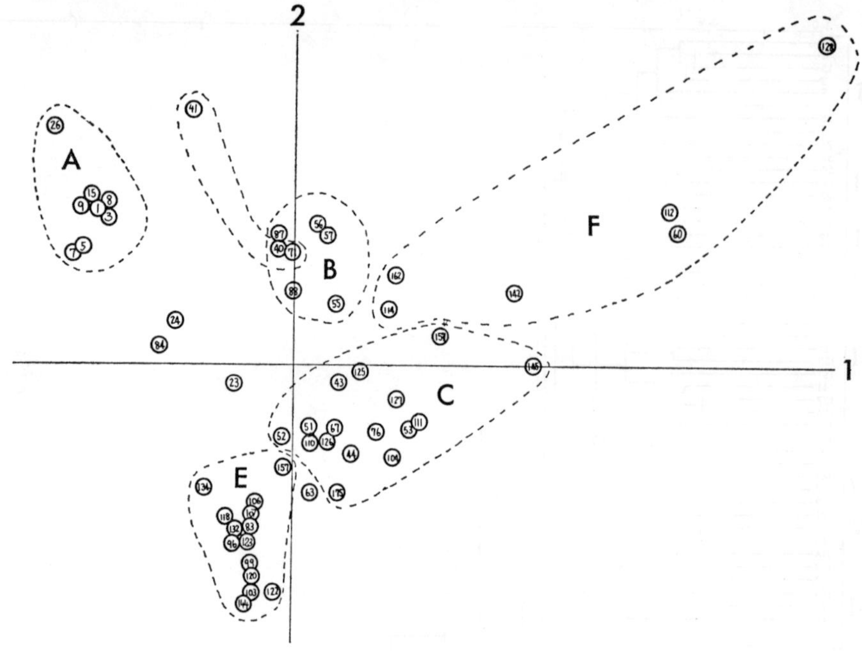

Text-fig. 13
Results of Q-mode Principal Component Analysis of fossil samples, plotted on axes 1 and 2.

the smaller area covered by recent ostracod samples, preservation bias, and transportation of dead carapaces. The last two will be discussed in a subsequent section.

Further differences result from the absence or underrepresentation of some species in the fossil or Recent assemblages. For instance, cluster 4 (text-fig. 11) is formed by species not present in the fossil ostracod assemblages. On the other hand, *Aurila* sp. and *Aurila cymba*, which are abundant in fossil samples, occur only in small numbers in a few samples in Tateyama Bay and form different clusters in text-figs. 10 and 11.

Biofacies of fossil ostracods: Results of the Q-mode cluster analysis are shown in text-fig. 12. Eight clusters (numbered using Roman numerals) can be discerned. Plots of samples on the 1, 2, and 4 axes (eigenvectors) obtained by PCA are shown in text-figs. 13 and 14. The first four axes account for 37% of the variance. Five clusters, marked by capital letters from A to F, are apparent on the 1 and 2 axes projection. On the 1 and 4 axes projection, the separation between clusters generally decreases but most components of the clusters delineated in text-fig. 14 still plot close together.

Comparison of text-figs. 12, 13 and 14 show that results obtained by the two methods are closely similar. Most components of clusters, joined at high value of correlation coefficient r in text-fig. 12, form clusters or plot close together in text-figs. 13 and 14. Samples joined to the cluster at low values of r in text-fig. 12 lie outside of well-defined clusters in text-figs. 13 and 14 or plot in different clusters when plotted on different axes.

Results of the CA and PCA were used to delineate groups of samples believed to

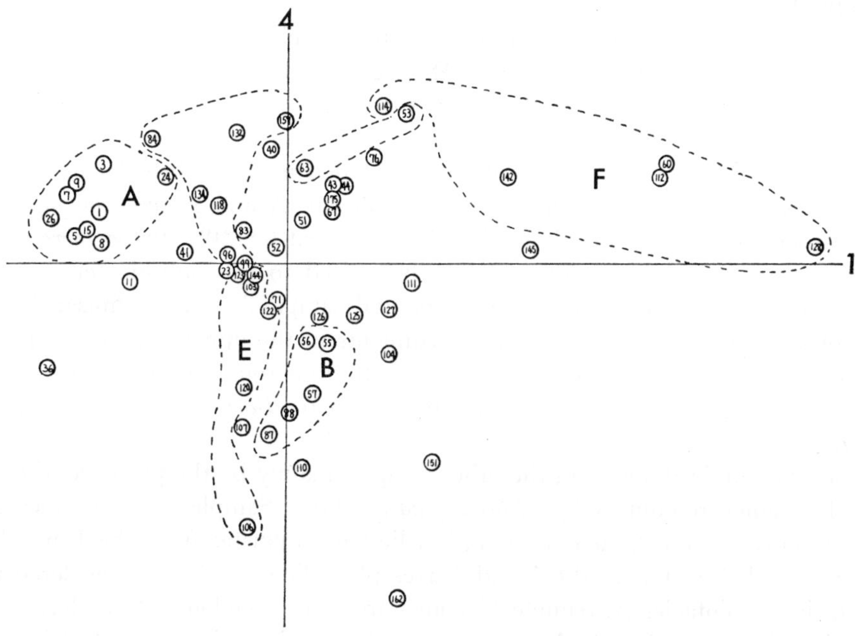

Text-fig. 14
Results of Q-mode Principal Component Analysis of fossil samples, plotted on axes 1 and 4.

represent ostracod biofacies, but they were not adhered to inflexibly. Various methods (discussed in a subsequent section) were used to distinguish in situ and transported components of the ostracod assemblage of each sample. Thus, together with information obtained from R-mode CA (text-figs. 10 and 11), the relative abundance of most common species in different samples, and information on the ecology of individual species and their values as environmental indicators (obtained from various sources) were used to confirm the assignment of samples to the biofacies and to decide the position of samples, which were put into different clusters by the CA and PCA methods.

A number of samples lying outside of PCA clusters join the clusters in text-fig. 12 at a low value of *r*. In order to facilitate the following discussion, an attempt has been made to minimize the number of such samples by assigning them to already existing clusters or by forming new clusters. The criteria used are detailed in the discussion of each biofacies.

As is apparent from text-figs. 12 to 14, the amount of variation among samples representing each biofacies is different in each biofacies (AL and SN biofacies), partly because samples available may not adequately represent the amount of variation within the biofacies. The amount of variation is also, to an extent, an artifact resulting from the reluctance to erect a large number of one or two sample biofacies.

Because the environment of the fossil biofacies is not known but only inferred, the biofacies were not named. Instead, each biofacies was assigned a one or two letter code, which (for mnemonic reasons) was derived from the names of the most abundant ostracods in the biofacies.

I. A Biofacies: This biofacies is represented by clusters I and II (text-fig. 12) and clusters A and B (text-figs. 13 and 14). It can be divided into two sub-biofacies.

(1) AL sub-biofacies: This sub-biofacies consists of samples 1, 8, 15, 3, 9, 7, 26, and 5, which form cluster I (text-fig. 12) and lie within cluster A (text-figs. 13 and 14). Samples 36, 24, 23, 11, and 84 are also included in this sub-biofacies. The sub-biofacies is characterized by dominance of *Aurila* sp. A and *Loxoconcha japonica*, a species association nearly coinciding with cluster 7 (text-fig. 10). *Ambocythere japonica* is not abundant, but its occurrence is nearly limited to this sub-biofacies.

Except for samples 23 and 24 from Numa and sample 36 from Kamisanakura, all the samples come from the coral-bearing sediments in the upper part of the Koyatsu valley. The sediment is silty sand to sandy silt with numerous coral fragments. Typical molluscs occurring with ostracods in this bio-facies are *Barbatia bicolorata* and *Lima sowerbyi*.

Samples 23 and 24 differ from the other samples mainly by the presence of *Cythere lutea* and a higher amount of *Spinileberis quadriaculeata*. Samples 11 and 36 are separated from cluster I on the dendrogram, but lie within cluster A on the 1 and 2 axes plot and near the cluster on the 1 and 4 axes plot. They could be considered to be sub-biofacies of biofacies A. Sample 11 comes from the conglomerate underlying the coral bed at Koyatsu. Sample 36 comes from the coral-bearing silty sand in Kamisanakura. Both samples 11 and 36 are remarkable for the abundance of *Parakrithella pseudadonta*. Sample 11 contains high amounts of *Callistocythere undata*, *Mutilus assimilis*, *Cytherois zosterae*, and *Cytherois* sp. 1. An increased abundance of *Pseudoaurila japonica* and *Xestoleberis hanaii* and the near absence of *Spinileberis quadriaculeata* characterize sample 36.

Sample 84 is exceptional, since the high proportion of *Spinileberis quadriaculeata* and *Aurila* sp. A is almost certainly due to selective removal by solution of ostracods with less resistant carapace.

(2) AA sub-biofacies: The sub-biofacies consists of samples 88, 55, 87, 56, and 57, which form cluster B in PCA and cluster II in CA. The close relationship between the AA and AL sub-biofacies is apparent both on the dendrogram and on the PCA plots. The major difference is that in the AA sub-biofacies *Aurila* sp. B becomes more abundant than *Aurila* sp. A; in most samples, *Aurila munechikai* and *Neonesidea oligodentata* are more numerous, whereas *Spinileberis quadriaculeata* is less common. Species commonly occurring in this sub-biofacies coincide with cluster 1 in text-fig. 10.

Samples of this sub-biofacies come from the center of Shiomi valley and the lower part of Koyatsu valley. The sediments range from massive silty sand to sand. Common molluscs include *Dosinella penicillata*, *Clementia papyracea*, *Panopea japonica*, and *Paphia undulata*.

II. N Biofacies: This biofacies consists of samples included in cluster III, most of which are also included in cluster C. In several respects the clusters are not clearly defined. Cluster C is closely related to cluster B (text-fig. 13) and consists of several clusters showing a rather wide dispersal. The unifying feature is the presence of a large number of *Nipponocythere bicarinata* and *Loxoconcha viva*, coinciding with species association of cluster 4.

Samples falling into this biofacies have a wide distribution; they occur throughout the Tateyama area and in the upper horizon in Tomoe River area. Sediments are predominantly silty sand to sandy silt. Molluscs commonly found are *Dosinella penicillata*, *Paphia undulata*, and *Ostrea denselamellosa*.

This biofacies can be subdivided into two sub-biofacies:

(1) NL sub-biofacies—This sub-biofacies is characterized by abundance of *Pontocythere subjaponica*, *Parakrithella pseudadonta*, and *Cythere lutea* and absence of *Keijella bisanensis*. It consists of samples 43, 67, 44, 175, 52, and 51 from the Tateyama area.

(2) KN sub-biofacies—This sub-biofacies stands appart due to the presence of *Keijella bisanensis* and is represented by samples 107, 126, 125, 151, 111, 127, 104, 110, and 145 from Tomoe River area.

III. KS Biofacies: This biofacies corresponds to cluster E shown in text-fig. 13 and cluster VI and VIII in text-fig. 12. It is closely related to the NL sub-biofacies and is separated mainly by the increased importance of *Keijella bisanensis*, *Spinileberis quadriaculeata*, and *Cytheromopha acupunctata* and the corresponding decrease in importance of *Nipponocythere bicarinata* and *Loxoconcha viva*.

As indicated by the CA, this biofacies can be split into two sub-biofacies corresponding to clusters VI and VIII:

(1) K sub-biofacies—K sub-biofacies corresponds to a well-defined cluster VI (text-fig. 12) and consists of samples 103, 122, 123, 120, 144, 106, and 83. Sample 76 includes species characteristic of the SN biofacies in addition to *Keijella bisanensis* and *Cytheromorpha acupunctata*. The latter two species appear to be in situ and the sample is therefore assigned to this sub-biofacies. In the PCA plot, samples of the K and S sub-biofacies lie very close together on the 1 and 2 axes projection (text-fig. 13) but are separated on the 1 and 4 axes projection (text-fig. 14).

The dominant species are *Keijella bisanensis* and *Cytheromorpha acupunctata*. Common species include *Echinocythereis bradyi*, *Spinileberis quadriaculeata*, and *Loxoconcha uranouchiensis*. This sub-biofacies is common in the lower silt and sandy silt of Tomoe River area and one sample occurs in silt in Tateyama area. Molluscs commonly found associated with this sub-biofacies include *Dosinella penicillata* and *Paphia undulata*.

(2) S sub-biofacies—As indicated by PCA in text-fig. 14, the separation between the S and K sub-biofacies is gradational, and as can be seen in text-fig. 12, this sub-biofacies does not form a good, tight cluster (VII) but consists of samples joined at a very low level of correlation. Examination of text-fig. 14 will show that, more than by the dominance of any one species, this sub-biofacies is characterized by an exceptionally high amount of *Keijella bisanensis* or *Spinileberis quadriaculeata* or both. It consists of samples 96, 99, 118, 120, 132, 134, 157, all of which fall within the part of cluster E above the number 1 axis in text-fig. 14. Samples 118 and 134 differ somewhat by the presence of a large proportion of *Loxoconcha japonica* or *Loxoconcha uranouchiensis* or both. Sample 59 was not included in the statistical analysis, but on the basis of its faunal composition (over 90% *Spinileberis quadriaculeata*), it was assigned to this sub-biofacies.

The S sub-biofacies is distributed in the lowest silt to silty clay in Tomoe River area, Chikura area, and in Maruyama area.

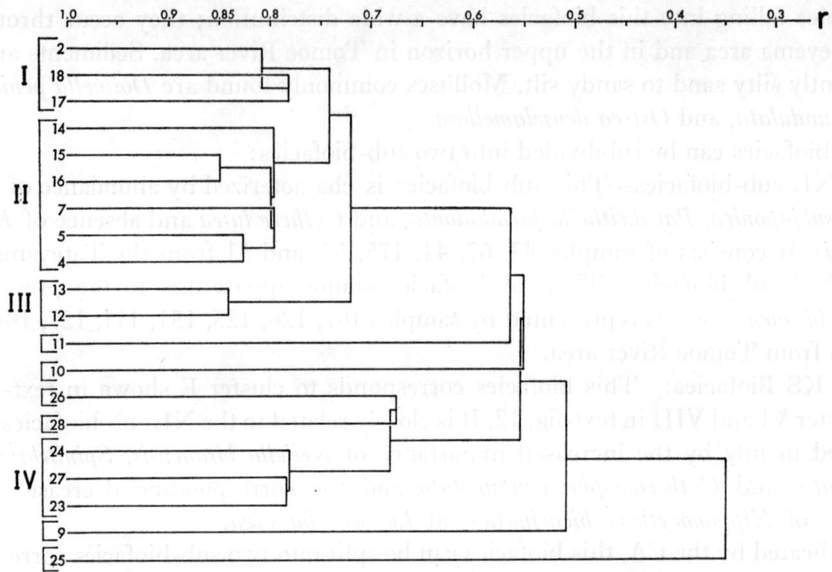

Text-fig. 15

Dendrogram of samples of recent ostracod samples obtained by Q-mode Cluster Analysis.

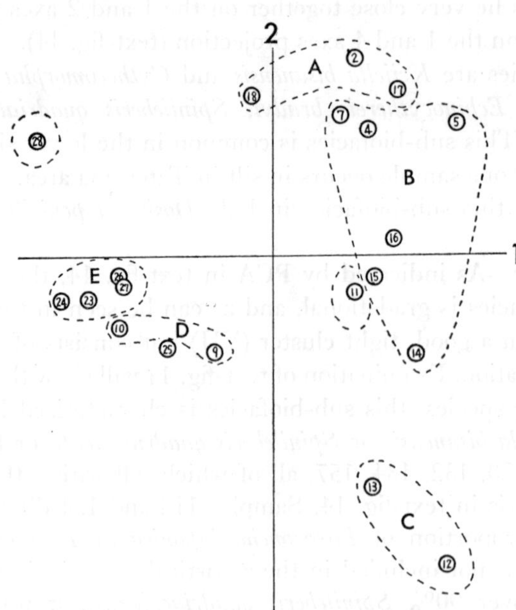

Text-fig. 16

Results of Q-mode Principal Component Analysis of recent samples, plotted on axes 1 and 2.

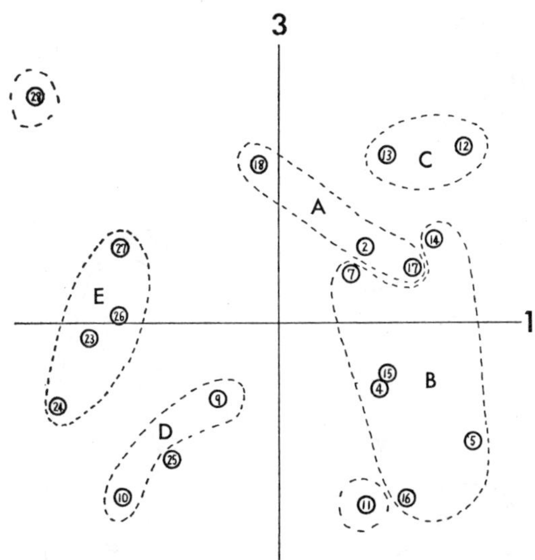

Text-fig. 17
Results of Q-mode Principal Component Analysis of recent samples, plotted on axes 1 and 3.

IV. SN biofacies: This biofacies consists of widely dispersed samples 114, 162, 142, 112, 60, and 128, which form a poorly defined cluster F. In CA the biofacies corresponds to cluster IV with more distantly related samples 114 and 128 added.

The SN biofacies is characterized by the presence of large numbers of *Schizocythere kishinouyei* and *Neonesidea oligodentata.* Other common species include *Pontocythere subjaponica, Cythere lutea, Aurila hataii, Xestoleberis hanaii* and other species of cluster 2, text-fig. 10. Sample 128 deviates only in the larger amount of the characteristic species *Neonesidea oligodentata* and *Schizocythere kishinouyei.* Sample 114 is included, based on the presence of a significant amount of *Schizocythere kishinouyei* and *Aurila hataii.*

Samples of this biofacies occur in well-sorted sand to silty sand in the uppermost horizon in Tomoe River area and in Chikura area.

V. LK biofacies: This biofacies corresponds to cluster V and consists of samples 40, 41, and 71. Samples 40 and 71 fall within cluster B in text-fig. 13 but are clearly separated in the projections on 1 and 3 axes (not illustrated).

The samples of this biofacies are characterized by the presence of a large amount of *Loxoconcha kattoi*, a species occurring only rarely in other biofacies (sample 114). The presence of *Ambocythere japonica* is also remarkable. All three samples come from the oyster reef in Tateyama area. The sediment is silty sand filling interstices between oysters (*Pretostrea imbricata*) and corals.

VI. PL biofacies: This biofacies is represented by only samples 63 and 53, which form cluster VII in text-fig. 12. In the plots of PCA, these two samples fall within cluster C and are partly separated only in the projection on the 1 and 4 axes (text-fig. 14).

Text-fig. 18
Distribution of ostracod biofacies in Tateyama Bay.

This biofacies resembles the NL sub-biofacies by the presence of *Pontocythere subjaponica*, *Cythere lutea*, and *Spinileberis quadriaculeata* but is distinguished from it by a much smaller content of *Nipponocythere bicarinata* and *Loxoconcha viva*. Samples 53 and 63 come from silty sand in the eastern part of Tateyama area.

Sample 61 could not be assigned to any of the above biofacies. In text-fig. 13 and 14 it plots within cluster C; but in text-fig. 12, it clearly lies outside of the corresponding cluster III. Sample 61 is set apart by the presence of 10% of *Loxoconcha* sp. and 7% of *Pontocythere sekiguchii* species encountered only very rarely. It comes from silty clay in Chikura area and the associated molluscs are *Dosinella penicillata* and *Scapharca inequivalvis*.

Biofacies of recent ostracods: Results of cluster analysis of samples of recent ostracods are shown in text-fig. 15. Results of PCA of recent ostracod samples are plotted on the first 3 axes, which account for 43% of the variance, and are shown in text-figs. 16 and 17. [Recent ostracod samples are identified by a number preceded by the letter T]

Biofacies were delineated following the same reasoning as used for fossil samples. Biofacies were given names based either on the type of sediment or location or both. As will be pointed out later, changes in ostracod fauna composition are gradational; in that sense, some biofacies boundaries are arbitrary. A map with inferred spatial distribution of biofacies within the bay is given in text-fig. 18. Biofacies boundaries were drawn to coincide with depth and sediment type distribution, which appear to be the major controlling factors of ostracod distribution.

I. Shallow Fine Sand Biofacies (SFS): This biofacies consists of samples T2, T18, and T17, which form cluster I (text-fig. 15) and plot within cluster A (text-figs. 16 and 17). The biofacies is characterized by large proportions of *Pontocythere subjaponica*, *Semicytherura miurensis*, and the presence of *Loxoconcha laeta*. It is distributed on a fine to medium sand bottom along the inner margin of the bay at depths from 3 to 10 m.

II. Central Bay Mud Biofacies (CBM): This biofacies is represented by samples T14, T15, T16, T7, T5, and T4, which form cluster II and plot as cluster B. The gradational relationship between this and the SFS biofacies is highlighted in text-figs. 16 and 17, where cluster B and A are not clearly separated. The biofacies is characterized by dominance of *Loxoconcha viva* and common occurrence of *Nipponocythere bicarinata*, *Krithe japonica*, and *Basslerites obai*. It is distributed on a fine muddy bottom in the central parts of the bay in depths ranging from 10 to 30 m.

The CBM biofacies could be divided into two sub-biofacies based on the relative abundance of *Krithe japonica*, *Basslerites obai*, *Spinileberis quadriaculeata*, and *Keijella bisanensis*. In samples T14, T15, and T16, *Basslerites obai* and *Krithe japonica* are more abundant than in samples T7, T5, and T4. Samples T5, T4, and T16 on the other hand, are characterized by the presence of *Spinileberis quadriaculeata* and *Keijella bisanensis*. Because samples T4, T5, T7 lie closer to the mouth of the Heguri River (and sample T16 near the mouth of the Shioiri River) than samples T15 and T14, it is possible that the faunal difference results from the influx of fresh water (perhaps periodic).

III. Deep Mud Biofacies (DM): This biofacies, represented by sample T11, further illustrates the gradational nature of faunal changes in Tateyama Bay. Its faunal composition is closely related to that of the CBM biofacies, but *Krithe japonica* and *Basslerites obai* become the dominant species over *Loxoconcha viva*. Sample T11 plots on the perimeter of cluster B in text-figs. 16 and 17, but it joins cluster II (text-fig. 15) at a rather low value of *r*. The DM biofacies is distributed on muddy bottom in the central part of the bay in depths exceeding 30 m.

IV. Deep Sand Biofacies (DS): This biofacies is represented by samples T12 and T13, forming cluster III and cluster C. It is similar to CBM biofacies but differs by the absence of *Nipponocythere bicarinata* and *Keijella bisanensis* and by smaller amounts of *Spinileberis quadriaculeata*, *Basslerites obai*, and *Krithe japonica*. The assemblage is dominated by *Aurila kiritsubo*, but since only dead carapaces showing effects of erosion are present, and since the usual habitat of *A. kiritsubo* is shallow subtidal, it is assumed that the carapaces of dead *A. kiritsubo* were transported into the area, possibly from the rocky coast of the island of Okinoshima, which lies to the southeast. The DS biofacies is distributed on a coarse, well-sorted sandy bottom of the outer part of the bay in depths from 20 to 40 m.

V. Shallow Coarse Sand Biofacies (SCS): This biofacies is represented by samples T9 and T10, which form a widely dispersed cluster D on text-fig. 16, and plot separately on the CA dendrogram in text-fig. 15. The uniting feature is the presence of *Cytherois zosterae* and *Platymicrocythere* sp., which are absent in other samples. *Semicytherura miurensis*, *S. tetragona*, and *S. sabula* n. sp. also occur in large quantities. The SCS biofacies is inferred to be distributed on the coarse sand bottom along the margins of the bay at depths from 5 to 10 m.

VI. Intertidal Rocky Biofacies (IR): This biofacies is represented by samples T24, T27 and T23, forming cluster IV, and samples T26 and T25. These samples plot within cluster E on text-figs. 16 and 17. This biofacies is characterized by dominance of *Xestoleberis hanaii* and *Paradoxostoma* spp. It is distributed on algal covered intertidal rocky shores of the bay.

VII. *Zostera* Biofacies (Z): This biofacies is represented only by sample T28. It is characterized by a large quantity of *Neonesidea oligodentata*, *Mutilus assimilis*, *Loxoconcha japonica*, *Xestoleberis hanaii* and absence of *Paradoxostoma* spp. It is found in a shallow embayment in the southwest part of the bay. The silty sand bottom is covered by growth of the sea grass *Zostera* sp.

Paleoecological interpretation of biofacies

I (1) AL sub-biofacies: *Aurila* sp. A and *Loxoconcha japonica* are the dominant species of the ostracod fauna living in *Zostera* sea grass community in central parts of Aburatsubo Bay (unpublished research by Hanai). Aburatsubo Bay is located on the southwest tip of Miura Peninsula, 50 km northwest-north of Tateyama. It is a bay 1 km long and from 150 to 200 m wide of the drowned valley type. *Zostera* grass grows in patches in shallow water (1 to 5 m) from the mouth of the bay up to about 600 m inland. In the inner part of the bay, an ostracod assemblage dominated by *Keijella bisanensis* is prevalent.

The AL sub-biofacies of Koyatsu valley can be interpreted as having originated in a similar sea grass community, where species from surrounding biofacies were added. *Spinileberis quadriaculeata*-dominated biofacies is commonly found in innermost part of bays, similar in position to *Keijella bisanensis* biofacies in Aburatsubo Bay. *Xestoleberis hanaii* is the dominant constituent of most intertidal rocky shore biofacies. Even though not preserved, the two biofacies probably existed alongside the *Zostera* community in Koyatsu and contributed to it. The compositional difference of samples 11, 36 and samples 23, 24 can be explained by the varying degree to which contributions from the nearby biofacies affected the final fossil assemblage.

I (2) AA sub-biofacies: *Pontocythere subjaponica* and *Parakrithella pseudadonta* are common in the SFS biofacies of Tateyama Bay. They also form a common component of the open coast assemblage in Hamanako estuary area (Ikeya and Hanai, 1980 MS). *Aurila munechikai* is most abundant in the mouth and outside of Uranouchi Bay (Ishizaki, 1968).

The increased abundance (as compared with the AL sub-biofacies) of these species and the concomitant decrease in abundance of *Spinileberis quadriaculeata* in the AA sub-biofacies indicates that even though it was formed in an environment similar to that of AL sub-biofacies, it was closer to the bay mouth. This interpretation agrees well with the actual location of AA sub-biofacies samples, which lie seaward from the samples of the AL sub-biofacies.

II (1) NL sub-biofacies: Composition of the NL sub-biofacies is very similar to the faunal composition of the SFS and CBM biofacies of Tateyama Bay and, in particular, samples T4, T5, and T7. The NL sub-biofacies contains large amounts of *Pontocythere subjaponica*, which characterizes the SFS biofacies, and *Nipponocythere bicarinata* and *Loxoconcha viva*, characteristic of the CBM biofacies. The most remarkable difference is the absence of *Basslerites obai* and *Krithe japonica* from the NL sub-biofacies. Near absence of *Cythere lutea*, which is common in the NL sub-biofacies, from Recent Tateyama Bay samples is also noticeable.

Both *Krithe japonica* and *Basslerites obai* become more abundant with increasing depth and distance from shore, and their absence from fossil samples could be explained by the gentler slope of the paleo-Tateyama Bay, allowing a clearer development and separation of the DM and CBM biofacies, which presently partly overlap in Tateyama Bay. *Cythere lutea* lives on soft algae and sea grasses, and the scarcity of suitable substrata in the present-day Tateyama Bay may explain the near absence of *Cythere lutea*.

The NL sub-biofacies can be inferred to have been present in the inner to central part of a widely open bay, similar to the present-day Tateyama Bay but with a more gently sloping bottom. Depth of deposition probably ranged from 10 to 15 m.

II (2) KN sub-biofacies: *Keijella bisanensis* is the dominant ostracod in the central and innermost part of Aburatsubo Bay. Ishizaki (1968) reported abundant *K. bisanensis* from the central parts of Uranouchi Bay in Kochi Prefecture, Shikoku. The Uranouchi Bay is a 10.5 km long, 1 km wide bay of the drowned valley type. Even though *K. bisanensis* is not restricted to such areas, its abundance there suggests that it prefers a low energy environment not strongly affected by inflow of open-sea water. *Nipponocythere bicarinata*, on the other hand, is found in the central parts of Tateyama Bay, directly affected by open-sea water.

The presence of approximately the same proportions of *Keijella bisanensis* and *Nipponocythere bicarinata* in the KN sub-biofacies suggests conditions where inflow of open-sea water is more restricted than in Tateyama Bay (or the influx of fresh water greater), but less so than in the central parts of Uranouchi Bay or inner parts of Aburatsubo Bay. This hypothesis fits well with the reconstructed outline of Tomoe Bay at relative sea levels above 20 m and the distribution of the KN sub-biofacies samples in the outer and central parts, but not in the inner part of the Tomoe River Valley.

III (1) K sub-biofacies: The dominance of *Keijella bisanensis* and *Cytheromorpha acupunctata*, both of which are found in abundance in inner parts of bays (Ishizaki, 1968; Ikeya and Hanai, 1980 MS) and the decreased amount of *Nipponocythere bicarinata*, which characterizes the central part of Tateyama Bay, suggest an inner bay environment with decreased inflow of open-sea water.

III (2) S sub-biofacies: Innermost parts of bays, Uranouchi Bay, Hamana-ko estuary, where inflow of open-sea water is very small, and where fresh water input by rivers may decrease salinity, are characterized by large numbers of *Spinileberis quadriaculeata* or *Keijella bisanensis*, and correspondingly smaller numbers of other species. S sub-biofacies is therefore interpreted as having developed in a marginal marine environment: the innermost part of a bay with a fresh water inflow.

IV. SN Biofacies: *Neonesidea oligodentata* lives on intertidal and shallow subtidal

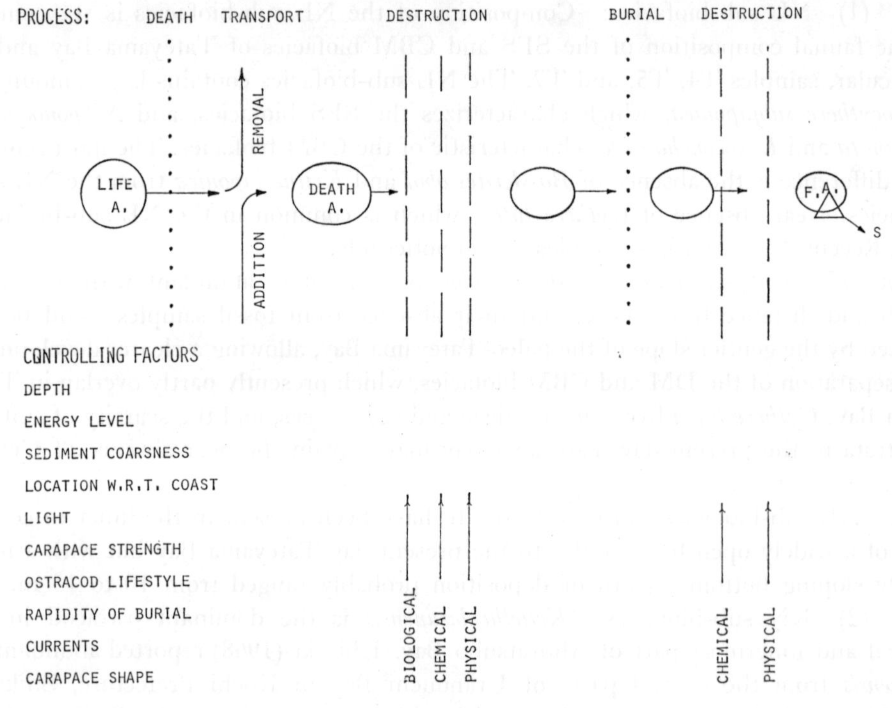

Text-fig. 19

Schematic representation of the live to fossil assemblage transition.

rocky shores covered with calcareous algae. Dead carapaces are commonly abundant in the adjacent sandy sediments. *Schizocythere kishinouyei* is found in the mouth and on the open coast outside of Uranouchi Bay (Ishizaki, 1968). *Pontocythere subjaponica* is found in abundance in sediments of the open coast adjacent to the Hamana-ko estuary (Ikeya and Hanai, 1980 MS.), and shallow parts of Tateyama Bay. The SN biofacies therefore developed in an open coast environment.

V. LK biofacies: A few empty carapaces of *Loxoconcha kattoi* were found in the intertidal rocky shore biofacies in Tateyama Bay (T27). In Uranouchi Bay, *L. kattoi* is common in the mouth and outside the bay at depths ranging from 10.3 to 25 m.

With the exception of *Loxoconcha kattoi*, *Ambocythere japonica*, and increased amount of *Xestoleberis hanaii*, the ostracod fauna of this biofacies resembles that of the NL biofacies. It is probable that it originated by mixing of fauna supported by the oyster reef and the fauna of the surrounding NL sub-biofacies.

VI. PL Biofacies: This biofacies closely resembles the SFS biofacies of Tateyama Bay. The environment of deposition was probably shallow (2 to 10 m) inner part of a widely opened bay. The PL biofacies differs from the SFS biofacies by the lack of *Semicytherura miurensis*, which probably resulted from post-depositional selective destruction of thin walled ostracods.

Life to fossil assemblage: In text-fig. 19, living assemblage is represented by the stippled circle at the far left. As time flows to the left, the assemblage appears to move

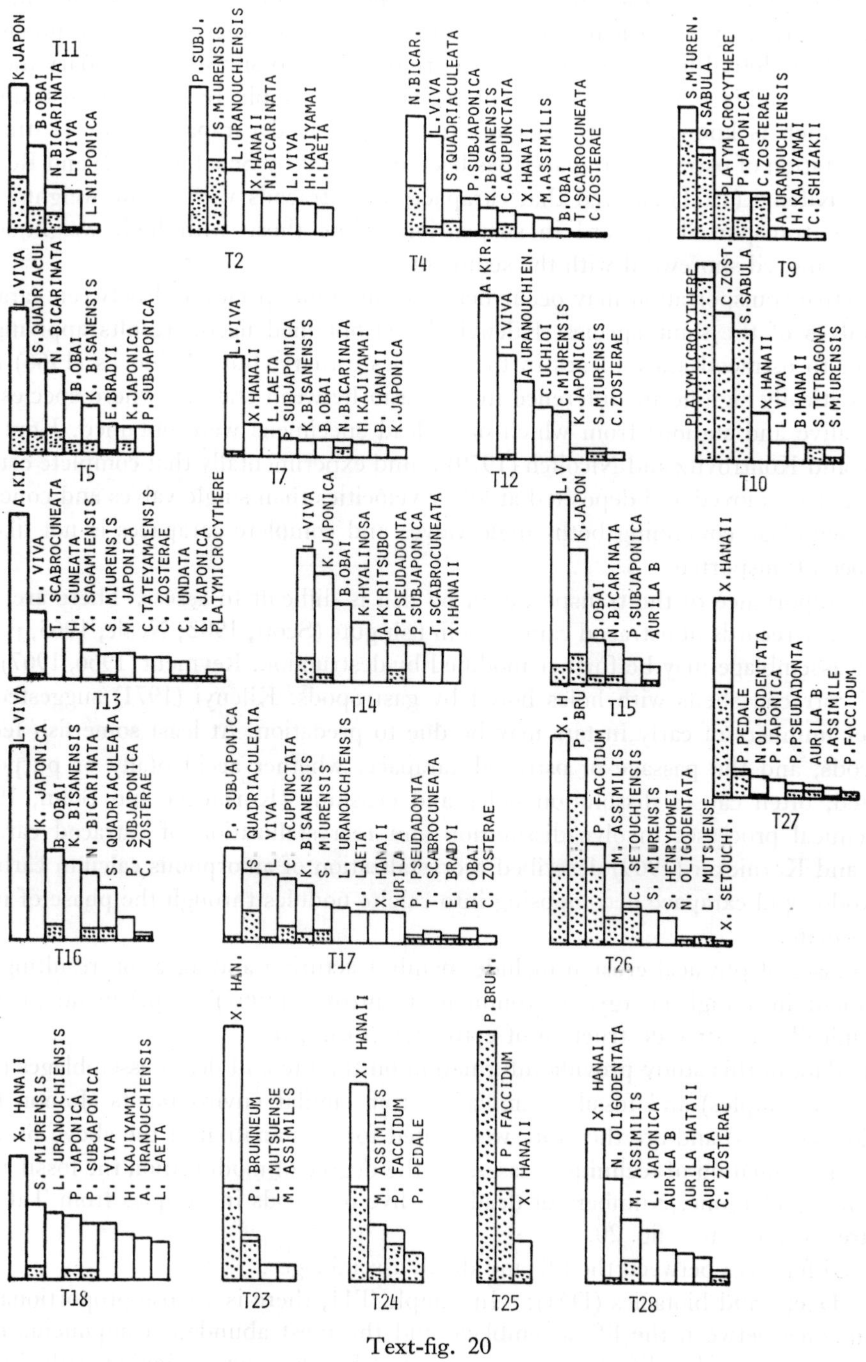

Text-fig. 20

Histogram showing abundance of live (dotted area) and dead (white area) ostracods in samples from Tateyama Bay.

to the right and undergoes many changes and modifications due to processes indicated on the diagram by vertical interrupted lines. Changes of the type of assemblage are indicated by dotted lines; the circle to the right of the fossilization boundary indicates the fossil assemblage, which is studied on the basis of a sample, represented by triangle S.

Not much is known about transportation of ostracod carapaces. The postmortem transportation of ostracod carapaces has been discussed by Schafer (1962) and Reyment (1966). Kilenyi (1971) showed that carapaces of ostracods with a low weight/surface ratio are transported seaward in suspension, while those with high weight/surface ratio are moved shoreward with the sediment.

Selective transportation may occur between different species and between carapaces and valves of the same species. Kilenyi (1971) obtained useful results applying this concept to ostracod assemblages in the Thames estuary, but Maddocks (1966) found no statistically significant difference in the ratio between stations where species were found alive and stations from which only dead specimens were obtained. Kontrovitz (1975) and Kontrovitz and Nicolich (1979) found experimentally that complete ostracod carapaces are moved and deposited at lower velocities than single valves and concluded that assemblage containing both single valves and complete carapaces is not likely to have been transported.

The importance of the transportation by fish is difficult to assess. There are, however, many records of ostracod carapaces in fish guts (Scott, 1902; Neale, 1965, p. 282).

The assemblage may be further modified by destruction. Reyment (1966, 1967) illustrated ostracod shells with holes bored by gastropods. Kilenyi (1971) suggested that smaller numbers of early instars may be due to predation. At least some fish feed on ostracods, and the passage of ostracod carapace, whether accidentally or purposedly ingested, often causes dissolution holes and crushing (Kornicker and Sohn, 1971).

Chemical processes involve dissolution and recrystallization of ostracod carapace. Sohn and Kornicker (1969) described recrystalization of amorphous calcium carbonate of myodocopid carapace decomposing into calcite nodules through the phase of monohydrocalcite.

Processes of physical erosion include mainly fracturing and abrasion resulting from movement in a high energy environment. Kontrovitz (1967)'s experiments prove the mechanically selective destruction of cytheracean carapace.

The data in this study provide information on the life and death assemblages (Tateyama Bay samples) and fossil assemblages from similar environments (Jomon transgression samples) and are therefore well suited for examination of the above processes as well as evaluation of techniques assessing the degree of modification the fossil assemblage has undergone. Numbers of dead and live ostracods in samples from Tateyama Bay are shown in text-fig. 20.

I. Differences between the life and death assemblages.

(1) Deep mud biofacies (DM): In sample T11, there is a close proportional correspondence between the life assemblage and the most abundant components of the death assemblage. The life assemblage cosists of 5 species and, together with the dead carapaces of these species, accounts for 95% of the assemblage. Modification of this assemblage is very small and consists of probable addition of species such as *Pontocythere subjaponica*, *Parakrithella pseudadonta*, and *Spinileberis quadriaculeata*.

(2) Central bay mud biofacies (CBM): Life to death assemblage correspondence in this biofacies ranges from very good (T4, T5) to very poor (T7, T16). Deviations are in part due to addition from outside the biofacies. *Xestoleberis hanaii* was found in the death assemblage of all the samples, yet live specimens were absent throughout the biofacies. *Pontocythere subjaponica* is also a common component of the death assemblage, but live specimens occur only in one sample (T4) in small quantity. *Pontocythere subjaponica* is, however, common in the life assemblage of the adjacent SFS biofacies, and clearly transport of dead carapaces over short distance is the reason.

Loxoconcha viva makes up the bulk of the death assemblage in T16 but is absent from the life assemblage of that sample. This may be partly accounted for by seasonal variation, because even though *Loxoconcha viva* is abundant in the death assemblages of other samples, it is typically present in the life assemblage in very small quantity. Its absence from T16 may be due to its small numbers at the time of the year, rather than absence at the sampling locality.

The unmodified death assemblages in CBM biofacies are altered by addition to a larger extent than the DM biofacies, but the change is not large enough to mask the nature of the unmodified death assemblage and life assemblage.

(3) Shallow fine sand biofacies (SFS): The death assemblage corresponds closely to the life assemblage in sample T17 and poorly in T2 and T18. *Xestoleberis hanaii*, *Loxoconcha uranouchiensis*, and *Loxoconcha viva* are present in the death assemblages of all three samples but are absent from the life assemblages and can be clearly accounted for by transport from outside. Significantly, T2 and T18 are located at lesser depth than T17, implying greater strength of physical agents of transport.

Addition by transport from outside is again the major process alering the assemblage. In T18 it is strong enough to mask the nature of the life assemblage.

(4) Deep sand biofacies (DS): The life to death assemblage correspondence is very poor in this biofacies. *Aurila kiritsubo*, which dominates the death assemblage of both samples is completely absent from the life assemblage. The ecology of *A. kiritsubo* (shallow subtidal) suggests that it was added to the assemblage by currents sweeping the nearby shallow rocky shores of Okinoshima. The presence of coarse sand at the depth of 30 m indicates furthermore the existence of strong currents, which may be responsible for adding various species.

The unmodified death assemblage of this biofacies is almost completely masked, owing to the presence of a strong current and proximity to a steep slope whose upper parts harbor a different biofacies.

(5) Shallow coarse sand biofacies (SCS): The proportions of species in the death assemblage of both samples closely resemble those of life assemblages, with the exception of some species which are abundant in the life assemblage but are rare or absent from the death assemblage (*Platymicrocythere* sp., *Semicytherura sabula* n. sp.). Furthermore, the number of live individuals greatly exceeds that of dead carapaces. These facts suggest that carapaces of dead ostracods are either being removed or destroyed and also that this process is selective. The near absence of *Cytherois zosterae* carapaces and the complete absence of *Platymicrocythere* sp. and *Semicytherura sabula* n. sp. from the death assemblages of the adjacent CBM biofacies seem to indicate that destruction is the dominant process.

The ostracods of this biofacies live in depths of less than 10 m in coarse sand that is constantly being moved by waves and currents. It is probably the high energy of the environment and associated physical destruction that is responsible for the under-representation of species in the death assemblage.

The modification processes in this biofacies do not completely obliterate the record of the life assemblage but are strong enough to significantly alter the composition of the unmodified death assemblage through selective destruction of carapaces of some species. Further, due to the destruction and resulting decreased abundance of carapaces of in situ species, carapaces added to the assemblage by transport may come to dominate the final death assemblage (*Xestoleberis hanaii* in T10).

(6) Intertidal rocky biofacies (IR): The correspondence between death and life assemblages is good. A remarkable exception is the poor representation in the death assemblage of *Paradoxostoma* spp., which are abundant or dominant in the life assemblage. This exception is striking when compared with the *Xestoleberis hanaii* live-to-dead ratios in the same samples. Clearly, destructive processes and transport are responsible. The abundance of dead *Xestoleberis hanaii* in this study and the near absence of *Paradoxostoma* spp. suggest that transport alone cannot account for the disparity and that destructive processes are prevalent.

(7) *Zostera* biofacies (Z): Addition to *Xestoleberis hanaii* from the adjacent IR biofacies changes the death assemblage. The absence of *Loxoconcha japonica*, *Aurila* sp. B, *Aurila* sp. A, and *Aurila hataii* from the life assemblage can be explained by the habitat of these species. They live on the leaves of the sea grass *Zostera* and generally off only after their death. The sample of the sand on which *Zostera* grows is therefore likely to contain dead carapaces of these species, even though they live in the area.

II. Differences between the death and fossil assemblages.

(1) Deep mud biofacies (DM): This biofacies is not found among the fossil samples. The absence of particular death assemblages from the fossil assemblages available does not necessarily mean that the death assemblage was destroyed during the fossilization process. In many instances, the absence may be accounted for by the fact that sediments deposited in an environment similar to that in which the assemblage is found at present are not represented among the samples available.

(2) Central bay mud biofacies (CBM): Fossil assemblage most closely resembling that of the CBM biofacies is found in the NL sub-biofacies. It is similar on account of *Nipponocythere bicarinata* and *Loxoconcha viva*, which range from dominant to common in both the fossil and recent assemblages. The two assemblages differ, however, by the absence of *Basslerites obai* and *Krithe japonica* from the fossil assemblage. Because the carapaces of both species are at least as thick as that of *Nipponocythere bicarinata*, which is abundant in the fossil assemblage, it is unlikely that the absence of *Basslerites obai* and *Krithe japonica* is due to selective destruction of their carapaces. Since there is no evidence of sorting in the fossil assemblage, their complete removal by currents is also not very probable. Both *Basslerites obai* and *Krithe japonica* are most abundant in sample T11 and decrease in abundance with decreasing depth and distance from shore. It is probable that the reasons for absence of these two species are ecological; they were probably absent from the biocoenosis of the fossil samples owing

to lesser depth of deposition, greater proximity to shore, and other ecological factors. Similarly, the presence of *Cythere lutea*, which was rare in the samples from Tateyama Bay, implies slightly different ecological conditions.

(3) Shallow fine sand biofacies (SFS): Fossil assemblages resembling that of the SFS biofacies are found in the PL biofacies and partly in the NL sub-biofacies (samples 43, 67). *Semicytherura miurensis* is dominant or very common in the Recent life and death assemblages but it is nearly absent from the fossil assemblages. This is not likely to be due to destruction prior to burial, since *Semicytherura miurensis* is well represented in the Recent death assemblages. Possible explanation is dissolution by acidic water percolating through the sediment subsequent to its exposure on land. This contention is supported by the eroded appearance of specimens of other species from these samples (53, 63) and absence of carapaces of thin-walled instars.

(4) Deep sand biofacies (DS): This biofacies is not found among the fossil samples; judging from sediment type, it was not sampled.

(5) Shallow coarse sand biofacies (SCS): This biofacies has no counterpart among the fossil samples; either the sediments were not preserved or not sampled because of lack of exposure. The near absence of a representative death assemblage in recent samples, coupled with the coarseness of the sediment which would facilitate circulation of ground water after uplift, makes it unlikely that fossil assemblage would be found even if the sediment of this biofacies were sampled.

(6) Intertidal rocky biofacies (IR): The death assemblage of this biofacies is dominated by *Xestoleberis hanaii* and includes a smaller proportion of *Paradoxostoma* spp., *Loxocorniculum mutsuense*, and *Neonesidea oligodentata*. Even though *Xestoleberis hanaii* is common in the fossil assemblages, no assemblage clearly dominated by *X. hanaii* and resembling the death assemblage of IR biofacies was found.

This is because the rocky intertidal zone is in times of regression and constant sea level a site of net erosion. Even at times of rapid transgression, the assemblage found in the thick growth of calcareous algae and local pockets of sediment in depressions is likely to be swept by currents and waves before it can be preserved by burial.

As suggested by the presence of *Xestoleberis hanaii* in death assemblages of the most recent samples, all components of the assemblage are not entirely destroyed, but some are preserved as extraneous elements, together with in situ species of other assemblages. Clearly, the shorter the distance from the rocky shore where the assemblage originated to the place of final burial, the better will the death assemblage be represented in the fossil assemblage. Under special conditions, such as a steep slope or cliff, the lateral distance between the living assemblage at the top and the burial zone at the foot of the slope can be of the order of a few meters. In such cases, most components of the IR biofacies assemblage will be well preserved along with the in situ biofacies assemblage of the burial zone. Such conditions probably existed at Koyatsu, and the fossil assemblage of the AL sub-biofacies there includes parts of transported IR biofacies assemblage (samples 5, 7, 8) and even includes some *Paradoxostoma* spp.

Carapaces of *Paradoxostoma* spp. and other thin-walled ostracods in the fossil assemblages are extensively bored by algae. It is uncertain whether the increased amount of light penetrating these transparent carapaces causes greater activity of boring algae. But even if an equal amount of boring is assumed for thin and thick carapace ostracods,

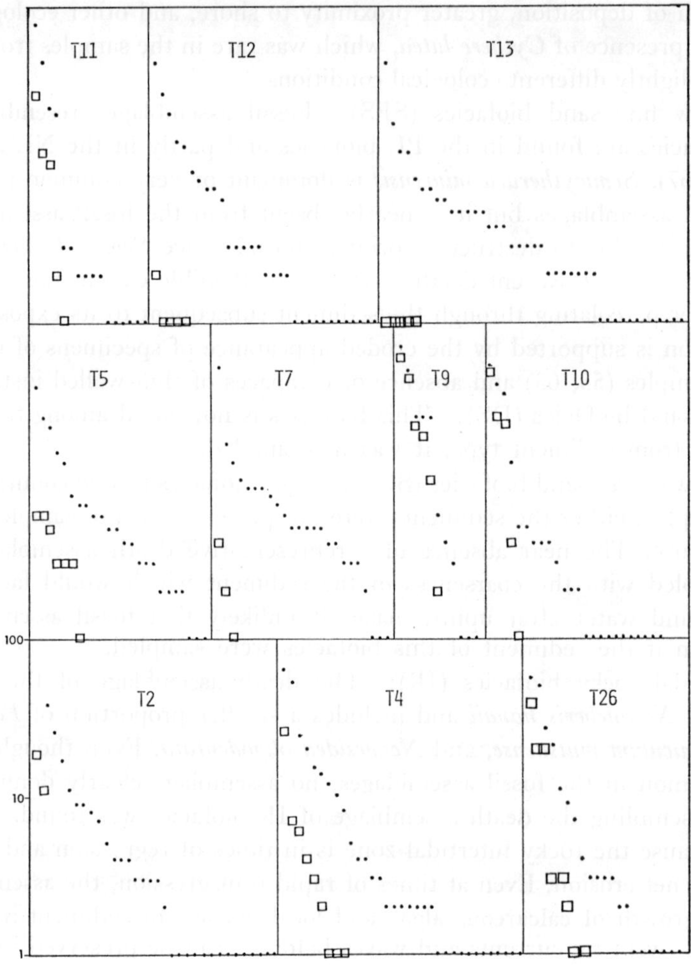

Text-fig. 21

Log-normal plot of abundance of live (squares) and total (dots) number of ostracods in samples from Tateyama Bay. Number of individuals is plotted on the ordinate, which is the log scale.

the effect is clearly much more detrimental to the thin-walled ones. This partly explains the rarity of *Paradoxostoma* in fossil assemblages and also in recent death assemblages, which are located far from the habitat of *Paradoxostoma*.

(7) *Zostera* biofacies (Z): The assemblage of this biofacies is similar to those of the AL and AA sub-biofacies. The only change is due to addition of IR elements as discussed above.

III. Recognition of modfied fossil assemblages: When only fossil assemblages are available, reconstruction of life assemblage becomes a much more difficult task. Several ways in which unmodified and altered fossil assemblage can be distinguished will be examined here, and the methods will be tested on the data available.

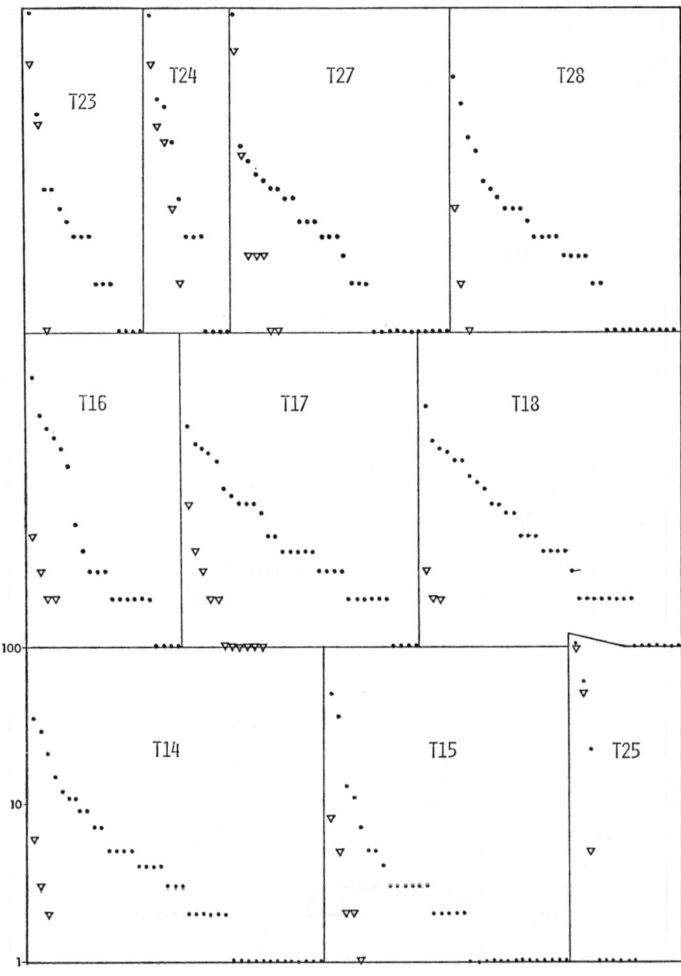

Text-fig. 22
Log-normal plot of the number of live and total ostracods present in samples from Tateyama Bay. Live ostracods are represented by triangles, total number by dots. Number of ostracods is plotted on the ordinate, which is the log scale.

(1) Community structure: Relative abundance of individuals of different species living in the same community can be approximated mathematically by the geometric series—a concept introduced by Motomura (1937) and known as Motomura's Law. Ikeya (1970) successfully used this relationship to distinguish modified and unmodified fossil assemblages of foraminifers. He found that in many life assemblages the relative frequency of individuals of species arranged by their abudnance rank follows the geometric series and, when plotted on log-normal paper, falls on a straight line. When Ikeya examined death assemblages, he found that in many cases the above relationship did not hold; when plotted on log-normal paper, a flexed or curved line was obtained. He explained the deviation as the result of the mixing of different assemblages, both

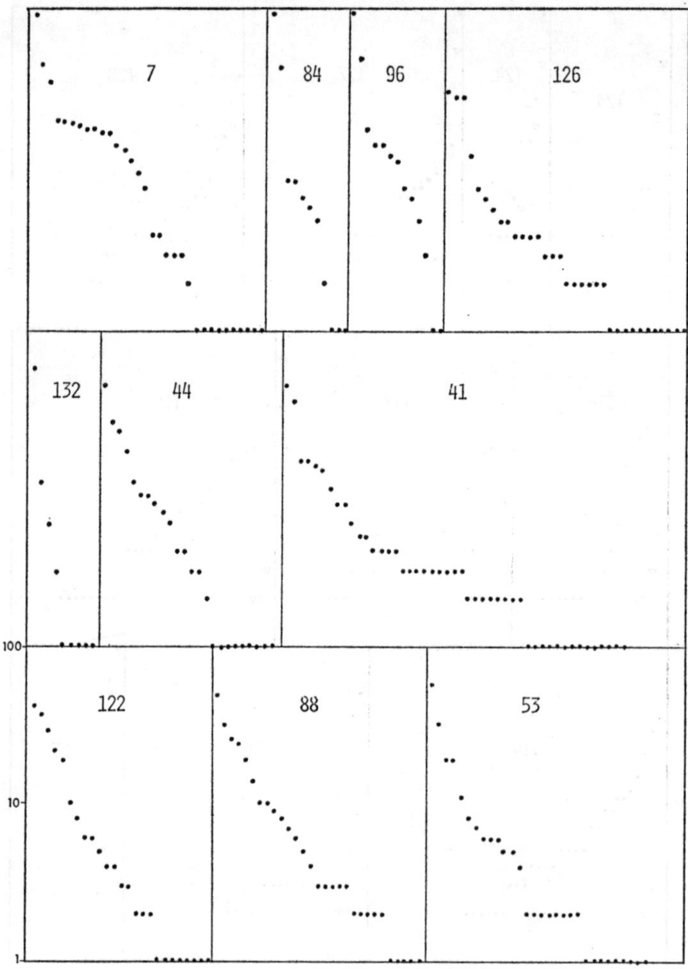

Text-fig. 23

Log-normal plot of the number of ostracods from selected Jomon transgression sediment samples. Number of individuals is plotted on the ordinate, which is the log scale.

in a temporal and geographic sense, and suggested that the degree of deviation from the geometric series demonstrates the amount of modification of the original life assemblage. Okada (1979) applied the same concept to fossil and recent ostracods.

Numbers of individuals within species from Tateyama Bay and selected Jomon transgression samples are plotted on log-normal scale in text-figs. 21, 22, and 23. In samples whose death assemblage was judged to be little modified, the points fall on or close to a steeply sloping straight line (samples T9, T10, T11, T24, T25, T26). In other samples (T13, T14, T17, T18), the numbers of individuals within a species plot on a line consisting of a short, steeply sloping part and a longer, more gently sloping part. According to the above theory, the species that plot on the steep segment would

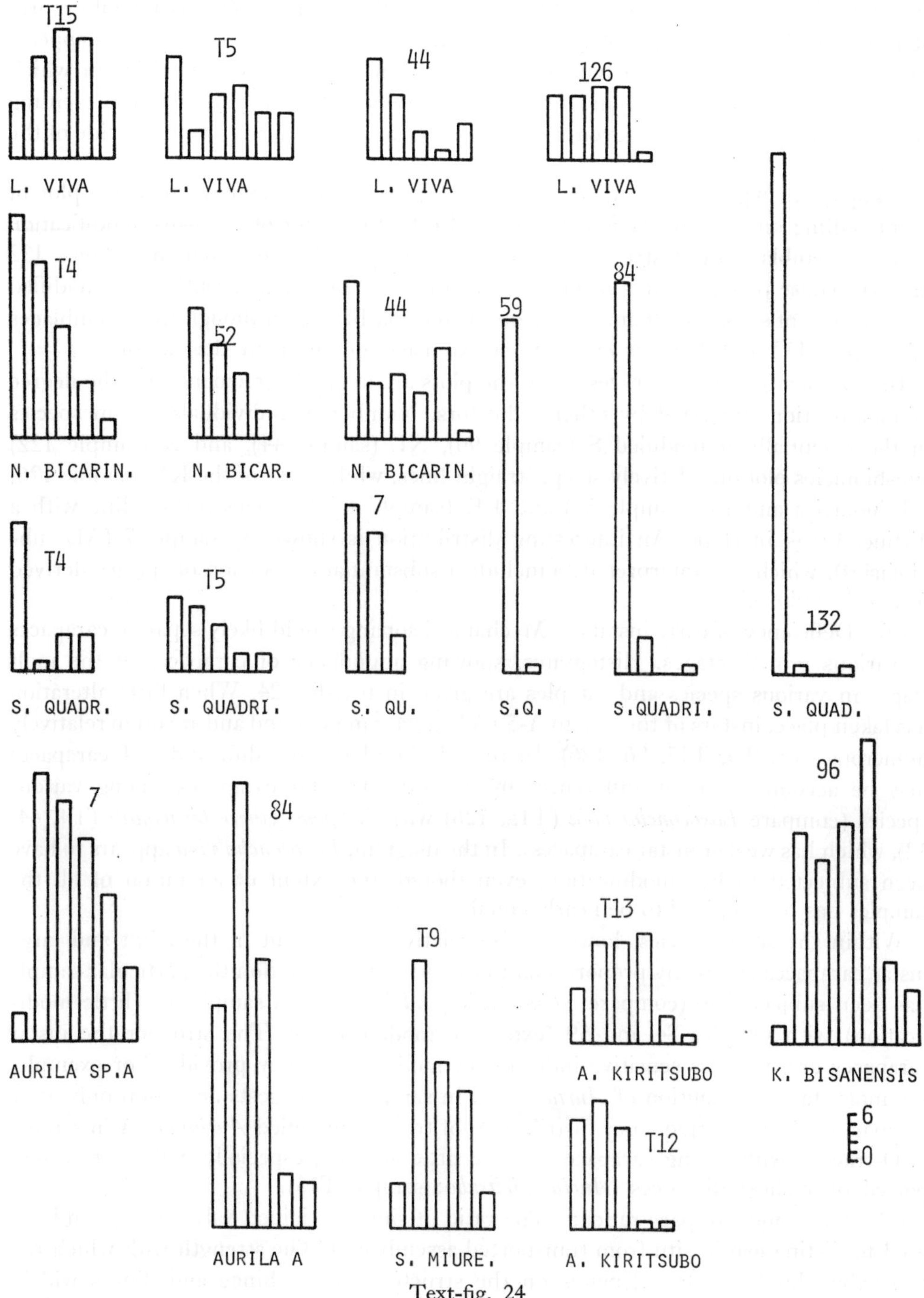

Text-fig. 24

Bar diagram showing the abundance of various growth stages of one species present in one sample. Selected samples and species from Tateyama Bay (prefixed with T) and Jomon transgression sediments are shown.

be the in situ species, while some of those lying on the gently sloping part would represent part of the assemblage not derived from the existing life assemblage. This appears to hold in most cases (T14) but does not apply in samples T12 and T13, in which *Aurila kiritsubo* is the most abundant species and thus plots on the segment representing the unmodified assemblage—even though it has clearly been derived from outside of the biofacies.

Comparison with fossil samples (text-fig. 23) brings forth another point: a plot of points falling on a steep, straight line may also be the result of extensive modification of the assemblage by destructive agents. This is well demonstrated in samples 132 and 84, whose plot compares favorably with the plots of the apparently little modified assemblages of samples T9 and T11 from Tateyama Bay, even though the assemblages of samples 132 and 84 show evidence of extensive alteration by dissolution.

In most samples, however, results of the plots are in good agreement with the degree of modification suggested by other indicators. Number of individuals within species in the essentially unmodified S (sample 96), NL (sample 44), and K (sample 122) sub-biofacies plot on relatively steep, straight lines, while those of the KN (sample 126) sub-biofacies and PL (sample 53) and LK (sample 41) biofacies form a line with a distinct break in slope. An interesting distribution is shown by sample 7 (AL sub-biofacies), which was interpreted to include a substantial proportion of species derived from other biofacies.

(2) Deficiency of early instars: Mechanical sorting would likely separate carapaces of various growth stages. Histograms, showing abundance of carapaces of the molt stages in various species and samples are given in text-fig. 24. When little alteration has taken place, instars of the A-3 to A-5 molt stage can be found and are often relatively numerous (T5, T9, T15, 96, 126). Increased abundance of adult and A-1 carapaces may be accounted for by differences in strength of instar carapaces among various species (compare *Loxoconcha viva* (T15, 126) with *Nipponocythere bicarinata* (T4, 44, 52), which has weaker instar carapaces). In the diagram, *Loxoconcha viva* appears to have been subjected to less modification, even though the extent of alteration of all the samples has been judged to be nearly equal.

Within the same species, however, the relative enrichment in the adult and large instar carapaces is directly proportional to the degree of alteration the particular sample has been subjected to (compare *Spinileberis quadriaculeata* in sample T5 (little modification) with samples 84 and 59 (extensive modification)). The stronger the instar carapaces are, the less sensitive indicator of modification they provide. For example, the molt stage distribution of *Aurila* sp. A in sample 84 suggests alteration only after comparison with the frequency distribution of truly unmodified *Aurila* sp. A in sample 7. Ostracods with strong carapaces will be little affected, especially if they are transported over short distances (*Aurila kiritsubo*: sample T13).

(3) Valve and carapace ratios: The ratio of single to joined valves has often been used to distinguish in situ from transported assemblages. The strength with which the two valves hold together depends on the structure of the hinge and differs widely depending on the ostracod species. Comparisons of ratios of single valves to whole carapaces are thus valid only when made within one species. Assemblage consisting only of complete carapaces could have been formed by rapid burial in situ or by trans-

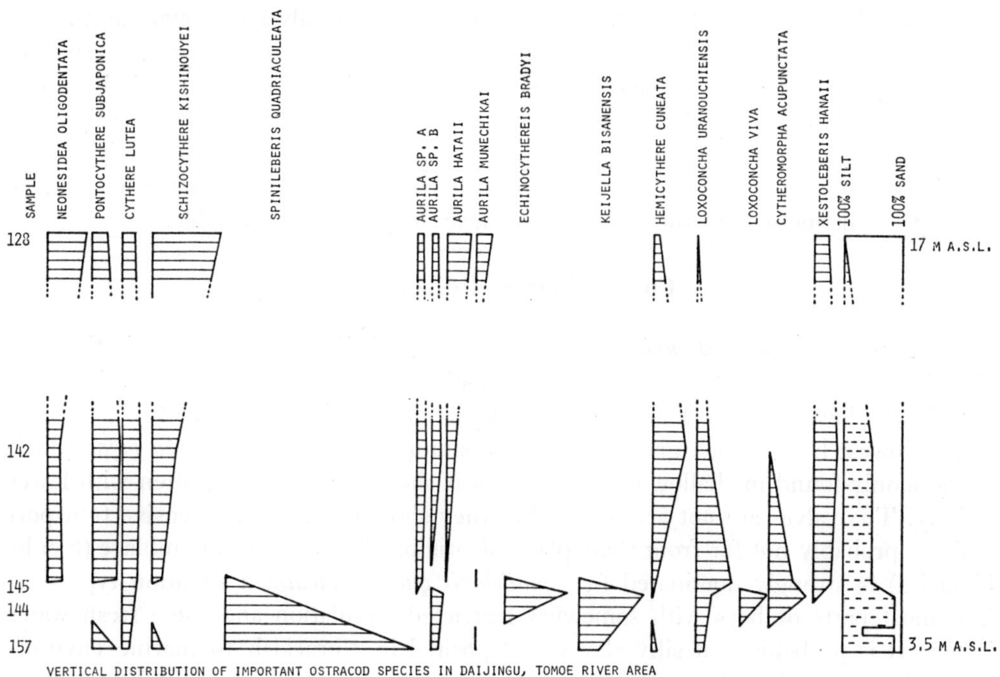

Text-fig. 25
Vertical changes in relative abundance of common ostracod species in Tomoe River Area.

port, in which case it would consist of carapaces of a species with similar threshold velocities.

The two valves of most ostracods differ only slightly in shape, size, and thickness, but in some species the difference is more pronounced (*Aurila* spp., *Neonedidea* spp.) and thus the transport characteristics as well as resistance to destruction are likely to be somewhat different. The ratio of left to right valves in a species characterized by marked lateral asymmetry could be used similarly to the single valve to carapace ratio to indicate the degree of modification of the assemblage, with the added advantage that it minimizes the effects of modification induced by sample preparation: valves may become separated during prolonged washing, but the ratio of left to right valves is not likely to change very much.

Ratios of left to right valves and single valves to complete carapaces of some species were examined. It appears that the left and right valve differences are too slight to serve as a useful indicator, unless the modification has been extensive and the number of valves available for study is large. Similarly, many ostracod carapaces are much too easy to separate (*Aurila* spp., *Loxoconcha viva*) to be useful in most cases. However, the carapace to single valve ratio may be, useful sometimes, as in the case of sample 132 and 59 where the high W/S ratio of *Spinileberis quadriaculeata* (10/48, 23/22) suggests that modification of the assemblage by size sorting has not taken place and that the alteration is probably due to dissolution.

The results indicate that the frequency distribution of valves of instars and adults is the best indicator of assemblage modification. The presence of abundant valves of earlier instars is the best measure of degree of modfication and the safest indicator of in situ origin of the species. Presence of only adult specimens does not necessarily mean that they are of allochthonous origin, since extensive in situ alteration might have destroyed the more fragile instar carapaces. In such cases, traces of dissolution may be of help in distinguishing the type of modification process responsible.

Conclusions and discussion

Interpretation of the past environment: I. Tomoe River Area: Changes in the relative abundance of the most common ostracods in sediments from 3 to 17 m above sea level in the lower part of Tomoe River valley are shown in text-fig. 25.

The lowest lying bluish green silty clay contains valves of *Crassostrea gigas* which is commonly found in shallow or intertidal near shore environments, often with lower salinity. The valves are not articulated but they show no signs of extensive transport and are probably not far from their place of origin. The ostracod assemblage (S sub-biofacies) is strongly dominated by *Spinileberis quadriaculeata*, a situation typical of innermost parts of bays with somewhat restricted circulation and some fresh water inflow. It is probable that silty clay was deposited in a brackish to marine environment, during the initial stages of the transgression, in a small incipient Tomoe Bay. Low sea level and proximity of shore is further indicated by the presence of peaty seams in the overlying sandy bed.

The peaty sand bed is overlain by one meter of silt to sandy silt. Presence of articulated *Ostrea denselamellosa*, a marine oyster, at the base of the bed indicates deeper water and at the same time completely marine conditions. The ostracod assemblage from this bed belongs to the K sub-biofacies. It is characterized by an increase in *Keijella bisanensis* and *Echinocythereis bradyi* and accompanying decrease of *Spinileberis quadriaculeata*. These changes also indicate a change toward more marine conditions.

The silty bed is abruptly overlain by a thin, organic matter rich bed. The lack of recognizable wood fragments, the nearly constant thickness of the bed, and its lateral continuity suggest that it is not a peat layer deposited during lowered sea level. It resembles the deposits of the dy biofacies (Ikeya and Hanai, 1980 MS) described from Hamana-ko estuary. Anoxic environment may develop in the inner parts of bays, where open-sea water circulation is not adequate, and which are too deep to be reached by wind generated waves. The presence of pyrite, indicating a reducing environment, as well as the absence of ostracods and other fossils, further supports this interpretation.

The sediment overlying the organic rich bed is substantially different from the sediment below it. It consists of 50% sand size particles. There are changes in the ostracod assemblage, but they are not as abrupt as one might expect. Species characteristic of muddy bottoms of bays, which dominated the assemblages below the organic-matter-rich bed, decrease in abundance and eventually disappear. The phasing out of *Spinileberis quadriaculeata*, *Echinocythereis bradyi*, and *Keijella bisanensis* is matched by the increasing importance of sandy open-coast species: *Neonesidea oligodentata*, *Pontocythere subjaponica*, *Schizocythere kishinouyei*, *Aurila hataii*, *Aurila munechikai*, and

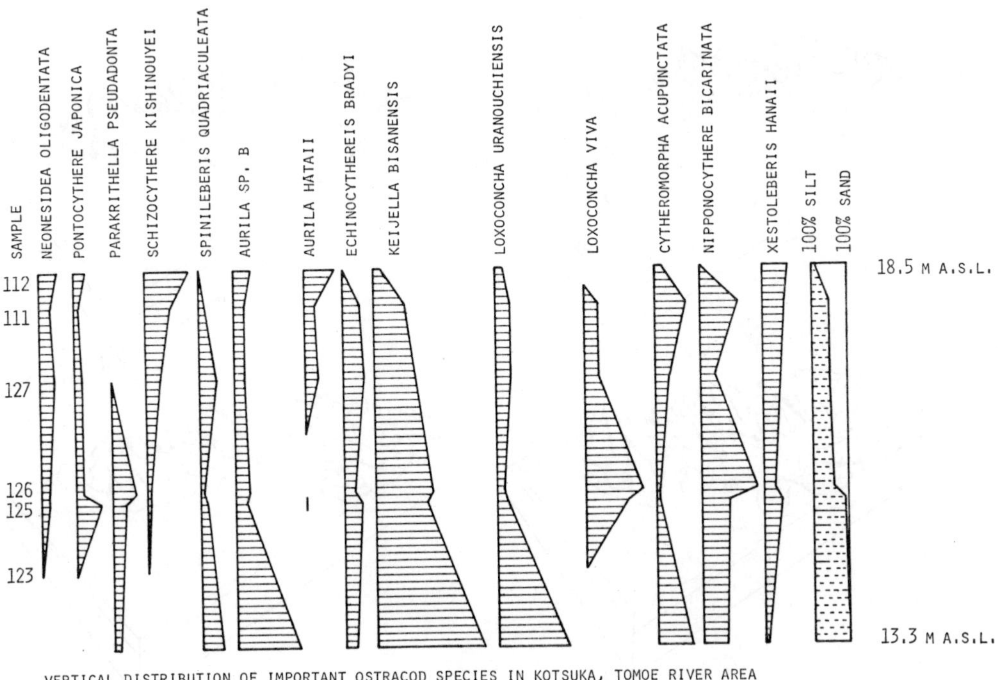

Text-fig. 26
Vertical changes in relative abundance of common ostracod species in Tomoe River Area near the village of Kotsuka.

Hemicytherura cuneata. The sediment overlying the organic-matter-rich layer is characterized by great valiability; it consists of alternating beds of varying thickness and composition. An overall trend is apparent; with increasing elevation, the sediment increases in coarseness and becomes better sorted. Ostracods characteristic of the open coast increase in importance and the final assemblage is dominated by *Schizocythere kishinouyei* and *Neonesidea oligodentata*. This is probably due to continued sea level rise, which resulted in the widening of the bay mouth and filling in by sediment, which kept the bottom within the reach of wave generated currents.

No C^{14} data are available for this section. Strong indications that the lowest exposed sediment was deposited in a brackish, shallow water environment during the initial stages of the transgression, together with a sea level of at least 18.5 m by 7,400 ($\pm$80) yr B.P., would indicate that the basal bed was laid down very early—perhaps around 10,000 to 9,000 yr B.P. This is further supported by the dating of a nearby bed at 10 to 11 m above sea level, dated at 7,690 ($\pm$190) yr B.P.

Text-fig. 26 shows similar changes in ostracod fauna in the upper part of the valley. The lowest S sub-biofacies is absent, probably due to lack of exposure. The lowest silty bed (at 13.3 m above sea level) with abundant *Dosinella penicillata* was dated at 7,630 ($\pm$70) yr B.P., making it contemporaneous with sediments at about 10 to 11 m above sea level in the lower part of the valley. The ostracod assemblage belongs to the K sub-biofacies and indicates a typical inner bay environment. Appearance and domi-

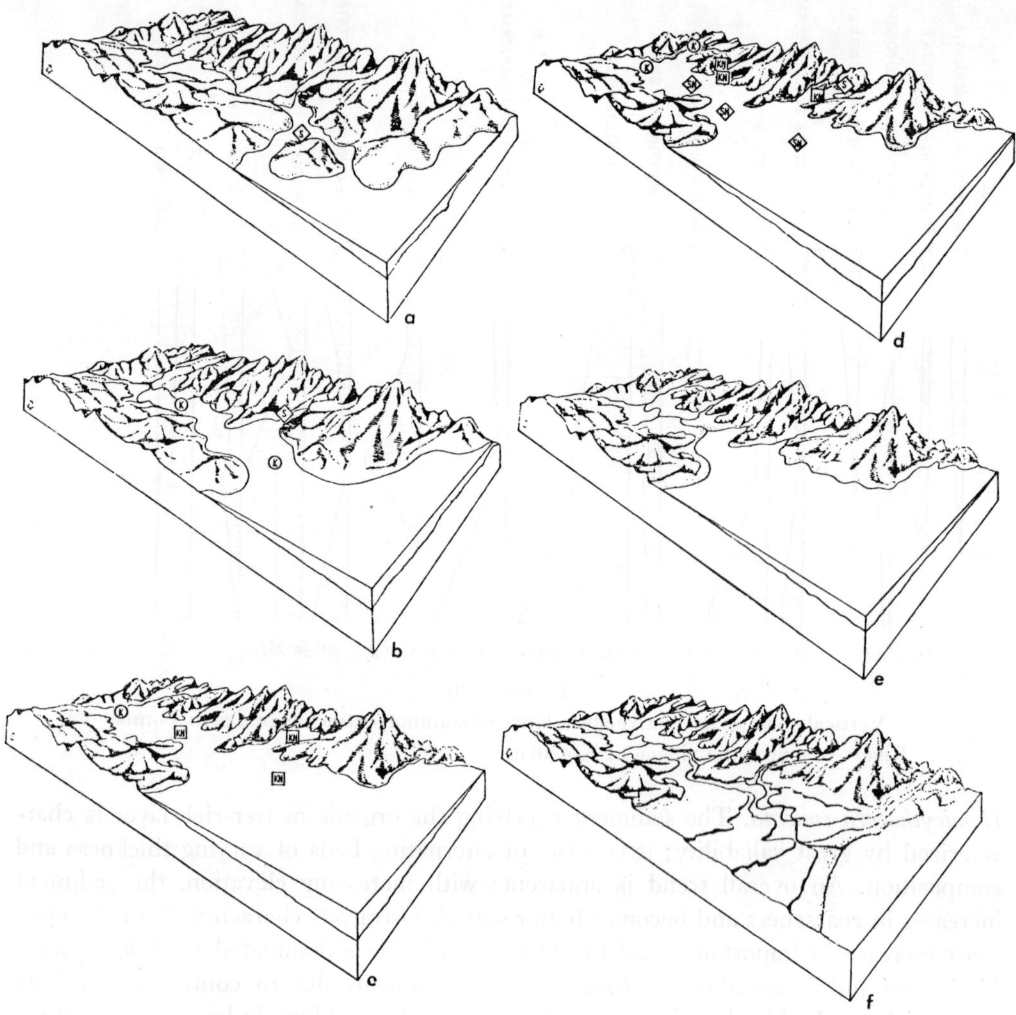

Text-fig. 27

Change of Tomoe Bay and biofacies distribution: (a) Initial Tomoe Bay and biofacies distribution in the early stage of the transgression, (b) Tomoe Bay and biofacies distribution around 8,000 yr B.P., (c) Biofacies distribution during period of high sea level around 7,000 yr B.P., (d) Biofacies distribution in gradually shallowing Tomoe Bay around 6,000 yr B.P., (e) Emergence of the Numa I terrace in Tomoe River area, (f) Present-day topography in Tomoe River area.

nance of the assemblage by *Loxoconcha viva* and *Nipponocythere bicarinata* at about 15 m above sea level, indicate a change in conditions, probably brought about by a further rise in sea level, which broadened the bay and improved access of open-sea water. The faunal change is accompanied by increased proportion of sand, indicating a higher energy environment. The amount of sand continues to increase until it reaches 90% at 20 m above sea level. *Keijella bisanensis*, *Loxoconcha viva*, and *Nipponocythere bicarinata* are gradually replaced as the dominant species. The new assemblage is dominated by *Schizocythere kishinouyei*, *Hemicytherura cuneata*, *Aurila hataii* and other species typical of an open coast environment. Comparison with the highest assemblage of the previous section is interesting. Even though important, *Schizocythere kishinouyei* and *Neonesidea oligodentata* do not dominate the assemblage to the extent of the upper beds in the lower part of the valley. Continued presence of species typical of the KN and K sub-biofacies also indicates that conditions were intermediate between the open coast and bay, a proposition fully supported by the outline of the reconstructed bay.

Summary of the development of Tomoe River area: Incipient bay was formed in the lower part of the valley about 9,000 yr B.P. (text-fig. 27a). Continued rise of sea level caused the shifting of the S sub-biofacies into inner parts of Tomoe River valley and its tributaries, while the K sub-biofacies replaced the S sub-biofacies in the lower parts of the bay (text-fig. 27b). Increased depth together with the presence of a barrier (possibly a sand bar) impaired circulation in the deeper water in the lower part of the bay, giving rise to an anoxic environment. Further rise in sea level or removal of the barrier or both restored normal marine conditions in the lower part of the bay where the KN sub-biofacies developed (text-fig. 27c). At the same time, the S sub-biofacies shifted further landward and was replaced by K sub-biofacies. Further rise in sea level widened the bay, facilitating circulation of open-sea water as well as movement of waves and currents, resulting in coarser sediment. K sub-biofacies was replaced by KN sub-biofacies and the S and K sub-biofacies moved to the inner parts of small bays. Shallowing, brought about by sediment infilling and possibly by partial lowering of sea level (uplift?) resulted in the development of a shallow open coast environment and NS biofacies in all, except the innermost, parts of the bay (text-fig. 27d). Uplift at about 5,500 yr B.P. resulted in the formation of the Numa I terrace and ended sedimentation in the larger part of the area (text-figs. 27e and f).

II. Tateyama city area: The oldest dated material is *Saxidomus purpuratus*, boring in the siltstone underlying the oyster reef near Nishigo. *S. purpuratus* lives in depths ranging from intertidal to 40 m, but is most common at depths of about 20 m. *Barnea manilensis inornata*, whose boreholes are found alongside those of *Saxidomus purpuratus*, is most common at shallower depths of about 4 m. At the above locality, its numbers are exceeded by those of *S. purpuratus* and it is therefore probable that water depth at this locality was already 10 to 20 m by 7,330 (±120) yr B.P. *Pretostrea imbricata*, which forms the oyster reef is known to live in depths ranging from 10 to 50 m, a fact that agrees well with the above conclusion.

Ostracod assemblages found in the interstitial sediment in the reef belong to the LK biofacies (text-fig. 28). The lower parts of the reef, where *Alveopora verrilliana* is the only coral present, are characterized by slightly higher proportions of *Spinileberis*

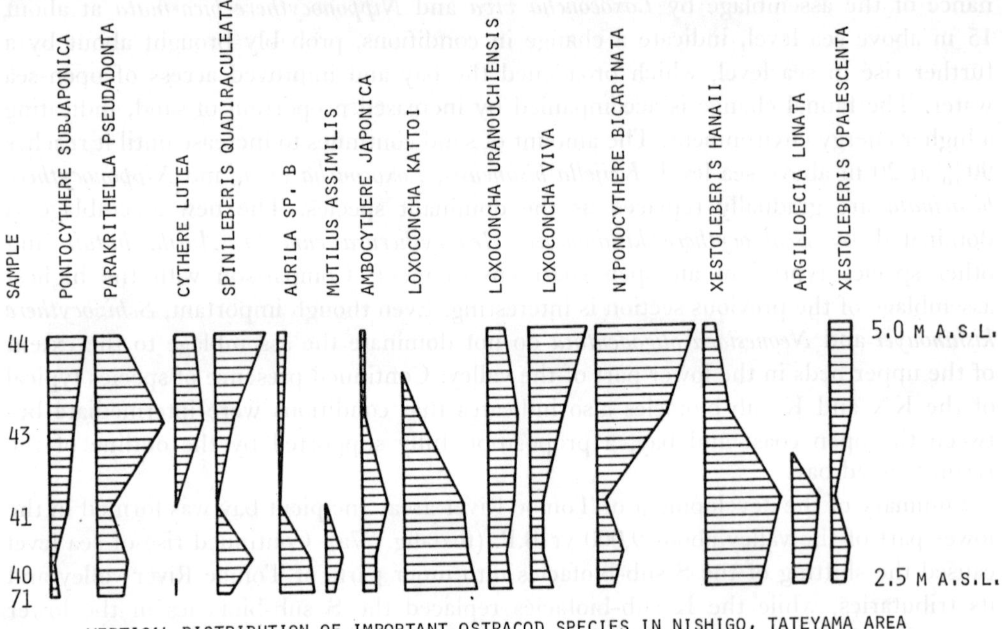

Text-fig. 28
Vertical changes in relative abundance of common ostracod species in and above the oyster reef in Nishigo, Tateyama Area.

quadriaculeata and *Nipponocythere bicarinata*. The decrease in the proportion of these species, typical of muddy substrata, coincides with the appearance of diverse coral fauna, dated 6,400 to 6,300 yr B.P., suggesting that absence of fine sediment or perhaps sufficient elevation above the surrounding muddy bottom, due to the growth of the reef, was responsible for the increased coral diversity. Increased abundance of *Ambocythere japonica*, a species found in coral-bearing sediments in the West Coast Valley area, also coincides with the coral rich zone in the oyster reef. *Xestoleberis hanaii*, characteristic of agitated hard substrate environments, is another species that is abundant in the sediment throughout the reef.

The cessation of the oyster reef growth coincides with an abrupt change of ostracod fauna. The assemblage from the overlying silty sand belongs to the NL sub-biofacies and is characterized by dominance of *Loxoconcha viva* and *Nipponocythere bicarinata*. *Spinileberis quadriaculeata*, *Cythere lutea*, *Pontocythere subjaponica*, and *Xestoleberis opalescenta* also increase in abundance. *Mutilus assimilis* and *Argilloecia lunata*, common in the reef sediment, are lacking; *Loxoconcha kattoi*, *Aurila* sp. B and *Xestoleberis hanaii* are reduced to insignificant proportions. *Parakrithella pseudadonta* becomes most abundant in the fragmented shell bed directly overlying the oyster reef, but decreases in abundance in the above sediments. Inferred age of the fragmented shell bed coincides with the time of emergence of the Numa I terrace indicated in other areas. It is possible that it originated by erosion of landward positioned muddy sediments as a result of the sea level drop.

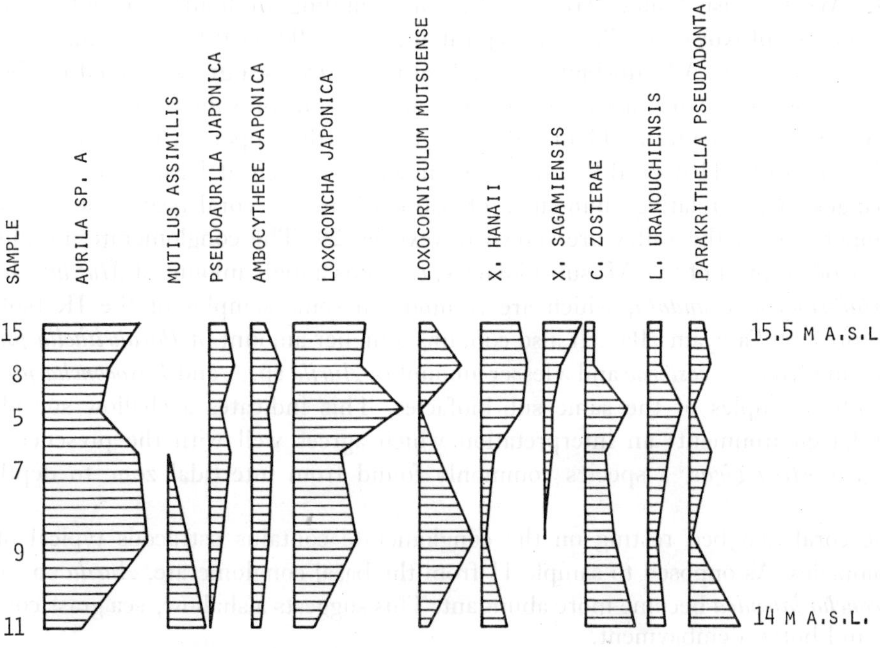

Text-fig. 29

Vertical changes in relative abundance of common ostracod species in the coral-bearing sediments in Koyatsu, West Coast Valley Area.

Spondylus barbatus from the top of the reef was dated at 5,800 yr B.P., while *Dosinella penicillata* from the directly overlying sediment was dated at 4,490 yr B.P. The 1,300-year gap is consistent with the interpretation of the reef as a local bathymetric high, and 1,300 years were needed for the surrounding sediment level to reach the top of the reef.

Most of the faunal changes can be explained by the absence of species which lived on the reef or on algae and sea grasses supported by it. The existence of NL sub-biofacies up to 4,440 yr B.P. indicates that water depth remained in the 10 to 20 m range. Younger sediments were, at this locality, removed by fluvial erosion.

Only one C^{14} date is available for the surrounding sediments. It suggests that the NL sub-biofacies, which occupies the central part of the Tateyama area, was likely deposited during the time interval from 7,000–8,000 to 4,400 yr B.P.

The PL biofacies, occurring in the western part of the area, appears to have been roughly contemporaneous with the sediments overlying the top of the reef; since its depth of depostion was probably 5 to 10 m, and its present location is 15 to 20 m above sea level, the bed overlying the oyster reef was probably deposited at depths of 15 to 30 m. This is in good agreement with the depth at which *Loxoconcha viva* and *Nipponocythere bicarinata* occur in the present Tateyama Bay.

III. West Coast Valley Area: Silty sand infilling in boulder conglomerate in the lower part of Koyatsu valley was deposited 7,740 (±90) yr B.P. It contains ostracods characteristic of the AA sub-biofacies, indicating *Zostera* sea grass covered the bottom in the area where open coast to bay mouth environment prevailed. Because *Crassostra gigas* from conglomerate at 14 m above sea level in the upper part of the valley was dated at 7,840 (±110) yr B.P., the depth must have been at least 13 m.

Changes in the relative abundance of ostracods in the coral-bearing sediments in the upper part of the valley are shown in text-fig. 29. The conglomerate, in addition to ostracods typical of the AL sub-biofacies, contains a high amount of *Mutilus assimilis* and *Callistocythere undata*, which are common in some samples of the IR biofacies in the present Tateyama Bay. It also contains a higher amount of *Parakrithella pseudadonta* and *Cytherois zosterae* and a lesser amount of *Aurila* sp. A and *Loxoconcha japonica* than other samples of the same sub-biofacies. This indicates a shallow subtidal to intertidal environment: an interpretation which agrees well with the presence of in situ *Crassostrea gigas*, a species commonly found from intertidal zone to depths of 5 m.

The coral-rich bed resting on the conglomerate contains ostracods typical of the AL biofacies. As opposed to sample 11 from the basal conglomerate, *Aurila* sp. A and *Loxoconcha japonica* become more abundant. This suggests a shallow, sea grass covered, silty sand bottom embayment.

An increase in the amount of *Spinileberis quadriaculeata* in the sediments overlying the in situ corals (sample 9) can be explained either by a lowering of sea level, resulting in a greater proximity to a fresh water source, or by a rise of sea level, which would result in a lesser agitation and allow deposition of finer sediment. The increase in *S. quadriaculeata* is not accompanied by a corresponding increase in *Mutilus assimilis* nor *Callistocythere undata*, species which characterize the intertidal to shallow subtidal environment of sample 11. It is probable that the change was caused by a rise of sea level. The subsequent decrease in the amount of *Spinileberis quadriaculeata* present, coupled with the increased proportion of *Xestoleberis hanaii* and *X. sagamiensis*, together with the continued presence of large numbers of *Aurila* sp. A and *Loxoconcha japonica*, might indicate further rise of sea level, which resulted in widening of the bay, better circulation of open-sea water, and development of intertidal rocky biofacies even along the inner margins of the bay. Shells from the top of the bed were dated at 6,720 (±60) yr B.P.

While corals were growing in Koyatsu, sediment containing ostracods of the AA sub-biofacies was being deposited in the central part of the neighboring valley of Shiomi, covering a conglomerate dated at 7,230 (±70) yr B.P. Ostracods of the AA sub-biofacies are found also in the overing sediments, along with in situ *Clementia papyracea*, *Paphia undulata*, *Panopea japonica*, and *Dosinella penicillata*, which has been dated at 6,370 (±165) yr B.P. The molluscan and ostracod assemblages suggest 10 to 15 m deep bay with sufficient influx of open coast sea water.

Corals also flourished in other bays—Numa, Kamisanakura—along the coast. All of them are found surrounded by sediment containing the AL ostracod sub-biofacies. Dated corals from Numa range from 6,160 (±120) yr B.P. to 5,365 (±170) yr B.P., and shells from the upper part of the coral bed at Kamisanakura were dated at 5,500

(±70) yr B.P.

Ostracod assemblage from Numa is rich in *Spinileberis quadriaculeata*, suggesting more inner bay environment, which is consistent with its location in the inner part of paleo-Tateyama Bay and in the center of Numa valley. The Kamisanakura ostracod assemblage (sample 36) is characterized by near absence of *Spinileberis quadriaculeata* and abundance of *Xestoleberis hanaii*, *Parakrithella pseudadonta*, and *Pseudoaurila japonica*, suggesting a shallower environment or greater lateral proximity to it.

To summarize, the transgressing sea reached at least 14 m above sea level by 7,840 (±110) yr B.P. It continued to rise and sometime before the formation of the Numa I terrace it reached at least 25 m, during which time molluscan borings in Koyatsu cliffs originated. Abundant hermatypic corals appeared slightly after 7,800 yr B.P. and continued to flourish at least until 5,500 yr B.P.

Sea level reconstruction: Relative changes of sea level over the past 10,000 years have determined the outline of the coast of the southern Boso Peninsula. The shape of the coastline extended profound influence on near shore sediment distribution and salinity, which are the two major factors controlling distribution of corals and ostracods. Knowledge of the relative sea level position is therefore crucial for understanding changes of coral and ostracod biofacies; moreover, information on distribution and succession of coral and ostracod biofacies helps to explain the movement of sea level.

The precision of reconstructed sea level depends on the precision of dating and on the certainty with which the position of the dated material for past sea level can be estimated. Under ideal conditions, where dated peat and marine sediment beds are available, sea level changes can be determined with a high degree of accuracy.

Reconstruction of past sea level changes in the southern Boso Peninsula is facilitated by the relative abundance of C^{14} dated material from Holocene sediments. Unfortunately, most of the material comes from sediments whose depth of deposition can not be easily estimated. There are only two dated peat deposits and one dating of beach rock. The approximate depth of deposition can be partially inferred from the fossil assemblage and sedimentary structures, but the resolution (in most cases, 5 to 10 m) is inherently far too coarse to detect most changes of sea level, considering that the maximum change in the area did not exceed 30 m.

The complications arising from the wide depth range of most potential depth indicators, such as marine fossils and bedding structures, are further aggravated by the sediments that were found in narrow bays, where environmental factors other than depth must have been often decisive. The information provided by terrace elevations also must be scrutinized carefully, becaue even though formed at the same time and at the same relative height of sea level, the rate of deposition and erosion may have been different in different localities, thus producing different elevations.

Finally, most outcrops are located in channels of rivers and streams and are at least seasonally below the water table. Contamination of the dated material is the probable cause of several spurious dates.

However, even in the absence of reliable precise indicators of sea level, minimum sea level curves for each area can be drawn based on the elevation of the highest available dated material in marine sediments for the given area and time. In most cases, the minimum sea level will lie several meters below the true sea level, but it gives the

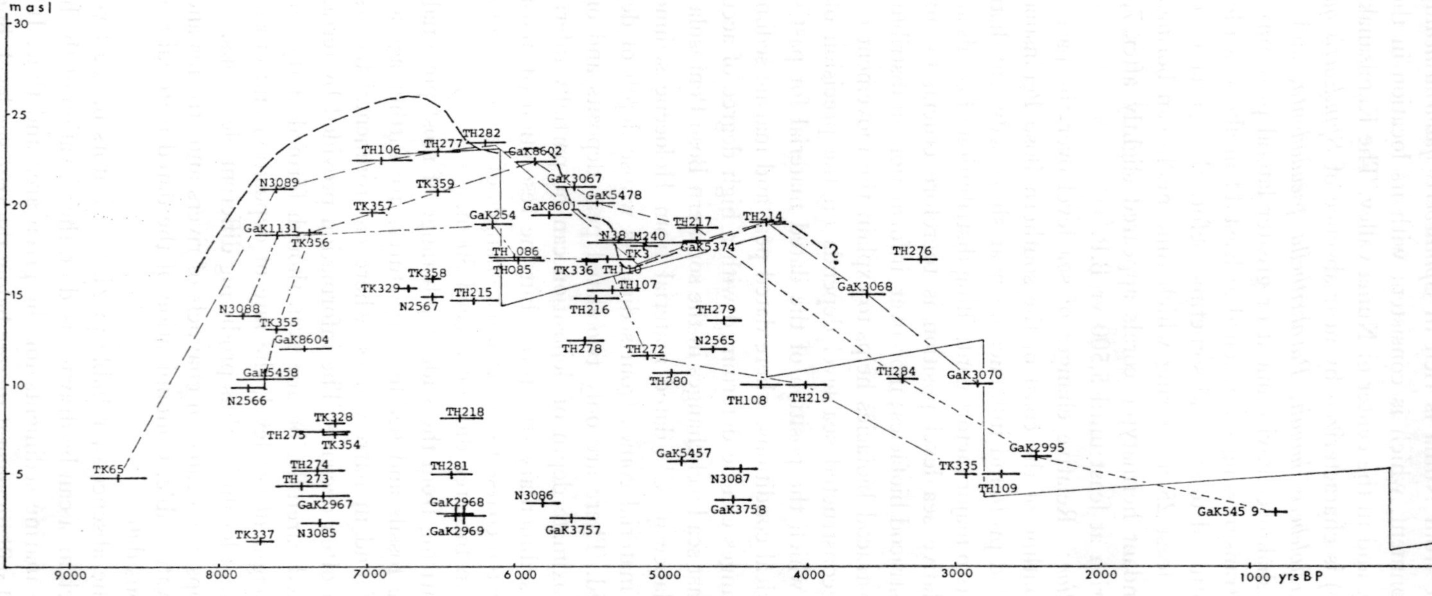

Text-fig. 30

Sea level changes in southern Boso Peninsula. Solid line shows the sea level curve proposed by Nakata et al. (1980); thick, interrupted line gives the sea level curve suggested by this study; thin, interrupted lines represent the minimum sea level curves in different study area.

absolute minimum value for given time and place.

The minimum sea level curves drawn for the three areas, West Coast Valley area, Tomoe River area, and Chikura area, resemble each other in shape (text-fig. 30). The curves rise rapidly and appear to reach a peak slightly before 6,000 yr B.P. No marine sediments dated younger than 5,480 ($\pm$110) yr B.P. are found above 20 m. Even though the curves have similar shapes, they differ in the absolute values of elevations; the curve for Chikura area is the highest (reaching 23 m); the curve for Tomoe River area is intermediate; and the curve for West Coast Valley area is the lowest, reaching only 19 m. Given the evidence of differential uplift during historical earthquakes, it is tempting to ascribe the differences to the cummulative effect of such events in the past. It is necessary to note, however, that there is an alternative explanation.

The depositional surface is seldom horizontal and often has different elevations in different areas. The elevation of the depositional surface, and thus dated material, depends on the elevation of the underlying bedrock and on the rate of deposition. Thus, material of practically the same age is found at widely differing elevations in the upper and lower parts of Koyatsu valley. Similarly, the higher elevation of the Numa I surface in Chikura and Tomoe River areas in comparison with the West Coast Valley area could be due entirely to higher rate of deposition. The watershed of both Tomoe and Seto Rivers far exceeds that of the streams in the West Coast Valley area. Removal of sediments by waves and currents would also be slower in the inner parts of protected bays (such as Tomoe Bay) resulting in formation of a higher depositional surface.

Accordingly, the variation in the elevation of terrace surfaces and material of the same age can be explained either by differential uplift or by differences in net deposition rate and location of dated samples. The presence of abundant bivalve borings up to 26 m in the Koyatsu valley and Tomoe River areas seems to indicate that the maximum sea level was the same in both areas, even though terrace surfaces are lower in the West Coast Valley area. It does not, however, prove conclusively that no differential uplift occurred. Uplift may have taken place, but evidence of sea level exceeding 26 m in Chikura area has either not been preserved or has not been found yet. In this sense there will always remain the possibility of discovering evidence of an even higher sea level in one of the areas, thus the question must be regarded as open.

The simpler theory, involving no net differential uplift, has been provisionally adopted because it appears sufficient to explain nearly all the data available at the moment.

The one exception is evidence of sea level at about 7 to 8 m above sea level around 7,240 ($\pm$80) yr B.P. in Maruyama area. The dated material comes from sediment which must have been deposited at a depth not exceeding a few meters. It is in direct contradiction with evidence from other areas, which indicates that sea level reached at least 21 to 22 m by this time. It is possible the material had been contaminated and thus yielded a younger age, but judging from the excellent preservation of the dated material and the low permeability of the sediment in which it was enclosed, contamination seems unlikely. The possibility of contamination cannot be rejected entirely, however, especially since the material comes from a river bank which is below the water table for at least part of the year.

Lesser uplift of the area in comparison with the southern part of the area is another way to explain the anomalous date. The Maruyama area experienced the least uplift during the Taisho earthquake (1 m in comparison with the maximum of 2 m in Tomoe River area). The question cannot be resolved, however, until additional data from the area become available.

Very little is known about the position of sea level prior to 8,000 yr B.P. The one clear indicator of a zero sea level, peaty seam at 4 m above sea level in Tomoe River area was not dated. Shells from 10 to 11 m above sea level from a nearby locality have yielded the age of 7,690 (±190) yr B.P. Assuming a similar rate of deposition as in the upper part of the Tomoe River valley, the age of the peat seam would be about 10,600 years.

In Tomoe River area, *Dosinella penicillata*, living in 0 to 20 m depth and found in silt at 13.3 m above sea level, was dated at 7,630 (±70) yr B.P. The associated ostracod assemblage dominated by *Keijella bisanensis* indicates moderate depth. The overlying sediments containing ostracod species *Keijella bisanensis* and *Nipponocythere bicarinata* suggest a rise of sea level. Presence of the *Keijella bisanensis* assemblage in sediments further upstream at 18.5 and 20.5 m above sea level, which were dated at 7,400 (±80) and 6,530 (±80) yr B.P., confirms a continuing rise of sea level after 7,600 yr B.P.

Crassostrea gigas from Koyatsu valley, dated at 7,840 (±110) yr B.P., is the oldest available date from the West Coast Valley and Tateyama areas. It lies at 14 m above sea level and indicates that sea level was at least that high by 7,840 yr B.P. *C. gigas* is commonly found in the intertidal zone but is present at a depth of several meters.

Coral and ostracod fossil assemblages in sediment overlying the conglomerate with *Crassostrea gigas* suggest an increase in depth. Several dates, spanning about 2,000 years, for corals and shells from coral beds in the West Coast Valley area come from sediments about 17 m above sea level. The highest coral is from 19 m in Numa, dated at 6,160 (±120) yr B.P. The youngest of these dates was obtained from shells in coral bed in Uesanakura, located at 17 m above sea level. The ostracod assemblage dominated by *Aurila* sp. and *Loxoconcha japonica*, along with high coral diversity, suggests a depth of 5 to 10 m. Abundant bivalve borings were found in the cliff in Koyatsu valley and Tomoe River as high as 26 m, suggesting that the sea level reached at least that level and perhaps a few meters above it. This is consistent with the interpretation of the depositional environment of the Koyatsu coral bed by Hamada (1963), who concluded that the coral probably lived in a depth of about 10 m. The interval of highest sea level was not long enough to result in the creation of a wave-cut terrace. *Crassostrea gigas* from 21 m above sea level in Chikura area was dated at 7,620 (±110) yr B.P., and a shell sample from 22.5 m above sea level was dated at 6,910 (±200) yr B.P., indicating that sea reached at least this level by that time. *C. gigas* at 23 m above sea level yielded the age of 6,530 (±140) yr B.P. The *C. gigas* is overlain by peat at 23.5 m, dated at 6,200 (±135) yr B.P., suggesting that sea level was below 23 m at that time. We may say that level was between 23 and 23.5 m above sea level during the interval from 6,500 to 6,200 yr B.P. Wood contained in marine sediments lying near the surface of the Numa I terrace at 20 m above sea level in Mera, on the coast adjacent to the Tomoe River, yielded the age of 5,480 (±110) yr B.P., indicating that sea level remained above 20 m at least up to 5,480 yr B.P. If differential uplift has occurred, the presence of a peat bed in Maruyama area colld be explained by lesser uplift in the northeast: a hypothesis which would be also supported by evidence of lower sea level there around 7,240 (±80) yr B.P. If no differential uplift has occurred, the peat bed would provide evidence of an abrupt lowering sea level around 5,500–5,400 yr B.P.

Presence of peat bed dated at 5,230 (±100) yr B.P. at about 18 m in Maruyama area indicates a lowering of sea level below 18 m. The absence of sediments younger than 5,480 (±110) yr B.P. from Numa I terrace indicates a fall in sea level. The maximum extent of the lowering is uncertain.

Nakata et al. (1980) suggested a lowering of sea level down to 14 m, based on unconformity they found in Tomoe River area (C^{14} dates TH 216 and TH 215). I was unable to locate and examine the outcrop of the unconformity. The brief description by Nakata et al. (1980) does not provide a strong indication of sea level lowered down to 14 m. The change in sediment type and erosion of underlying sediment (implied by the 1,000-year gap in the dates) could be accounted for by events taking place at depths of 8 to 10 m. The presence of shells dated at 4,740 (±100) yr B.P. and 4,280 (±130) yr B.P. in shallow sea sand (18 m above sea level) and in beach sand (18.9 m above sea level) in Chikura area suggests a new rise of sea level at least up to 18 m.

Dated material close to the surface of the Numa II terrace indicates that it emerged around 4,300 yr B.P. The movement of sea level during the interval between the emergence of the Numa I and II terraces is not clear.

Shells from cross-bedded sand, lying 0.3 m below the surface (at 10 m above sea level) of the Numa III terrace at Chikura were dated at 4,315 (±145) yr B.P. Shells from beach rock at the same elevation in a nearby locality yielded the age of 2,860 (±100) yr B.P. *Cyclina sinensis* from 10.3 m at Shirahama was dated at 3,360 (±100) yr B.P. Assuming that beach rock is the most precise indicator of sea level, the Numa III terrace emerged about 2,800 yr B.P.

The sea level was at least 6 m high at 2,440 (±100) yr B.P., as indicated by dated shells from shallow marine sand in Heisaura. As recently as 820 (±80) yr B.P., the level was above 2.8 m in the same area (C^{14} date Gak 5459). The Numa IV terrace emerged during the 1703 earthquake.

Two sea level curves for the southern Boso Peninsula have been published by Yokota (1978) and Nakata et al. (1980). The curves are in general agreement with each other and with the curve proposed in this study. Partly due to additional data and partly due to differences in interpretation the curve proposed here differs in several details (text-fig. 30).

New data have demonstrated that sea has reached high level about 2,000 years earlier than shown by Yokota. The wood from 4 to 5 m above sea level in Kawajiri River area is dated at 16,420 (±360) yr B.P. and lies in nonmarine sediments, which were probably deposited in the flood plain of Kawajiri River. It indicates only that sea level was below 4 to 5 m at that time.

The curve proposed by Nakata et al. (1980) is appealing in its simplicity and emphasis on the main mechanism of sea level change in the area. There are, however, several points where the curve disagrees with available data and other points where there are no data to support the curve.

For instance, the timing and extent of the drop of sea level at 6,100 yr B.P. is not convincingly demonstrated by the disparity in the TH 215 and TH 216 dates and is in clear contradiction with dates TH 086 and TH 085 from Numa, Gak 5478 date from Mera, and Gak 8602 date from Kotsuka, all of which demonstrate that sea level was still high at the time of its purported lowering to 14.5 m.

Lowering of sea level is, however, indicated by the peat bed in Maruyama prior to 5,230 yr B.P. (if no differential uplift occurred), following which it appears to have risen in accordance with Nakata's curve.

Further, there is no evidence of rise in sea level in the interval separating the emergence of Numa II and III terraces as shown in Nakata's curve. The sand from which sample TH 108 was obtained could have been deposited in a depth reaching several meters, but beach rock of sample Gak 3070 was most likely formed close or above the mean sea level.

The curves of Yokota (1978) and Nakata et al. (1980), representing the minimum sea level curve, are correct only if all the dated material through which the curve passes were deposited close to sea level—a supposition which is clearly not true. The curves therefore underestimate the maximum elevation which the sea level reached.

The curve proposed in this study suggests a sea level reaching at least 26 m. While supported by strong evidence of bivalve borings, the age and duration of this high stand of sea level has not been proven and therefore must be regarded with caution.

Presently available data seem to favor the lowering of sea level in several steps in the interval between the emergence of the Numa I and II terraces. The scarcity of data does not allow a reconstruction with confidence of the movement of relative sea level after the emergence of Numa II terrace.

Effect of the climatic change on ostracods: Climatic reconstructions based on fossil pollen indicate a warming up from around 9,000–8,000 yr B.P. and temperatures 1 to 2°C higher than at present in the interval from 8,000 to 4,900 yr B.P. (Kuroda and Hatanaka, 1979). The climatic optimum was reached around 6,000 yr B.P. The warm period was followed by an interval during which temperatures were several degrees lower than at present. Present-day climatic conditions extend back to about 1,500 yr B.P. (Yasuda, 1978).

The above paleoclimatic reconstruction agrees well with the result of the study of Holocene molluscs in southern Kanto by Matsushima (1978). He found that temperate molluscs appeared around 9,500–8,700 yr B.P. and were followed by molluscs of the tropical type around 6,500–6,000 yr B.P. The latter disappeared around 4,000 yr B.P.

The appearance and time of greatest diversity of hermatypic corals in the southern Boso Peninsula also coincides with the period of warm climate outlined above. While the appearance of both tropical molluscs and hermatypic corals is undoubtedly linked to the climatic amelioration, the connection between their disappearance or decreased diversity and the deterioration of climate is less clear cut. The suggestion by Matsushima (1979) that the disappearance of tropical molluscs was due to the disappearance of their habitat rather than due to climatic change deserves careful consideration. During the climatic optimum, the southern Boso Peninsula was characterized by a highly irregular coastline with drowned valley-type bays where hermatypic corals grew. The filling in of the bays, combined with tectonic uplift and eustatic lowering of sea level resulted in elimination of the drowned valley bay environment around 5,500–5,000 yr B.P. The effects of the climatic and physiographic changes on the corals are not easy to separate. However, because the present-day impoverished coral fauna of Tateyama Bay forms the northernmost limit of hermatypic corals in Japan, it seems unlikely that corals would have been able to survive the period of lower temperatures following the climatic optimum.

Knowledge of the present Japanese ostracod distribution indicates that ostracod fauna is not as strongly affected by small variations in temperature as corals are. The slight warming (several degrees centigrade) around 6,000 yr B.P. would therefore be unlikely to produce a substantial change in the composition of ostracod assemblages. If the warmer climate had an effect, it would probably be the appearance of moderate amounts of few subtropical species, which are not presently found in the area.

Examination of ostracod assemblages from fossil samples reveals that there is one species which might have appeared because of climatic change. *Ambocythere japonica* is present in most samples of the AL sub-biofacies and LK biofacies in amounts varying from 1 to 5%. All the samples come from sediments inferred to have been deposited during the climatic optimum and are associated with hermatypic corals.

Ambocythere is a predominantly tropical genus; all but two occurrences reported come from the Caribbean and Southeast Pacific (one species was reported from Chile [Hartmann, 1963]; the other from Japan). In Japan, live *Ambocythere japonica* has been found only in Shikoku (Ishizaki, 1968). It is absent from samples taken in Hamanako estuary (Ikeya and Hanai, 1980 MS.), and in Tateyama Bay (this study); unpublished research indicates that it is also absent from Aburatsubo Bay in Miura Peninsula.

Since the survey of recent ostracods in Japan is not complete and subsequent research may demonstrate presence of *Ambocythere japonica* in more northern latitudes, a definite conclusion can not be drawn at this point. The absence of *A. japonica* from other biofacies in sediments of the same age as those of the coral-bearing bed indicates that its presence is also controlled by environmental parameters other than temperature. However, its presence in two different biofacies in Jomon transgression sediments and its absence from Aburatsubo Bay, where the ostracod assemblage strongly resembles that of the AL sub-biofacies, warrant the speculation that the appearance and disappearance of *A. japonica* was the result of climatic changes.

Effect of the transgression and regression on ostracod assemblages: The sea level changes during the past 10,000 years have produced environmental variations which are reflected in lateral shifts of ostracod biofacies. Even though the main cause of the variations (sea level change) was essentially the same throught the southern Boso Peninsula, the environmental change and the accompanying changes in ostracod biofacies were different in the different study areas. This is principally because topographic and bathymetric characteristics of an area can maximize or minimize the overall environmental change caused by sea level variation.

Inspection of the outline of the present Tateyama Bay and that of the paleo-Tateyama Bay at the peak of the transgression demonstrates that the two are very similar. The same is true for a reconstructed bay at any sea level between 20 and 0 meters.

The sediment distribution in the paleo-Tateyama Bay also agrees well with that of the present Tateyama Bay; the central part is formed by silt and silty sand, while sand predominates in the peripheral parts.

Comparison between ostracod biofacies of the recent and paleo-Tateyama Bay also reveals good agreement; minor differences are ascribable to sampling and preservation bias and perhaps the gentler slope of the paleo-Tateyama Bay.

Along the western coast, the present-day and Jomon transgression costlines and sediment differ strongly. The present coast is characterized by a nearly straight coastline and well-sorted coarse to fine sandy sediment. In contrast, at the peak of the Jomon transgression, the western coast was formed by drowned valley bays filling with poorly sorted silty sand. Thus, sea level change in the West Caost Valley area had a substantially larger effect than in the Tateyama area. As the sea level rose, the bays reached farther inland, producing a greater number of different environments. Initially, the AA sub-biofacies occupied the small incipient bays, but as the sea level rose, a more quiet environment developed in the inner parts of the bay where the amount of agitation was small. AL sub-biofacies developed in this new environment. A continuing rise of sea level extended the bays farther inland where the inferred S sub-biofacies developed. As the sea level fell again, the environments of each biofacies disappeared in reverse order; the AA sub-biofacies was the last to disappear and still survives in some places (Z biofacies).

The greatest changes took place in the Tomoe River area, primarily due to the length and gentle slope of Tomoe River valley and its tributaries and to the comparatively large discharge of Tomoe River. The large discharge resulted in the formation of S sub-biofacies in the initial stages of transgression when the embayment was still relatively small. With a rising sea level, the S sub-biofacies gradually kept moving inland to stay at the head of the bay. As the water depth and width of the bay mouth increased, biofacies constituted by assemblages that were more open coast in character moved into the bay. The maximum variety of biofacies was reached at the peak of the transgression, when the S, K, and KN sub-biofacies and NS biofacies were all represented. The high rate of sedimentation, however, resulted in the shallowing of the bay and expansion of the open coast NS biofacies. The falling sea level accelerated this process and environments of the other biofacies gradually disappeared, the NS biofacies being the only one able to "retreat" with the receding sea.

Lack of sampling points does not allow reconstruction of the biofacies changes in the Chikura and Maruyama areas. It is probable that the changes taking place in Chikura area were very similar to those in Tomoe River area, because the two have similar topography and sediment distribution. In the Maruyama area the influx of fresh water and fluvial sediment seem to have a much bigger role than in Tateyama area.

Systematic description

Subclass Ostracoda Latreille, 1806
Order Podocopida Sars, 1866
Superfamily Cypridacea Baird, 1845
Family Pontocyprididae G. W. Müller, 1894
Genus *Argilloecia* Sars, 1866
Argilloecia lunata n. sp.
Pl. 8, figs. 3–7; text-fig. 31f.

Etymology.—From "lunatus" [Latin, "bent into a crecsent"].

Type.—Holotype, UMUT-CA 15001, left valve, male (Pl. 8, fig. 6), obtained from sample no. 7, from coral bed at Koyatsu, West Coast Valley area, AL sub-biofacies.

Diagnosis.—Species of the genus *Argilloecia* characterized by relatively broadly rounded posterior margin.

Description.—Elongate in lateral outline, highest at the middle. Dorsal margin convex, ventral margin concave at anterior third. Anterior margin broadly rounded, posterior margin narrowly rounded, extremely at lower third.

Sexual dimorphism moderate: in male, anterior margin obliquely truncated, meets dorsal margin with blunt cardinal angle, posterior margin more narrowly rounded.

Dorsal outline elongate ellipsoid, anterior and posterior margins pointed, widest at the middle. Right valve larger than left, overlaps left valve at second anterior fifth and posterior two-fifths of length.

Surface smooth, normal pores few, scattered.

Inner lamella broad, line of concrescence runs parallel to the margins, coinciding with the inner margin in ventral area. Anterior and posterior vestibule large. Marginal pore canals simple, straight, numerous in anterior end, few in ventral part, moderately numerous in posterior end. Several false radial pore canals occur in posterior end. Muscle scar large, consisting of three anterior, two posterior closely set scars. Hinge adont.

Dimensions.—Holotype, male LV; L, 0.54, H, 0.17; paratypes, female, LV, UMUT-CA 15002 (Pl. 8, fig. 3); L, 0.50, H, 0.18; male, LV, UMUT-CA 15003 (Pl. 8, fig. 4); L, 0.52, H, 0.19; female RV, UMUT-CA 15004 (Pl. 8, fig. 5); L, 0.55, H, 0.20; female RV, UMUT-CA 15005 (Pl. 8, fig. 7); L, 0.54, H, 0.19.

Remarks.—This is the first species of the genus *Argilloecia* reported from Japan.

Occurrence.—Common in LK biofacies and AL sub-biofacies, rare in AA, KN, K sub-biofacies and sample 162 of the SN biofacies. Absent in Tateyama Bay.

Genus *Propontocypris* Sylvester-Bradley, 1947
Subgenus *Propontocypris* Sylvester-Bradley, 1947
Propontocypris (*Propontocypris*) sp.
Text-fig. 32d.

Illustrated specimen.—A left valve, UMUT-CA 15042.

Remarks.—Assigned to the genus *Propontocypris* on the basis of typical muscle scar, consisting of three inclined rows of 5 scars, elongated subtriangular outline, and simple marginal pore canals.

Occurrence.—Rare in NL, AA sub-biofacies, and SN biofacies.

Family Candonidae Kaufmann, 1900
Subfamily Paracypridinae Sars, 1923
Genus *Paracypris* Sars, 1866
Paracypris sp.
Text-fig. 32e.

Illustrated specimen.—A right valve, UMUT-CA 15006.

Remarks.—The elongated, posteriorly tapering carapace, bifurcating marginal pore canals, and a muscle scar consisting of three anterior, two posterior and one upper scars characterize this species as belonging to the genus *Paracypris*. The species closely resembles *Pontocypris*? sp. 6, described by Maddocks (1966, p. 44, 45, figs. 32 J-Q) in outline, but differs in muscle scars and marginal pore canals.

Occurrence.—Rare to common in LK biofacies and NK sub-biofacies, rare in A, SN biofacies and K sub-biofacies.

Superfamily Cytheracea Barid, 1850
Family Leptocytheridae Hanai, 1957
Genus *Callistocythere* Ruggieri, 1953
Callistocythere littoralis group
Callistocythere numaensis n. sp.
Pl. 9, figs. 1–3; text-fig. 31a.

Etymology.—Named after the type locality of the Holocene coral-bearing sediments, village of Numa.

Type.—Holotype, UMUT-CA 15007, right valve (Pl. 9, fig 1), obtained from sample no. 7, from the coral-bearing sediments in Koyatsu, West Coast Valley area. The sample belongs to the AL sub-biofacies.

Diagnosis.—Species of the genus *Callistocythere* characterized by prominent sub-vertical ridge in the posterocentral area.

Description.—Carapace oblong, reniform, highest at anterior cardinal angle. Dorsal margin gently convex upward, merges with obliquely rounded anterior margin. Anterior margin crenulated at lower half, ventral margin strongly concave at anterior third. Posterior margin meets dorsal margin at a distinct cardinal angle, gently rounded below.

First anterior marginal ridge prominent, sinuous, running from the upper fifth of anterior margin to anterior sixth of ventral margin. Dorsal marginal ridge obscuring the dorsal margin, running from the eye tubercle to posterior fourth, where it bifurcates, one part joining the first posterior marginal ridge, the other running vertically to the posterocentral area. Posterior marginal ridge prominent, running from the posterior fourth of dorsal margin to the ventral sinuosity. First ventral margin starts at the ventral sinuosity and runs obliquely upwards to the posteroventral area, where it turns at a steep angle and continues to the central area. Second ventral marginal ridge running parallel to the ventral margin from anterior fourth to little past half, where it joins the first ventral ridge. Surface ornamented with coarse reticulation. Eye tubercle low, but large. Normal pores moderate in number, simple type, scattered.

Inner lamella moderately wide, radial pore canals simple.

Dimensions.—Holotype, RV; L, 0.43, H, 0.23; paratypes, LV, UMUT-CA 15008 (Pl. 9, fig. 2); L, 0.42, H, 0.23; LV, UMUT-CA 15009 (Pl. 9, fig. 3); L, 0.45, H, 0.24.

Remarks.—Development of the hingement and the short, unbranching radial pore canals seem to suggest that these are immature specimens; however, the width of the inner lamella indicates that the forms are adults.

Occurrence.—Common in AL sub-biofacies, rare in KN and K sub-biofacies.

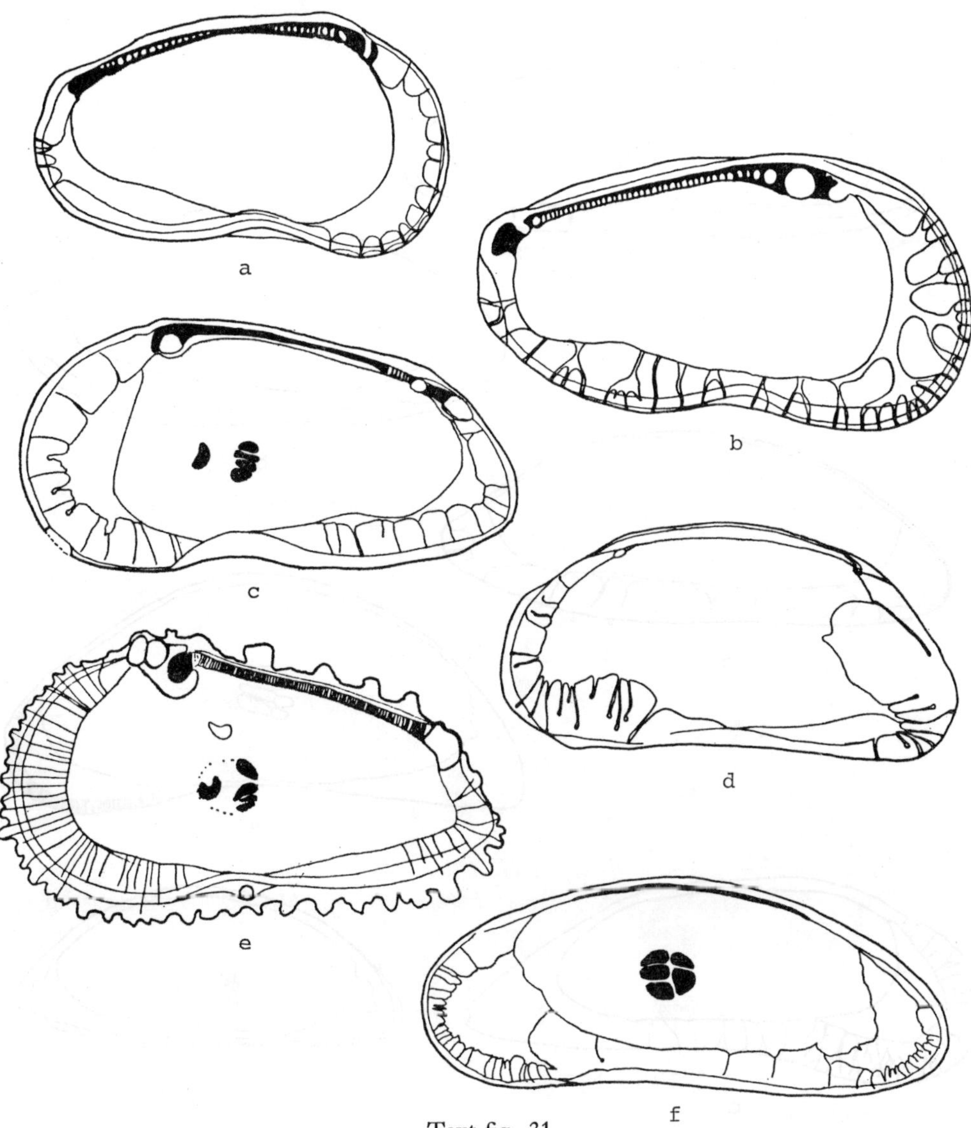

Text-fig. 31

a. Internal view of *Callistocythere numaensis* n. sp., left valve (UMUT-CA 15009). ×125.

b. Internal view of *Callistocythere tateyamaensis*, left valve, female (UMUT-CA 15012). ×125.

c. Internal view of *Nipponocythere hastata* n. sp., right valve, female (UMUT-CA 15034). ×160.

d. Internal view of *Semicytherura sabula* n. sp., right valve, female (UMUT-CA 15026). ×194.

e. Internal view of *Trachyleberis straba* n. sp., right valve, female (UMUT-CA 15021). ×100.

f. Internal view of *Argilloecia lunata* n. sp., right valve, female (UMUT-CA 15005). ×137.

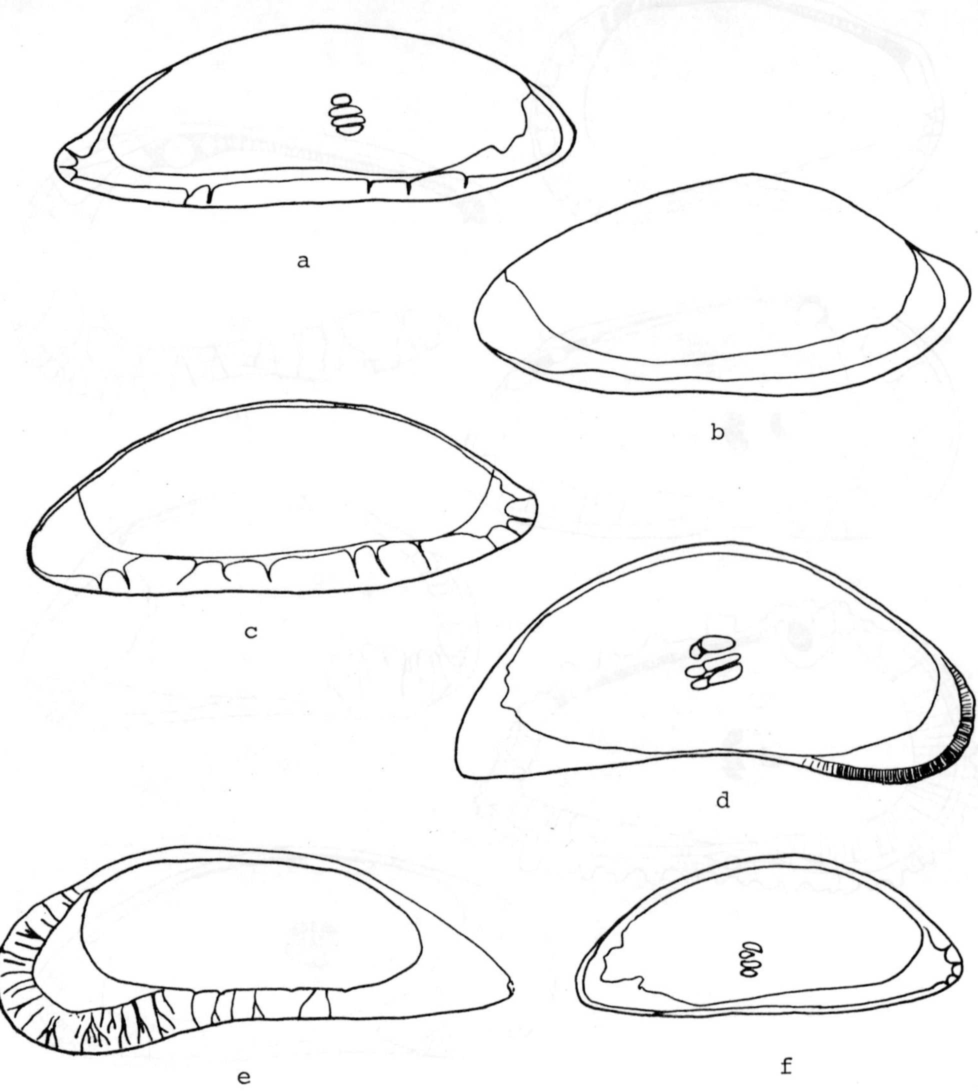

Text-fig. 32

a. Internal view of *Paradoxostoma* sp. 1, left valve (UMUT-CA 15036). ×64.

b. Internal view of *Paradoxostoma* sp. 3, right valve (UMUT-CA 15038). ×52.

c. Internal view of *Paradoxostoma* sp. 2, right valve (UMUT-CA 15037). ×49.

d. Internal view of *Propontocypris* sp., left valve (UMUT-CA 15042). ×44.

e. Internal view of *Paracypris* sp., right valve (UMUT-CA 15006). ×32.

f. Internal view of *Cytherois*? sp., right valve (UMUT-CA 15039). ×63.

Callistocythere tateyamaensis n. sp.
Pl. 9, figs. 4–9; text-fig. 31b.

Etymology.—Named after Tateyama City.

Type.—Holotype, UMUT-CA 15010, left valve, female (Pl. 9, fig. 5), obtained from sample no. 57, from the village of Shiomi in the West Coast Valley area. The sediment is coarse sand and the sample belongs to the AA sub-biofacies.

Diagnosis.—Species of the genus *Callistocythere* characterized by large depression developed between the first and second anterior marginal ridges, and reticulation of irregular outline.

Description.—Carapace thick, oblong, subreniform, highest at anterior cardinal angle. Anterior margin broadly obliquely rounded, dorsal margin gently arched. Posterior margin truncated above, meets dorsal margin with a distinct cardinal angle in the left valve, merges with dorsal margin in the right valve. Ventral margin slightly concave at the middle.

Surface sculptured with strong ridges and deep reticulation. First anterior ridge weak, runs from upper third of anterior margin to ventral sinuation. Second anterior ridge prominent, runs from anterior cardinal angle to lower third of anterior margin, where it joins the first anterior ridge. Third anterior ridge prominent, runs from anterior third of dorsal margin to anterior fifth of ventral margin, where it joins the first marginal ridge. Deep depression developed between the second and third marginal ridges and short, but very deep anteriorly oblique depression located behind the upper one third of the third marginal ridge.

Dorsal marginal ridge strong, obscuring the dorsal margin, running from anterior third of the dorsal margin to the posterior fourth, where it merges with the second posterior marginal ridge.

First posterior marginal ridge weak, running from posterodorsal cardinal angle to the ventral sinuation, obscures ventral margin behind the ventral sinuation. Second posterior marginal ridge very prominent, running from the posterior sixth of the dorsal margin, parallel to the posterior margin, and sloping obliquely downwards to the ventral sinuation. It runs parallel to the ventral margin and disappears a little anterior to the ventral sinuation. U-shaped depression present in dorsocentral area.

Anterocentral and posterocentral area ornamented with deep reticulations, commonly separated by thin, low partitions into two or more compartments. Eye tubercle obscure.

Inner lamella moderately wide along anterior and posteroventral margins, line of concrescence nearly coinciding with the inner margin. Anterior vetibule present but very narrow. Radial pore canals moderately numerous, repeatedly polyfurcate. Normal pores small, few, scattered.

Muscle scars located slightly below the center. Four adductor scars in a vertical row, the highest subtriangular, the middle two elongated, horizontal, the lowest circular. Selvage prominent. Snap-knob present in the right valve, corresponding snap-pit in the left. Hingement of the *Callistocythere littoralis* type.

Dimensions.—Holotype female LV; L, 0.55, H, 0.29; paratypes, male RV, UMUT-CA 15011 (Pl. 9, fig. 4); L, 0.54, H, 0.27; female LV, UMUT-CA 15012 (Pl. 9, fig. 6); L, 0.53, H, 0.26; female RV, UMUT-CA 15013 (Pl. 9, fig. 7); L, 0.58, H, 0.28; male LV, UMUT-CA 15043 (Pl. 9, fig. 8); L, 0.53, H, 0.28; female RV, UMUT-CA 15044 (Pl. 9, fig. 9); L, 0.53, H, 0.28.

Occurrence.—Rare to common in AA sub-biofacies, rare in AL and NL sub-biofacies. In Tateyama Bay, common to rare in DS biofacies.

Family Cytheridae Baird, 1850
Subfamily Schizocytherinae Mandelstam, 1960
Tribe Paijenborchellini Deroo, 1966
Genus *Neomonocertina* Kingma, 1948
Neomonoceratina sp.
Pl. 9, fig. 12.

Illustrated specimen.—A left valve, UMUT-CA 15014.

Remarks.—This species somewhat resembles *Neomonocertatina japonica* (Ishizaki, 1966) in carapace outline and surface ornamentation, but differs by possessing a dorsally located, downturned caudal process.

Occurrence.—One specimen was found in sample 61 and one in sample 114, both from the Chikura area.

Family Hemicytheridae Puri, 1953
Subfamily Hemicytherinae Puri, 1953
Tribe Hemicytherini Puri, 1953
Genus *Hemicythere* Sars, 1925
Hemicythere sp.
Pl. 9, fig. 15.

Illustrated specimen.—A right valve, UMUT-CA 15015.

Remarks.—This species somewhat resembles *Hemicythere*? *miii* (Ishizaki, 1969) but differs by absence of ridges, and surface ornamentation, consisting of fine reticulation in the marginal areas and punctation in the central area.

Occurrence.—One valve was found in sample 162, belonging to the SN biofacies.

Tribe Aurilini Puri, 1974
Genus *Aurila* Pokorný, 1955
Aurila sp. A
Pl. 9, figs. 10, 11.

Illustrated specimens.—A right valve, UMUT-CA 15016; a left valve, UMUT-CA 15017.

Remraks.—This species strongly resembles *Aurila cymba* (Brady, 1869) in lateral and posterior outline and ornamentation. It. differs by the absence of dorsal, dorsoposterior ridge.

Occurrence.—Abundant in AL sub-biofacies. Common in AA, KN sub-biofacies and SN biofacies, sample 61 and some samples of NL sub-biofacies. In Tateyama Bay, common in SFS and Z biofacies, rare in some samples of CBM and IR biofacies.

Aurila sp. B
Pl. 9, figs. 13, 14.

Illustrated specimens.—A right valve, UMUT-CA 15018; a left valve, UMUT-CA 15019.

Remarks.—This species resembles *Aurila* sp. A, but in posterior view *Aurila* sp. B is oval in outline, while *Aurila* sp. A is subtriangular. It further differs in possessing dorsal ridge ending in a tubercle in dorsoposterior area, while the ventral ridge is not as strongly developed as in *Aurila* sp. A. *Aurila* sp. B might be a sexual dimorph of *Aurila* sp. A but even though the two forms commonly occur together, their abundance is inversely proportional. Further, the differences in dorsal and ventral ridges become more prominent in immature forms.

Occurrence.—Abundant to common in A biofacies. Common in LK, LP, N, SN, and KS biofacies. In Tateyama Bay, common in some samples of the SFS, CBM, DM, IR, and Z biofacies. Rare live specimens found in samples 15 and 27 of the CBM and IR biofacies.

Family Trachyleberididae Sylvester-Bradley, 1948
Subfamily Trachyleberidinae Sylvester-Bradley, 1948
Tribe Trachyleberidini Sylvester-Bradley, 1948
Genus *Trachyleberis* Brady, 1898
Trachyleberis straba n. sp.
Pl. 8, figs. 1, 2; Pl. 9, figs. 16–18; text-fig. 31e.

Etymology.—From "strabus" [Latin, "squinting"]

Type.—Holotype, UMUT-CA 15020, left valve, female (Pl. 9, fig. 18), obtained from sample no. 107, from Tomoe River area near the village of Kotsuka. Sediment is silty sand, and the sample belongs to the KN biofacies.

Diagnosis.—Species of the genus *Trachyleberis* distinguished by the presence of a snap-knob and absence of eye tubercle.

Description.—Carapace thick, rather small for the genus. In lateral view elongate, subtrapezoidal, tapering toward posterior end. Highest at anterior cardinal angle. Anterior margin broadly rounded. Dorsal margin straight, ventral margin slightly sinuate at the middle. Posterior margin truncated, meets dorsal and ventral margins at a distinct angle. Anterior, ventral, and posterior margins ornamented with spines and tubercles. Carapace ornamented with tubercles, normal pore canals few, mostly located at the top of tubercles.

Anterior marginal ridge prominent, arising from area below anterior cardinal angle and running parallel to anterior margin to the upper half, where it disappears abruptly. A short, slightly sinuous ridges runs vertically from the anterior cardinal angle to a little below the beginning of the anterior marginal ridge.

Marginal infold narrow, line of concrescence coincides with inner margin; no vestibule. Radial pore canals narrow, slightly curved, numerous in anterior and posterior ends. Hingement holamphidont, central bar crenulate. Muscle scars situated in subcentral depression and composed of a V-shaped frontal scar and four horizontally elongated muscle scars in a vertical row.

Sexual dimorphism distinct, males longer and lower than females.

Remarks.—Even though this species strongly resembles *Trachyleberis scabrocuneata* in shape and ornamentation, the absence of an eye tubercle is remarkable. This is perhaps the only blind species of the genus.

Dimensions.—Holotype, female LV; L, 0.69, H, 0.40; paratypes, female RV, UMUT-CA 15021 (Pl. 9, fig. 17); L, 0.69, H, 0.39; male LV, UMUT-CA 15022 (Pl. 9, fig. 16); L, 0.77, H, 0.41.

Occurrence.—Rare to common in K and KN sub-biofacies.

Family Bythocytheridae Sars, 1926
Genus *Bythoceratina* Hornibrook, 1952
Bythoceratina sp.

Remarks.—This species resembles *Bythoceratina* sp. A described by Ishizaki (1968). It differs from *Bythoceratina hanaii* by near absence of median sulcus, lack of reticulation and less protruding, more broadly rounded posterodorsal margin. Only immature specimens were found.

Occurrence.—Rare in SN and N biofacies.

Family Cytheruridae G. W. Müller, 1894
Subfamily *Cytherurinae* G. W. Müller, 1894
Genus *Semicytherura* Wagner, 1957
Semicytherura sabula n. sp.
Pl. 8, figs. 8–14, 16, 17; text-fig. 31d.

Etymology.—From "sabulum" [Latin, "coarse sand"].

Type.—Holotype, UMUT-CA 15023, female carapace (Pl. 8, fig. 10), obtained from sample T10 taken in the Tateyama Bay at depth of about 10 m. Sediment was coarse to medium sand, and the sample belongs to the SCS biofacies.

Diagnosis.—Species of *Semicytherura* distinguished by its outline and surface ornamentation consisting of wrinkles in the anteroventral area.

Description.—Carapace small, thin, sexual dimorphism pronounced.

Female: Valves slightly asymmetrical, right valve higher than left. Valves reniform in lateral view, greatest height at the middle. Anterior margin broadly rounded, posterior margin more narrowly rounded, extremity at lower third. In right valve posterior margin slightly concave at upper half, meeting the dorsal margin at a distinct cardinal angle. In the left valve posterior margin merges with dorsal margin. Ventral margin gently concave at the middle. In dorsal view, carapace oblong oval, anterior pointed, posterior widely rounded, widest at the middle. Left valve overlaps the right at the anterior and posterior one-sixth; right valve overlaps left at the central two-thirds.

Male: Carapace elongated subrectangular in lateral view. Dorsal margin very gently arched, meets anterior margin with a distinct cardinal angle in left valve and merges with anterior margin in right valve. Anterior margin broadly rounded. Ventral margin gently concave at the middle. Posterior margin broadly rounded in left valve, more narrowly rounded in right valve. Extremity at slightly below middle. Posterior margin slightly concave in upper half in right valve, meeting dorsal margin at a small cardinal angle. In the left valve posterior margin merges with dorsal margin. In dorsal view, oblong, anterior narrowly pointed, posterior bluntly pointed, slightly compressed at middle, widest at posterior fourth.

Surface smooth except at anteroventral and posteroventral area, where it is sculptured with weak, wrinklelike, sinuous ridges parallel to margins. Normal pores few, scattered.

Infold and fused zone moderately wide: one-tenth length anteriorly, four-fifteenth length posteriorly in female, two-fifths in male. Radial pore canals moderate in number, mostly unbranched, often curved. False radial pore canals also present. Adductor muscle scars four in vertical row; frontal scars could not be observed. Hingement lophodont of *Cytherura* type.

Dimensions.—Holotype female carapace; L, 0.33, W, 0.10; paratypes, female LV, UMUT-CA 15024 (Pl. 8, fig. 8); L, 0.33, H, 0.16; male carapace, UMUT-CA 15025 (Pl. 8, fig. 9); L, 0.33, W, 0.09; female RV, UMUT-CA 15026 (Pl. 8, fig. 11); L, 0.33, H, 0.15; male RV, UMUT-CA 15027 (Pl. 8, fig. 12); L, 0.32, H, 0.13; male RV, UMUT-CA 15028 (Pl. 8, fig. 13); L, 0.31, H, 0.13; male carapace, UMUT-CA 15029 (Pl. 8, fig. 14); L, 0.33, H, 0.12; female RV, UMUT-CA 15040; L, 0.31, H, 0.15; female LV, UMUT-CA 15041; L, 0.35, H, 0.15.

Occurrence.—Live specimens abundant in SCS biofacies of Tateyama Bay.

Family Loxoconchidae Sars, 1925
Subfamily Loxoconchinae, Sars, 1925
Genus *Loxoconcha* Sars, 1866
Loxoconcha sp.
Pl. 9, figs. 19, 20.

Illustrated specimens.—A right valve, UMUT-CA 15030; a left valve, UMUT-CA 15031.

Remarks.—A species of *Loxoconcha* characterized by somewhat converging dorsal and ventral margins and surface ornamented with fine reticulation. Only immature specimens were found.

Occurrence.—Abundant in sample 61 and common in sample 76, both from Chikura area. Common to rare in KS, N, and LP biofacies. In Tateyama Bay one specimen was found in sample 13 of the DS biofacies.

Subfamily Cytheromorphinae Mandelstam, 1960

Genus *Nipponocythere* Ishizaki, 1971

Nipponocythere hastata n. sp.

Pl. 8, fig. 15, 18–20; text-fig. 31c.

Etymology.—From "hastatus" [Latin, "spear-shaped"].

Type.—Holotype, UMUT-CA 15032, right valve, female, (Pl. 8, fig. 19), obtained from sample no. 7 from the coral-bearing sediment in Koyatsu, West Coast Valley area. The sample belongs to the AL sub-biofacies.

Diagnosis.—Species of the genus *Nipponocythere* distinguished by strong sinuation of the ventral margin.

Description.—Carapace rather flat, in lateral view elongate subtrapezoidal in outline, highest at anterior cardinal angle. Anterior margin broadly, evenly rounded, dorsal margin straight, ventral margin concave slightly anterior to middle. Posterior end obliquely truncated, narrowly rounded at lower third. Longest at lower third. Sexual dimorphism present, male lower and longer than female.

Surface smooth in anterior, ventral and posterior areas, closely punctate in the central and dorsal areas.

Inner lamella wide anteriorly, narrower ventrally and posteriorly. Line of concrescence parallel to outer, subparallel to inner margin. Vestibule moderate anteriorly, small posteriorly. Radial pore canals few, evenly spaced, straight. False radial pore canals present. Normal pores of simple type with a low lip, moderately numerous, scattered.

Muscle scar consisting of a row of four elongate adductor scars and one inclined reniform anterior scar. Hingement Loxoconchinae gongylodont. In the right valve, anterior tooth is surrounded dorsally by a crescent-shaped extension of the median groove. Median groove narrow, smooth; first posterior tooth small, second large.

Dimensions.—Holotype female RV; L, 0.43, H, 0.20; paratypes, female LV, UMUT-CA 15033 (Pl. 8, fig. 15); L, 0.43, H, 0.20; female RV, UMUT-CA 15034 (Pl. 8, fig. 18); L, 0.41, H, 0.20.

Occurrence.—Rare to common in AL sub-biofacies and LK biofacies, rare in AA sub-biofacies. Absent in Tateyama Bay.

Family Paradoxostomatidae Brady and Norman, 1889

Subfamily Paradoxostomatinae Brady and Norman, 1889

Genus *Paradoxostoma* Fisher, 1855

Paradoxostoma sp. 1

Text-fig. 32a.

Illustrated specimen.—A left valve, UMUT-CA 15036.

Remarks.—Species of the genus *Paradoxostoma* characterized by prominent caudal projection at the lower third of the posterior margin.

Occurrence.—Rare to common in A biofacies. In Tateyama Bay one specimen found in sample 18 (SCS).

Paradoxostoma sp. 2

Text-fig. 32c.

Illustrated specimen.—A right valve, UMUT-CA 15037.

Remarks.—The present species resembles *Paradoxostoma lunatum* illustrated by Okubo (1976), but differs in more narrowly rounded anterior and posterior margins.

Occurrence.—One specimen was found in sample 60 and one in sample 61.

Paradoxostoma sp. 3

Text-fig. 32b.

Illustrated specimen.—A right valve, UMUT-CA 15038.

Remarks.—This species resembles *Paradoxostoma convexum* illustrated by Okubo (1977), but differs in more sharply angulated dorsal margin and slightly concave posterodorsal margin.

Occurrence.—Rare in LK, SN and A biofacies.

Genus *Cytherois* G. W. Müller, 1884

Cytherois? sp.

Text-fig. 32f.

Illustrated specimen.—A right valve, UMUT-CA 15039.

Remarks.—This species somewhat resembles *Cytherois zosterae* Schornikov, but differs in the much more acutely rounded anterior and posterior margins.

Occurrence.—Rare to common in A biofacies, rare in LK and SN biofacies. In Tateyama Bay common in sample 7 of the CBM biofacies.

Location of samples

West Coast Valley Area: Samples 1, 3, 5, 7, 9, 11, 15, 26, 84—Coral-bearing sediments croping out in the upper part of a valley lying south of Koyatsu village. The outcrop is located about 600 m south of the coastal road. Samples 23, 24—Coral-bearing sediments from the type locality of the coral bed in Numa. Samples were collected from the banks of a water reservoir about 300 m south of the Numa village. Sample 36—Coral-bearing sediments exposed along the banks of a nameless tributary of the Shioiri River joining it slightly upstream of Kamisanakura village. The outcrop is located about 700 m south of the Kamisanakura village. Samples 55, 56, 57—Samples collected from an outcrop along the banks of the little stream flowing through the Shiomi village. Outcrop is located about 300 m south of the coastal road. Samples 87, 88—Outcrop on the bank of a small stream flowing through the Koyatsu valley. The outcrop is located about 40 m downstream of the Ko bridge.

Tateyama Area: Samples 40, 41, 71—Oyster reef located near the village of Nishigo, about 100 m downstream of Nishigo bridge. Samples 43, 44—Sediments above the oyster reef described above. Samples 51, 175—Samples collected from outcrops along the Heguri River upstream of the Nishigo bridge. Samples 67, 52—Sediments exposed along the banks of Taki River, about 1 km upstream from its confluence with Heguri River. Samples 53, 63—Outcrops located near the village of Kokonoe. Sample 53 from a road ditch about 500 m south of Kokonoe station, sample 63 from a construction excavation about 300 m northwest of the station.

Tomoe River Area: Samples 59, 60—Sediments exposed along a small stream flowing through the Tateyama Golf course. The outcrop is located about 500 m upstream from the coastal road. Samples 142, 144, 145, 157—Banks of Tomoe River east of the village of Daijingu. Samples 96, 99, 103, 104, 151—Banks of Tomoe River tributary flowing into the Tomoe River from the south, about 200 m upstream of the Daijingu village. Samples 162, 128, 110,

Table 3
C^{14} ages of material collected in this study.

Code No.	Age	Location	Elevation a.s.l.	Material
TK-328	7,230±70	Tateyama, Shiomi 34°58′12″N, 139°48′49″E	8 m	*Chama reflexa*
TK-329	6,720±60	Tateyama, Koyatsu 34°58′03″N, 139°49′22″E	15.5 m	*Barbatia bicolorata*
TK-335	2,920±80	Tateyama, Kamisanekura 34°58′23″N, 139°52′07″E	5 m	*Dosinella penicillata*, *Macoma incongrua*
TK-336	5,500±70	Tateyama, Kamisanekura 34°58′05″N, 139°52′03″E	17 m	*Barbatia bicolorata*
TK-337	7,740±90	Tateyama, Koyatsu 34°58′25″N, 139°51′50″E	1.5 m	Shells
TK-354	7,240±80	Wada, Kaihotsu 35°01′22″N, 139°58′58″E	7.4 m	*Cyclina sinensis*
TK-355	7,630±70	Tateyama, Kotsuka, Daijingu 34°55′16″N, 139°51′25″E	13.3 m	*Dosinella penicillata*
TK-356	7,400±80	Tateyama, Tatsuoka 34°55′24″N, 139°51′50″E	18.5–18.8 m	*Scapharca broughtonii* *Paphia undulata*, *Dosinella penicillata*
TK-357	6,980±80	Shirahama, Ootsukuriba 34°55′06.5″N, 139°51′45″E	19.7 m	*Rapana venosa*, *Circe scripta*
TK-358	6,560±60	Tateyama, Daijingu, 34°54′56″N, 139°50′45″E	16 m	*Chama reflexa*, *Hyotissa hyotis imbricata*
TK-359	6,530±80	Tateyama, Tatsuoka 34°55′24″N, 139°51′50″E	20.8 m	*Saxidomus purpuratus*

Table 4
C^{14} ages in the southern Boso Peninsula (published).

Code No.	Age	Location	Reference
GaK-254	6,160±120	Tateyama, Okabu	M, H, K, Y, N
GaK-1131	7,620±150	Tateyama, Koyatsu	M, Y
GaK-2967	7,350±180	Tateyama, Heisaura	M, Y
GaK-2968	6,340±130	Tateyama, Nishigo	M, Y
GaK-2969	6,410±140	Tateyama, Nishigo	M, Y
GaK-2995	2,440±100	Tateyama, Heisaura	M, Y, N
GaK-3067	5,570±140	Chikura, Kawato	M, Y, Yo
GaK-3068	3,610±120	Chikura, Teraniwa	Y, Yo, N
GaK-3070	2,860±100	Chikura, Shiinokihara	Y, Yo, N
GaK-3757	6,600±150	Tateyama, Nishigo	M, Y
GaK-3758	4,490±120	Tateyama, Nishigo	M, Y
GaK-5374	4,740±100	Chikura, Seto	M, Y, N
GaK-5457	4,840±110	Tateyama, Nishigo	M
GaK-5458	7,690±190	Tateyama, Daijingu	M, Yo, N
GaK-5459	820± 80	Tateyama, Sakai	N
GaK-5477	16,420±360	Chikura, Enzooin	Yo
GaK-5478	5,480±110	Tateyama, Fura	M, Yo
GaK-8601	5,750±130	Nagaoka, Daijingu	M*
GaK-8602	5,840±140	Kotsuka, Daijingu	M*
GaK-8604	7,410±170	Tateyama, Takigawa R.	M*
M-240	5,100±400	Maruyama, Furukawa	M
N-38	5,290±140	Maruyama, Furukawa	M
N-2565	4,640± 90	Chikura, Baba	M, Yo, N
N-2566	7,810±110	Chikura, Baba	M, Yo, N
N-2567	6,560± 75	Minamihara	M, Yo
N-3085	7,330±120	Tateyama, Nishigo	M
N-3086	5,800±110	Tateyama, Nishigo	M
N-3087	4,440±110	Tateyama, Nishigo	M
N-3088	7,840±110	Tateyama, Koyatsu	M, N
N-3089	7,620±110	Chikura, Shinkawa	M
TH-085	5,970±180	Tateyama, Numa	M, O-76, N
TH-086	6,000±180	Tateyama, Okabe	M, O-76, N
TH-106	6,910±200	Chikura, Shinkawa	M, O-76, Yo, N
TH-107	5,325±170	Chikura, Shiinokihara	M, O-76, Yo, N
TH-108	4,315±145	Chikura, Shiinokihara	O-76, Yo, N
TH-109	2,685±130	Tateyama, Konuma	O-76, N
TH-110	5,365±170	Tateyama, Okabe	M, O-76, N
TH-111	285±100	Chikura, Ageshima	O-76
TH-213	3,640±125	Shirahama, Nemoto	O-78
TH-214	4,280±130	Chikura, W of Ch. Railway Stat.	O-78, N
TH-215	6,280±165	Tateyama, Daijingu	O-78, N
TH-216	5,440±155	Tateyama, Daijingu	O-78, N
TH-217	4,760±140	Tateyama, Nagaoka	O-78, N
TH-218	6,370±165	Tateyama, Shiomi	O-78, N

(Continue)

Code No.	Age	Location	Reference
TH-219	4,010±135	Tateyama, Shiomi	O-78, N
TH-272	5,090±120	Shiomi	N
TH-273	7,450±155	Chikura, Baba	N
TH-274	7,340±150	Chikura, Baba	N
TH-275	7,290±150	Chikura, Baba	N
TH-276	3,230±105	Chikura, Baba	N
TH-277	6,530±140	Chikura, Kawato	N
TH-278	5,510±130	Chikura, Teraniwa	N
TH-279	4,560±115	Chikura, Teraniwa	N
TH-280	4,920±120	Chikura, Shiinokihara	N
TH-281	6,430±140	Chikura, Shiinokihara	N
TH-282	6,200±135	Chikura, Kawato	N
TH-284	3,360±100	Shirahama, Harada	N
TK-3	5,230±100	Maruyama, Furukawa	M, Y, Yo
TK-7	7,870± 70	Tateyama, Koyatsu	Ko, M, Y, N
TK-65	8,680±190	Chikura, Teraniwa	M, Kb, Y, Yo

M=Matsushima, 1979; H=Hoshino, 1967; K=Kigoshi et al., 1969; Y=Yonekura, 1975; Yo=Yokota, 1979; O-76=Omoto, 1976; O-78=Omoto, 1978; Kb=Kobayashi et al., 1971; Ko=Konishi, 1967; M*=Matsushima, personal communication

107, 111, 112, 123, 125, 126, 127—Outcrops exposed along the Tomoe River and its small tributaries in the vicinity of Kotsuka village. Samples 120, 122—Banks of Tomoe River about 400 m upstream of Kotsuka village. Sample 106—Little stream flowing through the village of Otsukuriba.

Chikura Area: Samples 61, 114—Banks of a north flowing tributary of the Seto River, flowing through an area locally known as Baba. Sample 118—Bank of a small stream flowing into Seto River north of Kanzawa Village. The outcrop is located about 300 m north of the village. Sample 76—Outdrop located above the Seto River, about 100 m west of Ikubyo Center.

Maruyama Area: Sample 134—Outcrop along the banks of Maruyama River, about 300 m east of the railway bridge. Sample 132—Banks of Nuruishi River, about 300 m downstream of the village of Numa.

Acknowledgment

I would like to thank Drs. K. Chinzei, I. Hayami, T. Ozawa, T. Yamaguchi, Y. Okada, M. Yajima and Messrs. K. Abe, R. Matsui, R. Tabuki from the University of Tokyo and Mr. Y. Matsushima from the Kanagawa Pref. Museum. They have aided me in innumerable ways and, through discussions and suggestions both in the field and laboratory, helped to steer this work in the proper direction, and offered the much needed help in the last stage of preparation of this manuscript. Miss M. Kouchi, from the department of Anthropology of the University of Tokyo, patiently helped me with the computer programs. I am particularly grateful to my supervisors, Dr. T. Hanai and Dr. T. Hamada, whose numerous suggestions greatly improved the results of this work.

While conducting research, I was a recipient of the Kajima Foundation Scholarship and

Moriya Shogyo Scholarship. I would like to thank them deeply for their long and generous financial support.

Finally, I would like to thank my wife, Kumiko, for her patience and for drafting many of the figures in this report.

LATE PLEISTOCENE OSTRACODA FROM THE BOSO PENINSULA, CENTRAL JAPAN

Michiko Yajima
Tokyo Seitoku Gakuen, Oji, Tokyo 114

Abstract: This paper deals with ostracod fauna of the northern part of the Boso Peninsula. These ostracods lived in the Paleo-Tokyo Bay during four periods of high sea level accompanying interglacial stages (ca. 300,000–100,000 yr B. P.), as represented by the deposition of the Yabu, Kiyokawa, Kamiiwahashi, and Kioroshi Formations, which are the middle and upper parts of the Shimosa Group. Each formation shows a cyclic tendency reflecting a cycle of glacioeustatic sea-level changes. The Paleo-Tokyo Bay may have been open toward the east, however, during the maximum high sea-level phase, warm water could have entered the bay directly from the south.

Ostracods representing 121 species were identified. Most of the modern species of ostracods living in the shallow sea of the temperate zone on the Pacific coast of Japan are represented. Four assemblages were recognized: subtidal sand, warm water sand, shallow water mud, and brackish water clay. Warm and cold water influences are represented by the displacement of several species within one assemblage.

One genus, *Robustaurila*, and 14 species are described as new: *Neocytherideis aoi*, *Eucythere yugao*, *Munseyella oborozukiyo*, *Callistocythere hotaru*, *Schizocythere asagao*, *Campylocythereis? ukifune*, *Aurila kiritsubo*, *Pseudocythere frydli*, *Eucytehrura utsusemi*, *Semicytherura wakamurasaki*, *Kangarina hayamii*, *Loxoconcha hanachirusato*, *L. tamakazura*, and *Xestoleberis suetsumuhana*. The following genera are reported from Japan for the first time: *Paracypris* (Candonidae), *Eucythere* (Eucytheridae), *Stigmatocythere*, *Rocaleberis*, *Carinovalva*, *Robertsonites*, *Australimoosella*, *Campylocythereis* (Trachyleberididae), *Pseudocythere* (Bythocytheridae), and *Kangarina* (Cytheruridae).

Introdcution

Upper Pleistocene sediments of the northern part of the Boso Peninsula consist of dominantly marine strata which reflect cyclic changes of the sedimentary environment and contain well-traceable marker-tephras and well-preserved fossils. They represent the last stage of the filling-up of the middle and southern parts of the Paleo-Tokyo Bay, which opened toward the east and, at the time of high sea level phase, perhaps also toward the south. The sediments preserve eventful records of faunal changes resulting from sea level fluctuations during glacial and interglacial periods. In this paper, ostracods which lived during four interglacial stages of high sea level represented by the deposition of Yabu, Kiyokawa, Kamiiwahashi, and Kioroshi Formations are described. These formations correspond to the middle and upper parts of the Shimosa Group.

The geology and paleontology of this area has been studied intensively since the latter half of the last century. From the paleontological point of view, the following

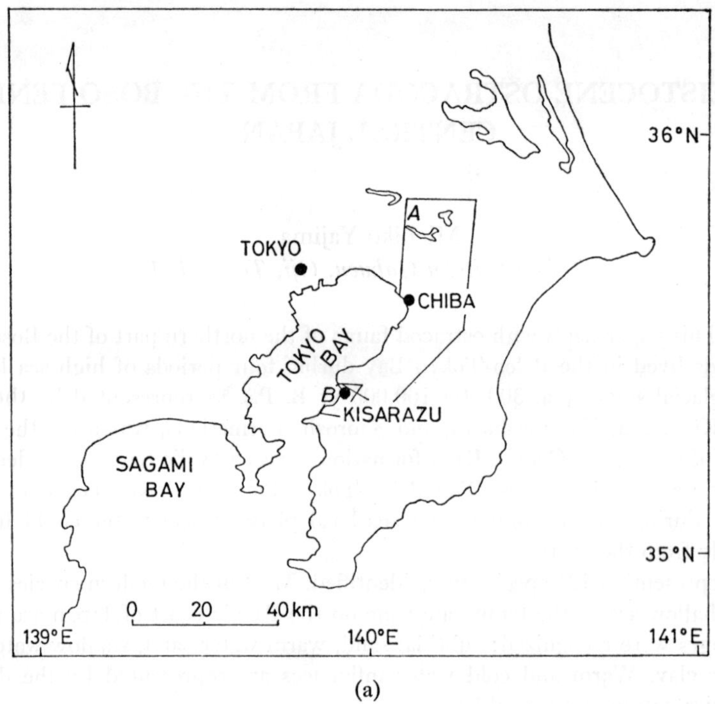

(a)

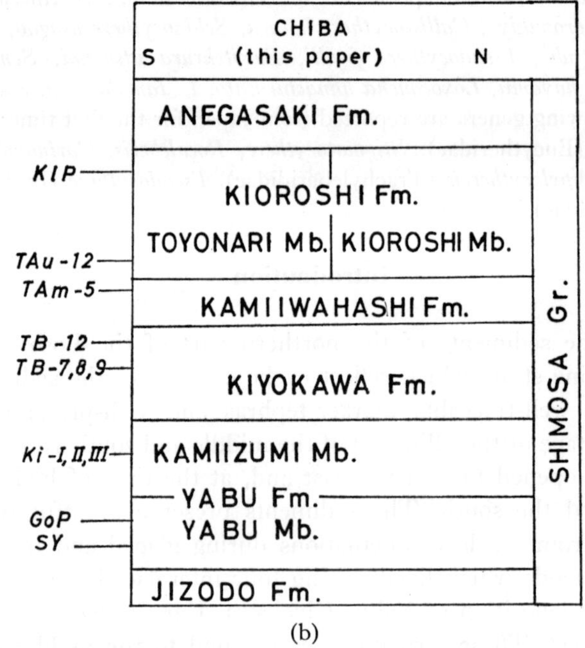

(b)

Text-fig. 1 Index map and stratigraphic units.

1–a A: the surveyed area; B: Kisarazu area (Yajima, 1978).

1–b Italics show marker-tephras.

three types of studies are of special interest: (1) reports of fossil occurrence at the several famous localities, Ikebe (1936), Oinomikado (1937), Oyama and Ishiyama (1968), Ohara (1968), and Horiguchi and Ohara (1972); (2) discussions on changes of paleoenvironments based on areal geological surveys, Kozima (1958–1966) and Aoki et al. (1962–1971); (3) tephrochronological studies to find the exact interrelation between local depositional sequences, Sugihara (1970) and Sugihara et al. (1978).

Recently, ostracods from the Kisarazu area (A of text-fig. 1) were discussed in relation to the paleonenvironment reconstructed from detailed field observations (Yajima, 1978) of a reconnaissance study on the Japanese Upper Pleistocene warm-water ostracod fauna.

As the first step of the present study, the stratigraphy of the ostracod-bearing horizons was described to restore the physiographic environments of deposition. Data were obtained through field mapping and observation of sediments with special reference to the occurrence of megafossils. Each formation shows a cyclic tendency of sedimentation reflecting a cycle of glacioeustatic sea level changes.

As the second step, effort has been made to update the taxonomy of the Pleistocene ostracods of the Yabu to Kioroshi Formations. In many species, classification at the specific level remains in confusion. Examples can be seen in the species of *Aurila* and *Loxoconcha*. For a complete solution of the problem, the study of live material is often more effective than that of Pleistocene fossils. Published and unpublished information on the modern species was referred to as much as possible for species identification.

Several hundred new genera of ostracods have been proposed since the publication of the Treatise in 1961. Most of these are cytheracean genera and some attempts have been made to classify this large number of cytheracean taxon (Hartmann and Puri, 1974; Liebau, 1975; Gründel, 1973, 1976). Such attempts of classification were to a large extent successful, but many controversial questions still remain. For Japanese ostracods, many species of the Trachyleberididae and Hemicytheridae have still been assigned to inappropriate genera. This paper attempts to clarify this confusion and to classify Japanese trachyleberidids and hemicytherids into the proper genera.

Grouping of genera into tribes, especially into the two families mentioned above, raises another problem. Tribe by tribe or subfamily by subfamily revisions are now becoming popular among ostracod taxonomists. However, appropriate tribes to accommodate some genera which occur in Japan have not yet been found. Although hasty assignment may result in confusion, it is not practical to wait till satisfactory tribe-by-tribe classification is available. Thus, the resultant family group classification used in this paper is quite tentative.

As the third step, the Pleistocene ostracods of the area were analyzed ecologically. The Pleistocene ostracods described in this paper seem to include most of the modern species of ostracods living in the shallow sea of the temperate zone on the Pacific coast of Japan. Therefore, the information on the ecology of modern species found in the checklist of the Japanese ostracods (Hanai et al., 1977), ecological study of ostracods of Hamana-ko Estuary (Ikeya and Hanai, 1980 MS) and unpublished data on ostracod ecology of Aburatsubo Cove, Misaki, was especially useful in the analysis. Time and space distribution of the Pleistocene ostracods of this area may be explained by interfingering of ostracod fauna of warm-temperate and cold-temperate water in this area.

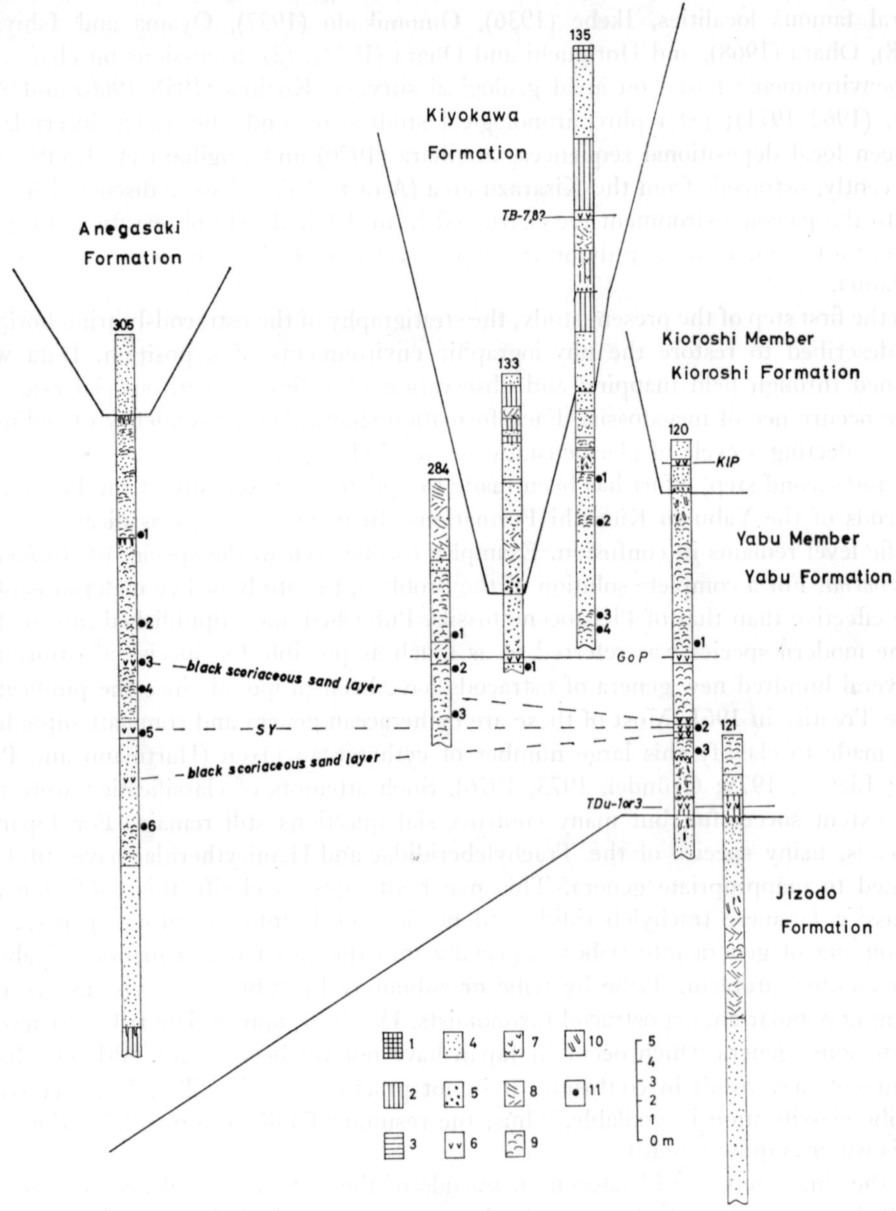

Text-fig. 2

Columnar sections of Yabu Member of Yabu Formation.

(1) clay, (2) silt, (3) alternation of silt with sand, (4) sand, (5) gravel, (6) marker-tephra, (7) pumice tuff, (8) cross-lamination, (9) shells, (10) type 1 burrows, (11) sampling horizon of ostracods.

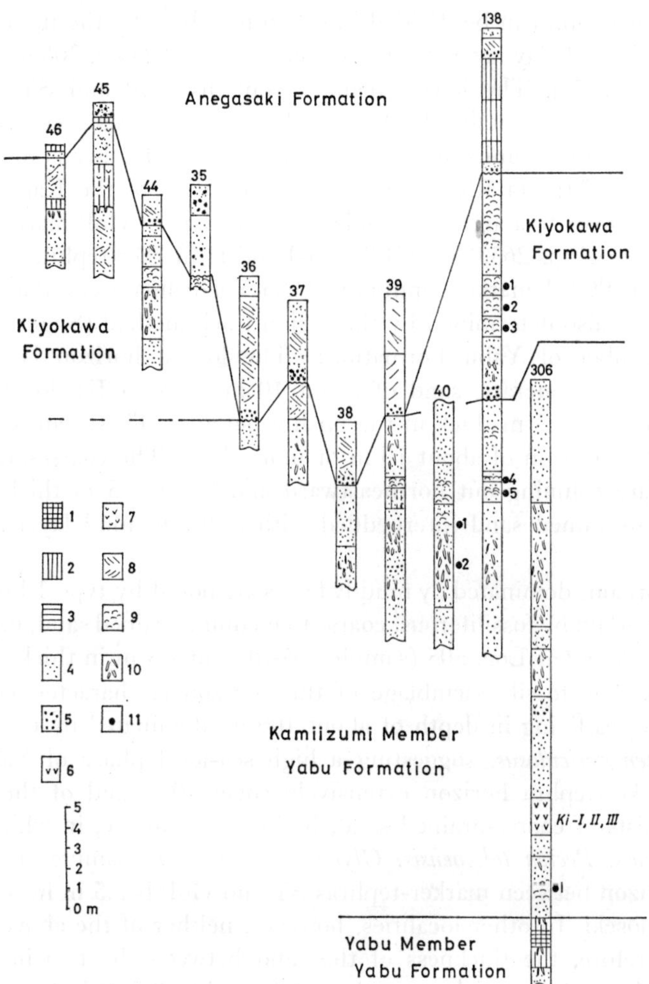

Text-fig. 3

Columnar sections of Kamiizumi Member of Yabu Formation. See explanation of legend of text-fig. 2.

Four assemblages are recognizable; subtidal sand, warm water sand, shallow water mud, and brackish water clay. Warm and cold water influences may be represented by the displacement of certain species within one assemblage.

Description of lithology at sampling localities

Yabu Formation: The Yabu Formation lies unconformably on the uppermost, shallow water silty fine-grained sand of the Jizodo Formation (Loc. 120 of text-figs. 2, 10), and is overlain by gravel-bearing, coarse-grained sand of the Kiyokawa Formation (Loc. 138 of text-figs. 4, 10) and by a 30 cm thick gravel bed of the Kioroshi Formation (Loc. 120).

The Yabu Formation can be divided into two members by the upper surface of an extensively distributed clay layer, approximately 2 m thick (Loc. 306 of text-figs. 3, 10) (Sugihara et al., 1978). The lower member contains GoP and SY marker-tephras and the upper member contains Ki-I, II, III marker-tephras, hereinafter called the Yabu Member and the Kamiizumi Member. The Yabu Member is distributed extensively south of the Murata River, while the distribution of the Kamiizumi Member is restricted to a small area around Kamiizumi. The date of GoP marker-tephra determined by fission tracks is 263,000 yr B.P.; Ki-I and II marker-tephras can be identified with TC1-3 and TC1-4 marker-tephras in Oiso-Yokohama area. The date of TC1-4 is 243,000 yr B.P. also determined by the fission track method (Sugihara et al., 1978).

I. Yabu Member of Yabu Formation: The lowest horizon is represented by nonfossiliferous conglomeratic sand of about 10 m thick in Jizodo area and coarse-grained sand of about 27 m thick in the Nishiyatsu area. These sands are overlain by a mud dominated horizon of about 10 m or more thick. The coarse-grained sand and mud horizons are thinning out northeastward and become 3 m thick, yellow-brown fine- to medium-grained sand interbedded with a 40 cm mud layer at Loc. 120 of Sematanoseki.

The lower horizons dominated by muddy facies are bored by type 1 burrows (Yajima, 1978) and are overlain by fossiliferous, coarse to medium-grained sand, not well stratified and 10 m in thickness at Loc. 305 (sample 305-6), and 1.3 m in thickness at Loc. 120 (sample 120-3). The fossil assemblage of this horizon is characterized by abundant warm water species living in depth of about 100 m, dominated by *Glycymeris pilsbryi* and *Cryptopecten vesiculosus*, suggesting a high sea-level phase of Yabu cycle.

The SY marker-tephra horizon extensively covers the sand of the high sea-level phase and consists of coarse-grained sand, besides SY pumice, in which certain cold-temperate species, *Pecten tokyoensis*. *Glycymeris yessoensis* (samples 120-2 and 305-5) occur. The horizon between marker-tephras SY and GoP is 3.5 m in thickness at Loc. 120 of Sematanoseki. In other localities, however, neither of the above two tephras is detectable; therefore, the thickness of the sand between the two marker-tephras is uncertain, but it tends to thicken southward (samples 305-4, 3, 2, samples 284-3, 2). The fossil assemblage of this sand is characterized by abundance of articulated *Pecten tokyoensis* with the convex right valve upward. GoP horizon is actually a coarse-grained sand mixed with GoP pumice (sample 133-1).

Grey fine- to medium-grained, well-sorted sand covers GoP marker-tephra and contains numerous disarticulated valves of *Mactra chinensis* and *Glycymeris yessoensis*, lying convex side upward and suggesting shallow water under cold-temperate submarine climate. Thickness of the sand is 8.5 m at Loc. 120 and tends to thicken toward Yabu area (samples 120-1, 135-1–4, 284-1 and possibly sample 305-1). This sand is overlain by silt in the Yabu area and by sandy silt in the Mino area south of Kisarazu. The Yabu Member terminates with this silt of about 1.5 m in thickness (Sugihara et al., 1978).

Sampling localities of the Yabu Member are given below. Local popular names of localities are given in parentheses after locality numbers.

Loc. 305 (Atebi-nishi): a cliff, 1.5 km S 85°E of Makuta station of Japanese National Railway (J.N.R.), Mariyatsu, Fukita-machi, Kisarazu-shi (Lat. 35°21′47″N, Long. 140°04′46″

E) (Loc. 6 of Sugihara et al., 1978). Yabu Member of Yabu Formation (32 m+) and Anegasaki Formation (4 m).
Sample 305-6, black scoriaceous coarse-grained sand, 4 m below SY pumice marker-tephra, with abundant shell fragments of *Crassostrea gigas*, *Glycymeris pilsbryi*, *Dentalium* sp. and *Echinarachnius mirabilis*.
Sample 305–5, SY marker-tephra (80 cm in thickness), 17 m above the upper surface of mud which is exposed at the lower horizon of the column.
Sample 305-4, medium-grained sand, 2 m above SY, with numerous articulate valves of *Pecten tokyoensis* lying right valve upward.
Sample 305-3, black scoriaceous sand, 3 m above SY, with abundant barnacles and bryozoans.
Sample 305-2, medium-grained sand, 5.5 m above SY, with abundant *Panopea japonica* in life position and scattered *Pecten albicans* and *Mactra chinensis*.
Sample 305-1, shell lenses with pebbles, 9 m above SY, with abundant *Glycymeris yessoensis*, *Umbonium costatum*, and *Solen krusensterni*. Abundant burrows have their openings at the base of this horizon.

Loc. 284 (Yamakura): an exposure, 5 km E of the Yamakura-ko Lake, Ichihara-shi (Lat. 35°29′23″N, Long. 140°11′56″E) (Yumukura of Aoki and Baba, 1971; Loc. 51 of Sugihara et al., 1978). Yabu Member of Yabu Formation (13 m+).
Sample 284-3, fine-grained sand, 2.5 m below GoP marker-tephra, with abundant shell fragments of *Pecton tokyoensis*, *Mactra chinensis*, and *Glycymeris yessoensis*.
Sample 284-2, fine-grained sand just below GoP, with *Glycymeris yessoensis* and *Mactra chinensis* lying convex side upward.
Sample 284-1, fine-grained sand, 40 cm above GoP, with *Mactra chinensis*, *Pecten tokyoensis*, and *Glycymeris yessoensis*.

Loc. 133 (Higashikuniyoshi): a cliff, 200 m W of Higashikuniyoshi Post Office, Ichihara-shi (Lat. 35°30′10″N, Long. 140°13′00″E). Yabu Member of Yabu Formation (3.7 m+) and Kiyokawa Formation (10.9 m+).
Sample 133-1, coarse-grained sand with abundant *Glycymeris yessoensis*, containing GoP pumice, 3 m below the unconformity between Yabu Formation and Kiyokawa Formation.

Loc. 135 (Takakura): a cliff, 3.5 km S 08°E of Honda railway station (J.N.R.), Ichihara-shi (Lat. 35°30′44″N, Long. 140°13′18″E) (Loc. 332 of Kozima, 1962, 1963; Loc. 7b of Hanai, 1970; Takakura of Aoki and Baba, 1971). Yabu Member of Yabu Formation (13 m+) and Kiyokawa Formation (16.3 m+).
Sample 134-4, coarse-grained sand, 11.1 m below the unconformity between sand of Yabu Formation and basal conglomerate of Kiyokawa Formation, with abundant *Pecten tokyoensis*.
Sample 135-3, fine-grained sand, 11 m below the unconformity, with abundant *Pecten tokyoensis*.
Sample 135-2, medium-grained sand, 6 m below the unconformity, with *Glycymeris yessoensis*, *Pecten tokyoensis*, *P. albicans* occurring commonly.
Sample 135-1, medium-grained sand, 4.15 m below the unconformity, with *Glycymeris yessoensis* and *Pecten tokyoensis* occurring commonly.

Loc. 120 (Sematanoseki): a cliff, along the Murata River, 2.4 km S 35°E of Honda railway station (J.N.R.), Ochi-shimoshinden, Ichihara-shi (Lat. 35°31′30″N, Long. 140°13′48″E) (Loc. 4 of Aoki et al., 1962; Loc. 327 of Kozima, 1962, 1963; Semata of Ohara, 1968; Loc. 54 of Sugihara et al., 1978). Jizodo Formation (13 m+), Yabu Member of Yabu Formation (16 m), and Kioroshi Member of Kioroshi Formation (2 m).
Sample 120-3, coarse-grained sand, 3.2 m above the base of Yabu Formation, with abundant pumice particles and shell fragments of *Glycymeris pilsbryi* and pteropods.
Sample 120-2, 40 cm thick, coarse-grained sand with coarse-grained SY pumice, 4.3 m above

the base of Yabu Formation, lying between two black layers each of which is 20 cm thick. *Pecten tokyoensis* and *Glycymeris yessoensis* are abundant.

Sample 120-1, grey fine-grained, well-sorted sand, 1 m above the base of GoP pumice tuff, 40–60 cm thick. Abundant convex side upward *Mactra chinensis* and *Glycymeris yessoensis* from several layers.

II. Kamiizumi Member of Yabu Formation: The deposition of the Kamiizumi Member of the Yabu Formation begins with silty sand which fills up type 1 burrows (Yajima, 1978), penetrating into the uppermost silt of the Yabu Member of Yabu Formation at Loc. 306. The Kamiizumi Member, about 20 m thick, distributed near Kamiizumi is composed of lower silty sand and upper well-sorted, fine-grained sand with Ki-I, II, III marker-tephras in its basal part.

The lower grey silty sand is 3 m thick. It contains numerous molluscan fossils: *Pecten albicans*, *Chlamys farreri*, *Paphia naganumana*, and *Siphonalia fusoides*. In the lower part (sample 306-1), they occur as scattered masses of irregularly packed shells. In the upper part, *Pecten albicans* is usually articulated, and *P. tokyoensis* occurs rarely. The grain size becomes coarser upwards. Among the molluscan species, *P. albicans*, *Chlamys farreri* and some others are warm sea inhabitants in depths of about 20 to 50 m.

The upper well-sorted, fine-grained sand with many tubelike burrows and molluscan fossils lies over the lower silty sand. Tuff layers of Ki-I, II, III marker-tephras composed of angular pumice, showing graded bedding, are found 1 m above the base of the upper sand. Tube-like burrows in the upper sand are cone shaped, about 1 cm in diameter and 10 cm in length, oriented with their lower axes vertical to the bedding plane. Molluscan fossils occur in two horizons. The lower fossiliferous horizon, about 4 m thick, is characterized by the abundance of shallow, warm-temperate water molluscs such as *Umbonium costatum*, *Tapes japonicus*, *Siphonalia fusoides*, and *Solen krusensterni* (samples 40-2, 1 and samples 138-5, 4). Fossils of the upper fossiliferous horizon dominated by *Mactra chinensis* are completely dissolved at loc. 138.

Sampling localities of this member are as follows.

Loc. 306 (Kawaharai): a cliff, 100 m E of Kawaharai elementary school, Sodegaura-machi, Kimitsu-gun (Lat. 35°24′30″N, Long. 140°04′20″E) (Loc. 30 of Sugihara et al., 1978). Yabu Member (3.5 m+) and Kamiizumi Member (26 m+) of Yabu Formation.

Sample 306-1, bluish grey silty sand, 2.5 m below Ki-I, II, III marker-tephras, with abundant *Pecten albicans*, *Chalmys farreri*, and *Siphonalia fusoides* which occurs as scattered masses of irregularly packed shells.

Loc. 40 (Kamiizumi): an exposure, 2.75 km N 15° E of Higashiyokota railway station (J.N.R.), Sodegaura-machi, Kimitsu-gun (Lat. 35°24′30″N, Long. 140°02′56E). Kamiizumi Member of Yabu Formation (12 m+).

Sample 40-2, fine-grained sand, 2 m above 2 m thick, medium-grained sand with abundant short vertical burrows (10 cm in length and 0.5 cm in diameter). *Mactra chinensis* and *Umbonium costatum* are abundant.

Sample 40-1, fine-grained sand, 4 m above 2 m thick, medium-grained sand. Besides burrows, *Umbonium costatum* and other shell fragments are abundant.

Loc. 138 (Nagayoshi): a cliff, 2.75 km N 40° E of Higashiyokota railway station (J.N.R.), Sodegaura-machi, Kimitsu-gun (Lat. 35°24′12″N, Long. 140°03′20″E) (Nagayoshi of Aoki and Baba, 1971). Kamiizumi Member of Yabu Formation (11 m) and Anegasaki Formation (7 m).

Sample 138-5, fine-grained grey sand, 4 m below the unconformity between the Kamiizumi

Member of the Yabu Formation and the Kiyokawa Formation, with abundant shell layers and irregular shell lenses containing *Mactra chinensis*, *Umbonium costatum* and *Pecten albicans*. Sample 138-4, lens of shell fragments consisting of *Batillaria zonalis*, *Crassostrea gigas*, *Solen krusensterni*, 3.5 m below the unconformity.

Kiyokawa Formation: The gravel of the Kiyokawa Formation about 3 m in thickness unconformably overlies medium-grained sand of the Kamiizumi Member of the Yabu Formation in the Kamiizumi area (Loc. 138 of text-figs. 4, 11), while it unconformably overlies medium-grained sand (Loc. 135), sand with trace fossils made by isopods of the genus *Exciolana* (Kikuchi, 1972) (Loc. 126), and probable nonmarine clay (Loc. 286) of the Yabu Member of the Yabu Formation in the Semata area.

In the Kamiizumi area, basal gravel contains type 2 burrows (Yajima, 1978) in the sand matrix in the lower half and shells of *Pecten albicans*, *P. tokyoensis*, and *Mactra chinensis* in the upper half. Light brown medium-grained thick sand, which overlies the basal gravel, includes abundant shells of *Pecten albicans*, *P. tokyoensis*, and *Mactra chinensis* in the lower part (sample 138-3), sporadic *Paphia naganumana* and *Fulvia mutica* in the middle part (sample 138-2). *Mactra chinensis* forms layers in the upper part (sample 138-1); in the uppermost part, *Mactra chinensis* and *Umbonium costatum* are compactly packed. Type 2 burrows (Yajima, 1978) also occur in the uppermost part.

Black coarse-grained sand exposed in the Otabe area (Loc. 290) may be correlated with a horizon higher than the light brown medium-grained thick sand described above. The black coarse-grained sand includes *Mactra chinensis*, *Glycymeris yessoensis* and *Umbonium costatum* (samples 290-1, 2, 3). It is overlain by black clay, which contains oyster banks near Otabe (Loc. 111); the black clay is overlain by the medium-grained sand, including TB-7 (?) marker-tephra.

Near Takinokuchi (Locs. 308, 310-315), silt with *Dosinia japonica* in living position, which may be correlated with black clay, is overlain by fine-grained sand of about 7 m in thickness with many layers of molluscan fossils consisting mainly of cold-temperate water species and characterized by the dominance of *Mactra chinensis*, which lie convex side upward and form several layers in the sand. The fossiliferous sand becomes darker and coarser in the upper 3 m and contains fragments of *Mactra chinensis* and solitary valves of *Glycymeris yessoensis*. The fossiliferous sand is overlain by coarse-grained sand, about 2.5 m thick, with small lenticular bodies of shell fragments. A 50 cm thick tuff layer which may be correlated to the TB-8 marker-tephra covers extensively the coarse-grained sand (Loc. 311).

Near Nakatakane (Loc. 82), grey medium-grained, well-sorted sand, 10 m+ thick, characterized by type 1 burrows (Yajima, 1978), abudance of *Glycymeris yessoensis* (sample 82-2), and trace fossils produced by isopods of genus *Exilolana* (Kikuchi, 1972) is overlain by brackish water silt, 6 m in thickness, including *Batillaria zonalis* and plant fragments. TB-7 marker tephra is found in this silt. TB-7, 8, 9 marker-tephras are usually interbedded in the middle part of the silt bed, cover a wide area, and provide a good marker dividing the Kiyokawa Formation into upper and lower parts. The date of 220,000 yr B.P. has been obtained for TB-7 marker-tephra in the Oiso-Yokohama area by the fission track method (Sugihara et al., 1978).

The lowest horizon of the upper part of the Kiyokawa Formation covers directly the TB-7 marker-tephra in the Nakatakane area (Loc. 82), where silty sand 1–2 m in

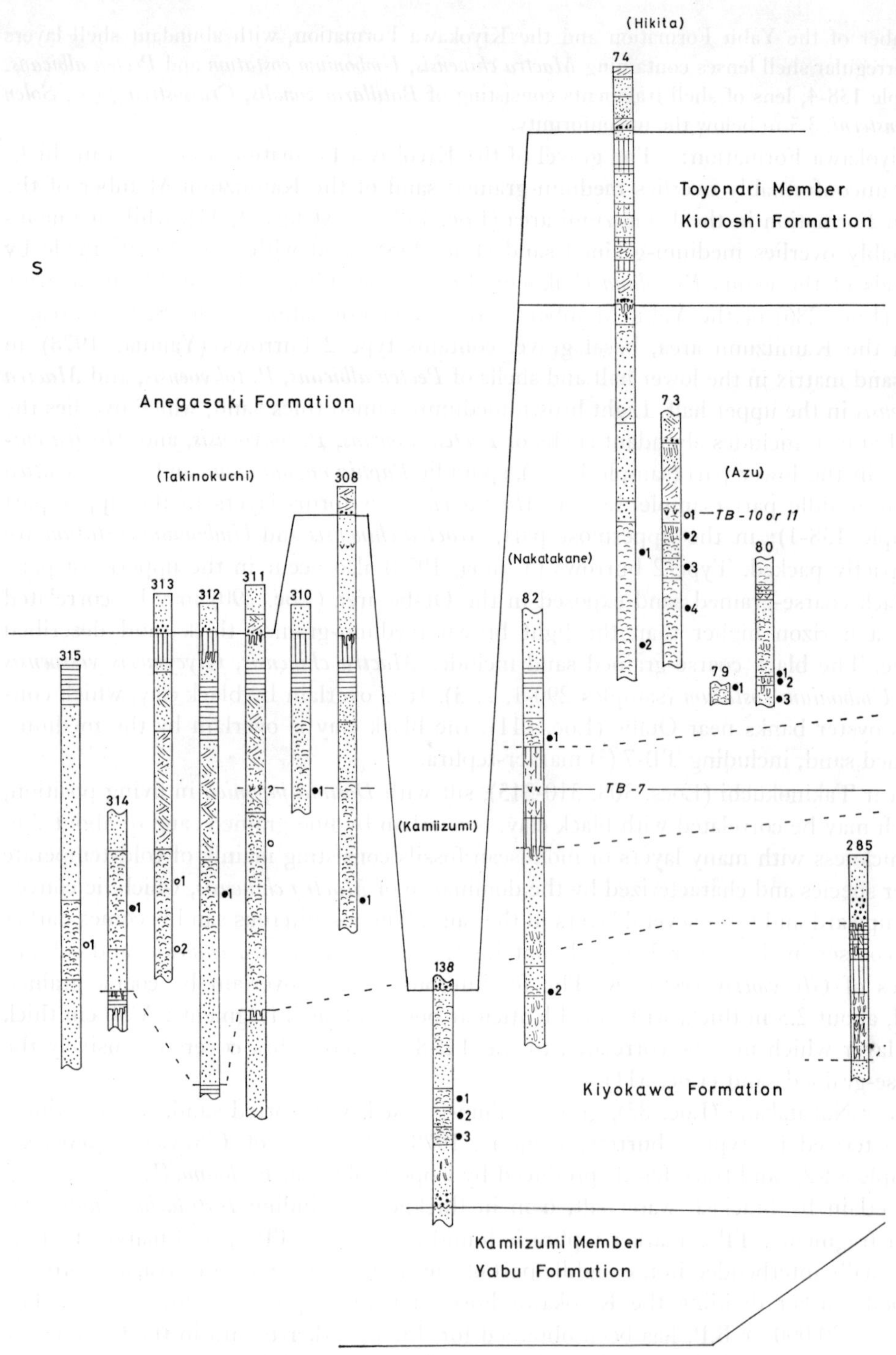

Text-fig. 4
Columnar sections of Kiyokawa Formation. (1) clay, (2) silt, (3) alternation of silt with sand, (4) sand, (5) gravel, (6) peat, (7) marker-tephra, (8) pumice

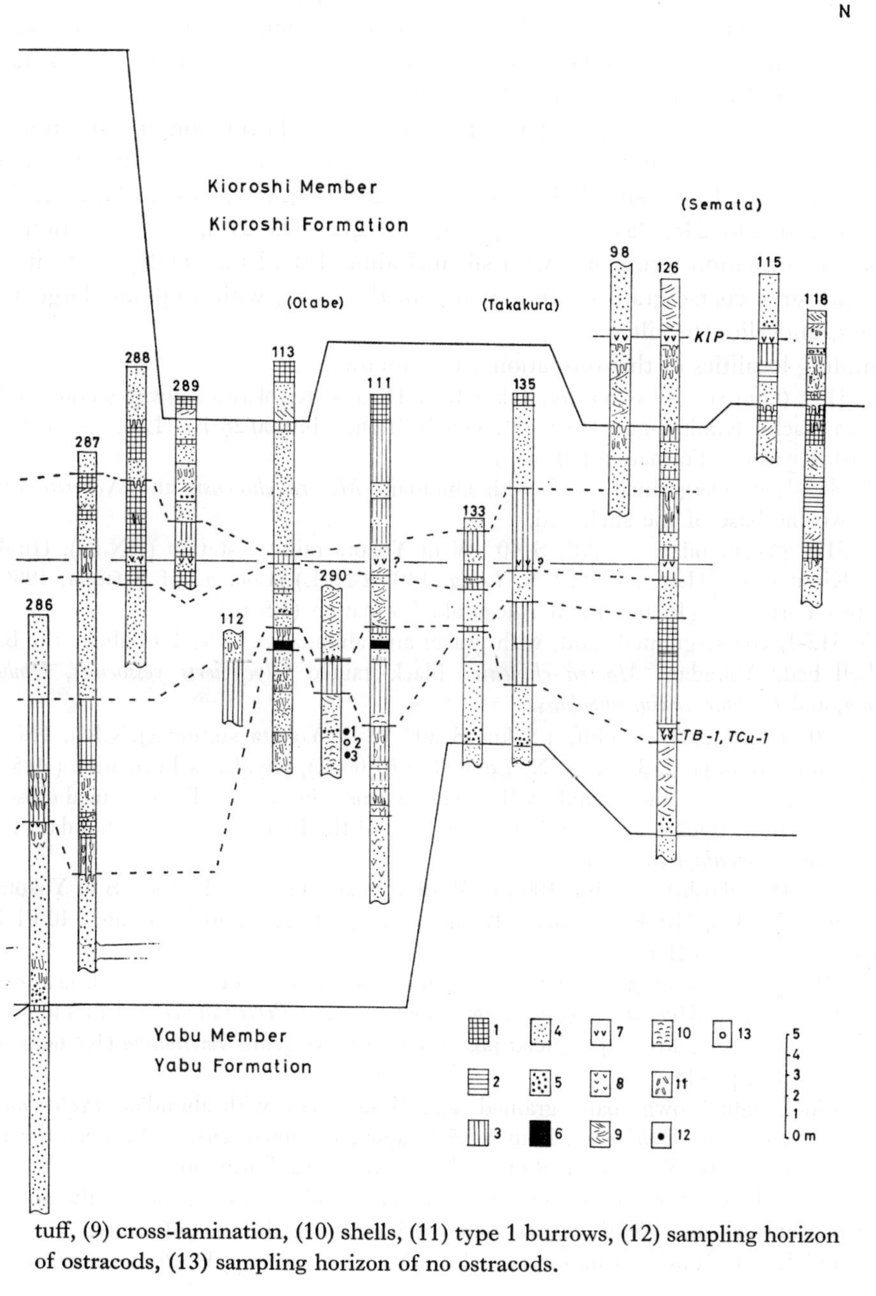

tuff, (9) cross-lamination, (10) shells, (11) type 1 burrows, (12) sampling horizon of ostracods, (13) sampling horizon of no ostracods.

thickness overlies the silt and includes various warm-temperate water molluscs: *Pectan albicans*, *Glycmeris vestita*, *Siphonalia fusoides*, and *Pectan tokyoensis* (sample 82-1).

The upper part of the Kiyokawa Formation, perhaps a horizon higher than the warm-temperate silty sand, is exposed around Hikita (Locs. 73 and 74). It is about 15 m thick and is composed of fine-grained, well-sorted sand including current valves of *Mactra chinensis* which also lie with their convex side upward and form several layers in sand. Shell bed (samples 74-1, 2, and 73-2, 3, 4) is overlain by TB-10 or 11 (?) marker-tephra (sample 73-1). In Azu area (Locs. 79 and 80) grey medium-grained well-sorted sand contains abundant *Glycymeris yessoensis* and *Mactra chinensis* (sample 79-1) and type 1 burrows (samples 80-1, 2, 3).

Various horizons of the upper part of the Kiyokawa Formation are unconformably overlain by the Kioroshi Formation in the northern areas around Imbanuma Lake. In the vicinity of Imbanuma Lake (Locs. 205, 208, 210, 214, 215, 218–220, 223, 234, 236–238, 240, 246–248, 251–253, 264, 265, 271, 272, 293–296), the deposition of the Kiyokawa Formation terminates with silt including TB-12 marker-tephra in its lower part; dark grey coarse-grained sand, 3 m+ in thickness, with long and large type 1 burrows, underlies the silt.

Sampling localities of this formation are as follows.

Loc. 315 (Otorii): an exposure, 1 km E of Higashi-Kiyokawa railway station (J.N.R.), Hirakawa-machi, Kimitsu-gun (Lat. 35°23′06″N, Long. 140°00′28″E) (Loc. 57 of Kozima, 1966a, b). Kiyokawa Formation (20 m+).

Sample 315-1, coarse-grained sand, with abundant *Mactra chinensis* and *Neverita didyma*, 2 m above the base of the shell bed.

Loc. 312 (Yoshinoda): a cliff, S 40° W of Yokota railway station (J.N.R.), Hirakawa-machi, Kimitsu-gun (Lat. 35°22′24″N, Long. 140°01′20″E) (Loc. 52 of Kozima, 1966a, b). Kiyokawa Formation (16 m+) and Anegasaki Formation (1.5 m).

Sample 312-1, coarse-grained sand, with gravel and abundant shells, 1 m above the base of the shell bed. Abundant *Mactra chinensis*, black-stained *Glycymeris yessoensis*, *Umbonium costatum*, and *Echinarachnius mirabilis*.

Loc. 310 (Yarimizu): a cliff, 1.5 km S 30° W of Yokota station (J.N.R.), Hirakawa-machi, Kimitsu-gun (Lat. 35°22′30″N, Long. 140°01′00″E), Kiyokawa Formation (10.5 m+).

Sample 310-1, coarse-grained sand, with shells in lenses in grey medium-grained cross-laminated sand 6 m in thickness, 7.5 m below the base of the bluish-grey silt. Abundant *Mactra chinensis* and *Umbonium costatum*.

Loc. 308 (Uchikoshi): a cliff, 100 m SW of Uchikoshi shrine, 1.25 km S of Yokota railway station (J.N.R.), Hirakawa-machi, Kimitsu-gun (Lat. 35°22′30″N, Long. 140°01′10″E). Kiyokawa Formation (21 m+).

Sample 308-1, grey medium-grained sand, 1 m above the base of 7.5 m thick layered shell sand, with abundant *Mactra chinensis*, *Tapes japonicus*, and *Pecten albicans*. In its upper part, the sand contains trace fossils produced possibly by isopod genus *Excilolana* (Kikuchi, 1978).

Loc. 138. See p. 148.

Sample 138-3, light brown coarse-grained sand (1 m thick), with abundant *Pecten albicans*, *P. tokyoensis* and *Mactra chinensis*. Found 3.5 m above the unconformity between the Kamiizumi Member of the Yabu Formation and the Kiyokawa Formation.

Sample 138-2, light brown coarse-grained sand (1 m thick), with sporadically distributed *Paphia naganumana* and *Fulvia mutica* found 4.5 m above the unconformity.

Sample 138-1, light brown medium-grained sand, convex side up *Mactra chinensis* in layers,

5.5 m above the unconformity.

Loc. 82 (Nakatakane, Kazato): a cliff, 1.9 km N 30° W Umatate railway station (Kominato Line), Ichihara-shi (Lat. 35°26′02″N, Long. 140°06′25″E) (Nakatakane of Aoki and Baba, 1971; Loc. 39 of Sugihara et al., 1978). Kiyokawa Formation (22 m+).
Sample 82-2, grey medium-grained, well-sorted sand with sporadically distributed *Glycymeris yessoensis* found 10.5 m below TB-7.
Sample 82-1, bluish grey fine-grained sand, 2 m above TB-7, with abundant *Pecten albicans*, *P. tokyoensis*, *Glycymeris vestita*, and *Dosinia japonica.*

Loc. 74 (Hikita): a cliff, 3 km S 85° W of Kazusamitsumata railway station (Kominato Line), Ichihara-shi (Lat. 35°27′30″N, Long. 140°05′50″E) (Hikita of Horiguchi and Ohara, 1972). Kiyokawa Formation (19 m+), Toyonari Member of Kioroshi Formation (3 m).
Sample 74-2, fine-grained sand, 17 m below the unconformity between Kiyokawa Formation and Kioroshi Formation, with abundant *Mactra chinensis* lying convex side upward and *Umbonium costatum.*
Sample 74-1, coquinalike medium-grained sand, 12.5 m below the unconformity, with abundant *Mactra chinensis* and *Glycymeris yessoensis.*

Loc. 73 (Hikita-higashi): a cliff, 2.5 km S 85° W of Kazusamitsumata railway station (Kominato Line), Ichihara-shi (Lat. 35°27′40″N, Long. 140°06′05″E). Kiyokawa Formation (13 m+).
Sample 73-4, pumiceous fine-grained sand, 4.2 m below a pumiceous layer (TB-10 or 11?), with *Umbonium costatum* and *Anomia lischkei.*
Sample 73-3, medium-grained sand, 2.2 m below a pumiceous layer (TB-10 or 11?), with abundant *Mactra chinensis*, *Glycymeris yessoensis*, and *Umbonium costatum.*
Sample 73-2, medium-grained sand, 60 cm below a pumiceous layer (TB-10 or 11?), with abundant burrows (0.8 cm in diameter and 2 cm in length) and *Mactra chinensis* lying convex side upward.

Loc. 79 (Azu-kita): a small exposure, 1 km N 60° W of Kazusayamada railway station (Kominato Line), Ichihara-shi (Lat. 35°27′20″N, Long. 140°09′40″E). Kiyokawa Formation (115 cm).
Sample 79-1, grey well-sorted medium-grained sand, 80 cm above the base of the exposure, with *Glycymeris yessoensis* and *Mactra chinensis.*

Loc. 80 (Azu): an exposure, along the Yoro River, I km S 70° W of Kazusayamada railway station (Kominato Line), Ichihara-shi (Lat. 35°26′48″N, Long. 140°07′08″E) (Azu of Aoki and Baba, 1971). Kiyokawa Formation (7 m+).
Sample 80-3, grey fine-grained sand with type 1 burrows and abundant shell fragments, 35 cm below the upper surface of the sand where burrows have their openings and 65 cm below 10 cm thick mud layer.
Sample 80-1, grey fine-grained sand with abundant molluscs, 15 cm above mud layer, 4 m below shell bed bearing *Pecten tokyoensis.*

Loc. 290 (Otabe): an exposure, 300 m SE of Otabe hamlet, Ichihara-shi (Lat. 35°29′11″N, Long. 140°08′34″E). Kiyokawa Formation (9 m+).
Sample 290-3, dark coarse-grained sand with abundant *Mactra chinensis*, *Glycymeris yessoensis*, and *Umbonium costatum*, 2.6 m below 1 m thick dark clay.
Sample 290-1, gravel bed, 2 m below the dark clay. Shell lens 1 m above the gravel bed contains molds of *Mactra chinensis*, *Glycymeris yessoensis*, and *Spisula sachalinensis.*

Kamiiwahashi Formation: The Kamiiwahashi Formation, about 16.5 m maximum thickness (Loc. 203), lies unconformably on the Kiyokawa Formation and once extensively covered the entire area surveyed. However, in the area south of the Murata River,

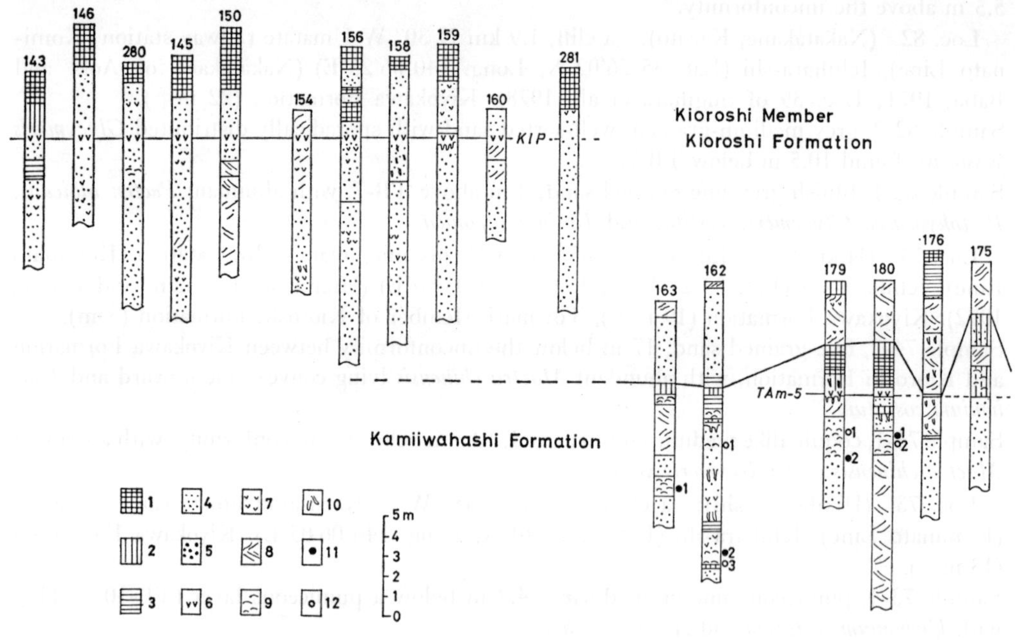

Text-fig. 5
Columnar sections of Kiyokawa Formation, Kamiiwahashi Formation, and Kioroshi Member of Kioroshi Formation.

its distribution is now restricted to a few scattered small areas, mainly because of the erosion at the base of the Kioroshi Formation cutting deeply the Kamiiwahashi and Kiyokawa Formations and reaching even the Yabu Formation in this area.

The deposition of the Kamiiwahashi Formation begins with medium-grained sand filling type 1 burrows (Yajima, 1978) which penetrates the silt of the Kiyokawa Formation. The medium-grained sand includes *Panopea japonica* in living position in its basal part (samples 203-3, 246-1, 247-1, 211-1, 218-1, 264-1). Allochthonous molluscan fossils of cold-temperate water inhabitants, *Solen krusensterni*, *Glycymeris yessoensis*, *Mactra chinensis*, and *Diplodonta usta*, dominate in its lower 4 m (samples 174-3, 170-3, 203-2, 1, 207-1, 209-1, 211-1, 214-1, 215-1, 219-1, 238-2, 234-1, 259-1, 256-1, 253-1, 240-1, 251-4, 279-4). In the upper half, the medium-grained sand includes type 2 burrows (Yajima, 1978) and shell lenses of faunal assemblage similar to that of the lower half (samples 163-1, 162-2, 179-2, 180-1, 174-2, 170-1, 2, 276-1, 202-1, 198-1, 2). The upper part of the Kamiiwahashi Formation consists of brackish water clay (Locs. 179 to 269), which includes TAm-5 marker-tephra in its basal part. The clay contains *Batillaria zonalis*, *Anadara broughtoni*, and *Raeta yokohamensis* in some places

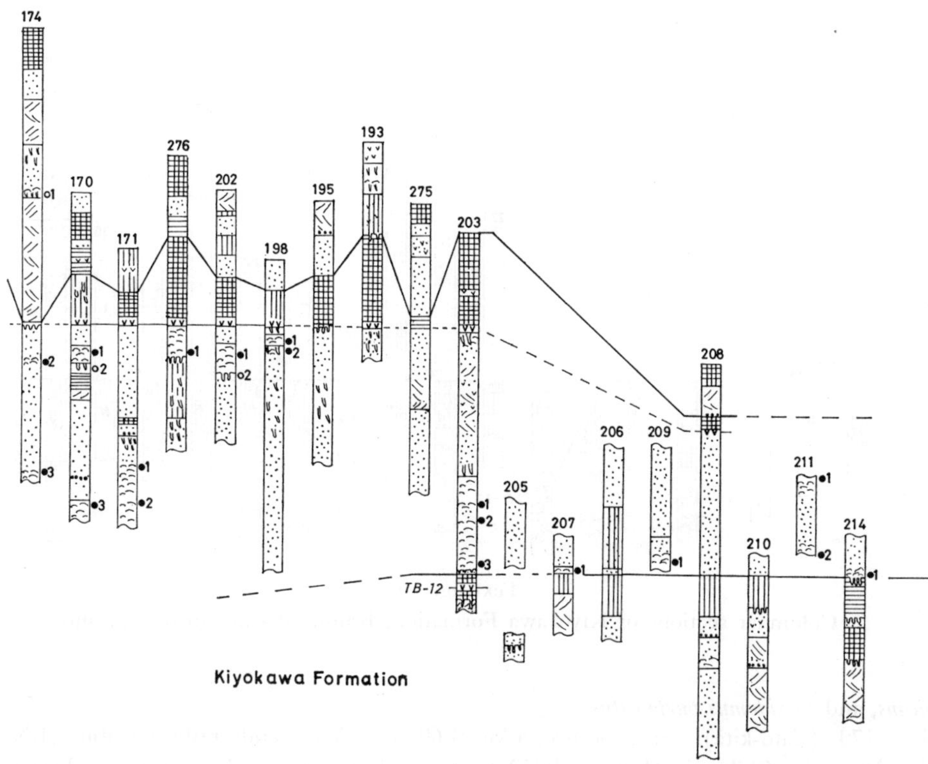

(1) clay, (2) silt, (3) alternation of silt with sand, (4) sand, (5) gravel, (6) marker-tephra, (7) pumice tuff, (8) cross-lamination, (9) shells, (10) type 1 burrows, (11) sampling horizon of ostracods, (12) sampling horizon of no ostracods.

(sample 279-3). Around Narita (Locs. 224-230), the Kamiiwahashi Formation terminates either with silt including *Dosinia japonica* in living position, abundant *Raeta yokohamensis*, plant fragments (sample 224–2), or sandy silt with water-strewn *Mactra chinensis* and *Raeta yokohamensis* (samples 228-1, 2, 3).

Sampling localities of this formation are as follows.

Loc. 163 (Ikazuchi): a small exposure, 50 m E of Ikazuchi-machi, Chiba-shi (Lat. 35° 34′33″N, Long. 140°13′23″E). Kamiiwahashi Formation (7 m+) and Kioroshi Member of Kioroshi Formation (4 m).

Sample 163-1, grey coarse-grained sand, 5 m below the unconformity between Kamiiwahashi Formation and Kioroshi Formation, with abundant *Glycymeris yessoensis* and *Mactra chinensis* lying convex side upward.

Loc. 162 (Ikazuchi-kita): an exposure, along the road halfway between Ikazuchi-machi and Kawai-machi, Chiba-shi (Lat. 35°34′52″N, Long. 140°13′22″E). Kamiiwahashi Formation (10 m+) and Kioroshi Member of Kioroshi Formation (4.5 m).

Sample 162-2, shell lens in grey coarse- to medium-grained sand, 8.5 m below the upper surface of top mud of Kamiiwahashi Formation, with abundant *Glycymeris yessoensis*, *Pecten*

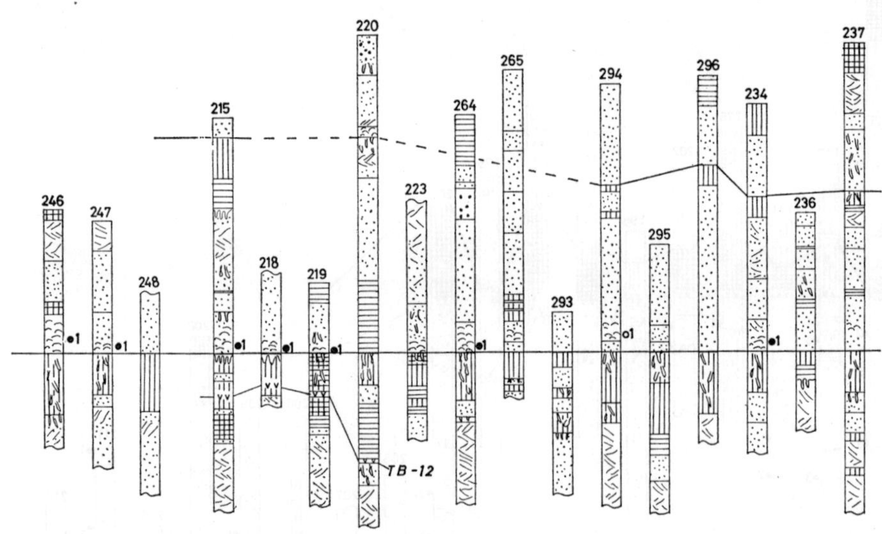

Text-fig. 6
Columnar sections of Kiyokawa Formation, Kamiiwahashi Formation, and

albicans, and *Saxidomus purpuratus*.

Loc. 179 (Yato-kita): an exposure, 5 km S 60° E of Yotsukaido railway station (J.N.R.), Shimoda-machi, Chiba-shi (Lat. 35°38′32″N, Long. 140°13′13″E). Kamiiwahashi Formation (8 m+) and Kioroshi Member of Kioroshi Formation (3 m).
Sample 179-2, medium-grained sand, 3.2 m below TAm-5, with abundant *Mactra chinensis* and articulated *Glycymeris yessoensis*.

Loc. 180 (Oushi): an exposure, along the road halfway between Sakato and Uchida, southern part of Sakura-shi (Lat. 35°38′58″N, Long. 140°13′46″E). Kamiiwahashi Formation (13 m+) and Kioroshi Member of Kioroshi Formation (3 m).
Sample 180-1, coarse-grained sand, 1.8 m below TAm-5, with abundant shell fragments.

Loc. 174 (Iwatomi-minami): an exposure, 500 m S of Iwatomi-machi Post Office, Iwatomi-machi, Chiba-shi (Lat. 35°38′34″N, Long. 140°14′30″E) (Loc. 405 of Kozima, 1962, 1963). Kamiiwahashi Formation (7.5 m+) and Kioroshi Member of Kioroshi Formation (14.5 m).
Sample 174-3, light brown medium-grained, well-sorted sand, 7.2 m below the upper surface of top clay of Kamiiwahashi Formation, with *Solen krusensterni*.
Sample 174-2, black coarse-grained sand, 1.8 m below the unconformity, with abundant *Mactra chinensis*, *Glycymeris yessoensis* and *Dosinia japonica*.

Loc. 170 (Enokida-minami): an exposure, 700 m N of Iwatomi-machi Post Office, Sakura-shi (Lat. 35°39′24″N, Long. 140°14′28″E) (Loc. 410 of Kozima, 1962, 1963). Kamiiwahashi Formation (12.5 m+) and Kioroshi Member of Kioroshi Formation (3.3 m).
Sample 170-3, medium-grained sand, 8.8 m below TAm-5, with abundant *Glycymeris yessoensis*, *Spisula sachalinensis*, and shell fragments.
Sample 170-1, coarse-grained, poorly sorted sand, 1.35 m below TAm-5, with *Mactra chi-*

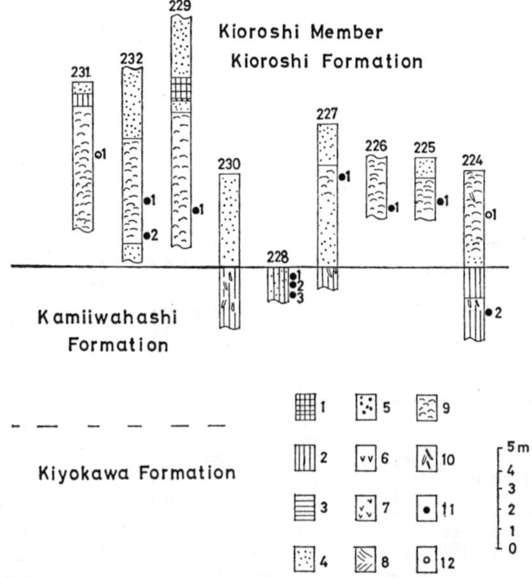

Kioroshi Member of Kioroshi Formation. See explanation of legend of text-fig. 5.

nensis lying convex side upward.

Loc. 171 (Enokida): a small exposure, 800 m N of Iwatomi-machi Post Office, Sakura-shi (Lat. 35°39′28″N, Long. 140°14′25″E) (Loc. 410 of Kozima, 1962, 1963). Kamiiwahashi Formation (11.5 m+) and Kioroshi Member of Kioroshi Formation (2.2 m).
Sample 171-2, grey medium-grained sand, 8.9 m below TAm-5, with shell fragments of *Mactra chinensis* and *Glycymeris yessoensis.*
Sample 171-1, medium- to coarse-grained sand, 6.9 m below TAm-5, with *Mactra chinensis, Glycymeris yessoensis,* and *Pecten albicans.*

Loc. 276 (Godo-minami): an exposure, 600 m S 10° W of Godo police station, 500 m S 45° E of Loc. 202, Sakura-shi (Lat. 35°40′09″N, Long. 140°14′20″E) (Loc. 6 of Oinomikado, 1937; Loc. 414 of Kozima, 1962, 1963). Kiyokawa Formation (4.6 m+), Kamiiwahashi Formation (6.8 m) and Kioroshi Member of Kioroshi Formation (4 m).
Sample 276-1, coarse-grained sand, 1 m below yellow sand (possibly TAm-5), with abundant *Mactra chinensis, Pecten albicans* lying convex side upward, and *Tonna luteostoma.*

Loc. 202 (Godo): an exposure, 4 km S 06° E of Sakura railway station (J.N.R.), Sakura-shi (Lat. 35°40′16″N, Long. 140°14′00″E) (Loc. 5 of Oinomikado, 1937; Loc. 415 of Kozima, 1962, 1963). Kamiiwahashi Formation (8 m+) and Kioroshi Member of Kioroshi Formation (4.2 m).
Sample 202-1, grey medium-grained sand, 1.5 m below TAm-5, with *Mactra chinensis, Tonna luteostoma,* and *Pecten albicans.*

Loc. 198 (Mukai): an exposure, 1 km S 05° E of Monoi railway station (J.N.R.), Yotsukaido-machi, Imba-gun (Lat. 35°40′25″N, Long. 140°12′17″E). Kamiiwahashi Formation (1.5 m).
Sample 198-2, light brown medium-grained sand, 1.3 m below TAm-5, with shell fragments

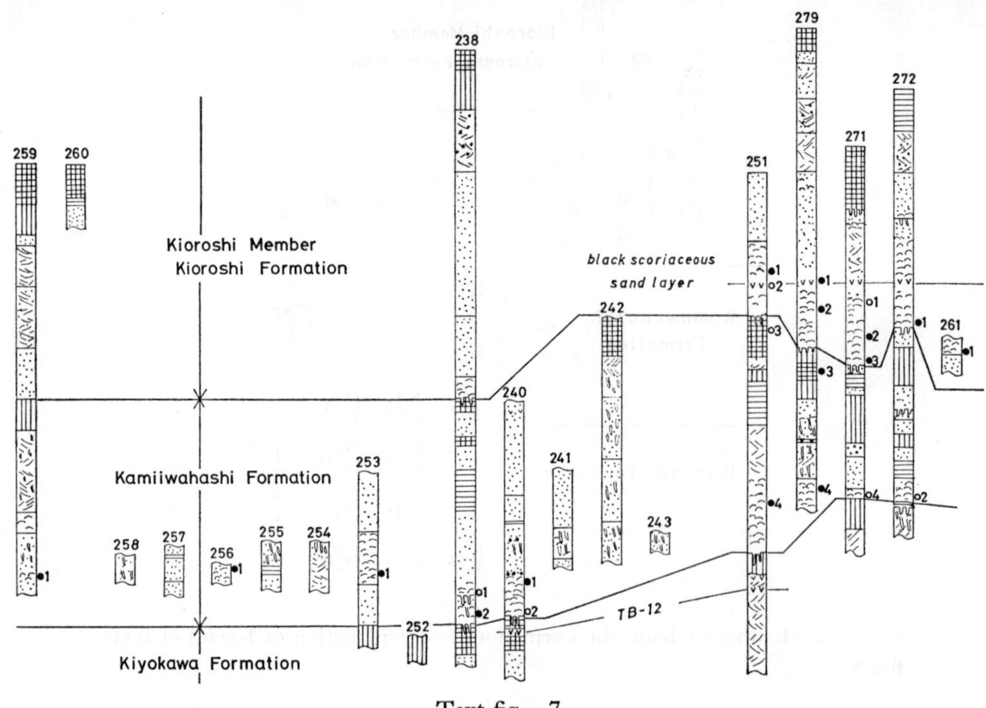

Text-fig. 7
Columnar sections of Kiyokawa Formation, Kamiiwahashi Formation, and

of *Mactra chinensis* and silt burrows.
Sample 198-1, coarse-grained sand, 80 cm below TAm-5, with abundant *Mactra chinensis* and *Echinarachnius mirabilis*.

Loc. 203 (Yamanosaki): an exposure, 500 m N 45° W of Keisei-Sakura railway station (Keisei Line), Sakura-shi (Lat. 35°44′32″N, Long. 140°13′42″E) (Loc. 254 of Kozima, 1959). Kiyokawa Formation (2 m+) and Kamiiwahashi Formation (16.5 m).
Sample 203-3, grey to dark grey medium-grained sand, 20 cm above the unconformity between Kiyokawa Formation and Kamiiwahashi Formation, with *Solen krusensterni*, *Glycymeris yessoensis*, and *Panopea japonica*.
Sample 203-2, fine-grained sand, 2.5 m above the unconformity, with *Panopea japonica* in living position and abundant *Mactra chinensis*.
Sample 203-1, shell bed of medium-grained sand, 3.3 m above the unconformity, with abundant small *Glycymeris yessoensis* which are compactly packed.

Loc. 207 (Iwana): a small exposure, 1.25 km N 05° E of Keisei-Sakura railway station (Keisei Line), Sakura-shi (Lat. 35°44′00″N, Long. 140°14′05″E). Kiyokawa Formation (3 m+) and Kamiiwahashi Formation (1.8 m+).
Sample 207-1, shell lens in grey medium-grained sand, immediately above the unconformity between Kiyokawa Formation and Kamiiwahashi Formation, with *Glycymeris yessoensis* and *Solen krusensterni*.

Loc. 209 (Iwana-kita): a small exposure, 1.75 km N 10° W of Keisei-Sakura railway station (Keisei Line), Sakura-shi (Lat. 35°44′17″N, Long. 140°14′50″E). Kamiiwahashi Formation (6.5 m+).

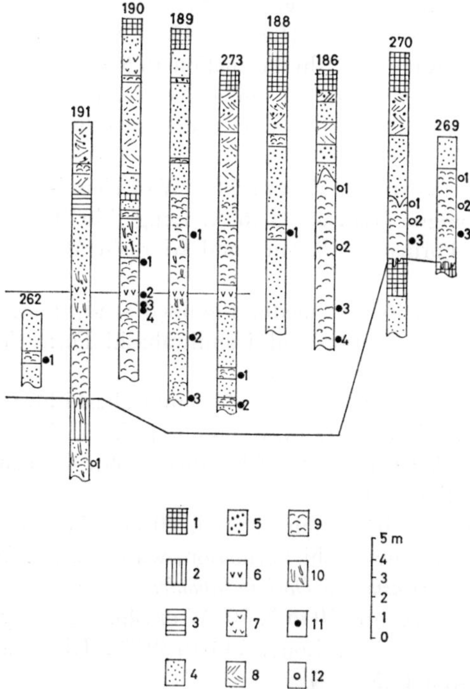

Kioroshi Member of Kioroshi Formation. See explanation of legend of text-fig. 5.

Sample 209-1, shell lens in light brown medium-grained sand, the basal part of the upper shell lens, 50 cm thick, with *Solen krusensterni*, *Glycymeris yessoensis* and *Pecten albicans*.

Loc. 211 (Osakura): a small exposure, 550 m N 20° W of Osakura railway station (Keisei Line), Sakura-shi (Lat. 35°43′05″N, Long. 140°15′05″E). Kamiiwahashi Formation (7 m+).
Sample 211-2, medium-grained sand with *Panopea japonica* in living position and abundant *Solen krusensterni* and *Glycymeris yessoensis*, 4 m below boundary between fossiliferous sand and light brown laminated fine- to medium-grained sand.
Sample 211-1, well-sorted, coarse-grained sand with *Glycymeris yessoensis* and abundant *Mactra chinensis*, 1 m below the boundary between fossiliferous sand and light brown laminated fine- to medium-grained sand.

Loc. 214 (Hamajuku): a small exposure, 550 m N 25° E of Osakura railway station (Keisei Line), Sakura-shi (Lat. 35°43′54″N, Long. 140°15′21″E). Kiyokawa Formation (7 m+) and Kamiiwahashi Formation (2 m+).
Sample 214-1, shell lens in grey coarse-grained well-sorted sand, 50 cm above the unconformity between Kiyokawa Formation and Kamiiwahashi Formation, with *Solen krusensterni* and *Mactra chinensis*.

Loc. 246 (Heta): an exposure, 2.25 km N 40° W of Keisei-Shisui railway station (Keisei Line), Imba-mura, Imba-gun (Lat. 35°44′57″N, Long. 140°15′24″E). Kiyokawa Formation (4.4 m+) and Kamiiwahashi Formation (7 m).
Sample 246-1, medium-grained sand, 70 cm above the unconformity between Kiyokawa Formation and Kamiiwahashi Formation, with *Panopea japonica* in living position and abundant *Glycymeris yessoensis*, *Mactra chinensis*, and *Solen krusensterni*.

Loc. 247 (Hirago): a small exposure, 2.9 km N 65° W of Sogosando railway station (Keisei Line), Imba-mura, Imba-gun (Lat. 35°45′23″N, Long. 140°15′12″E). Kiyokawa Formation (5.6 m+) and Kamiiwahashi Formation (6.5 m).
Sample 247-1, medium-grained sand, 20 cm above the unconformity between Kiyokawa Formation and Kamiiwahashi Formation, with abundant *Glycymeris yessoensis*, *Solen krusensterni*, and *Mactra chinensis*.

Loc. 215 (Shinbori): an exposure, 700 m S 60° W of Keisei-Shisui railway station (Keisei Line), Shisui-machi, Imba-gun (Lat. 35°43′46″N, Long. 140°16′00″E). Kiyokawa Formation (7.6 m+) Kamiiwahashi Formation (8.5 m), and Kioroshi Member of Kioroshi Formation (1 m).
Sample 215-1, shell lens in coarse- to medium-grained sand, immediately above the unconformity between Kiyokawa Formation and Kamiiwahashi Formation, with *Glycymeris yessoensis*, *Mactra chinensis*, and *Dosinia japonica*.

Loc. 218 (Kamiiwahashi): type locality of Kamiiwahashi Formation, along the highway route 51, 450 m N 70° W of Shisui railway station (J.N.R.), Shisui-machi, Imba-gun (Lat. 35°43′46″N, Long. 140°16′29″E) (Loc. 212 of Kozima, 1959). Kiyokawa Formation (2.5 m+) and Kamiiwahashi Formation (4 m+).
Sample 218-1, light brown medium-grained sand, 20 cm above the unconformity between Kiyokawa Formation and Kamiiwahashi Formation, with *Panopea japonica* in living position and abundant *Glycymeris yessoensis* and *Dosinia japonica*.

Loc. 219 (Shisui): an exposure, 50 m N 70° W of Shisui railway station (J.N.R.), Shisui-machi, Imba-gun (Lat. 35°43′43″N, Long. 140°16′39″E). Kiyokawa Formation (7.5 m+) and Kamiiwahashi Formation (3.5 m+).
Sample 219-1, grey coarse-grained well-sorted sand, immediately above the unconformity between Kiyokawa Formation and Kamiiwahashi Formation, with abundant *Glycymeris yessoensis* and *Solen krusensterni*.

Loc. 264 (Owashi): a small exposure, 1 km N 35° E of Keisei-shisui railway station (Keisei Line), Shisui-machi, Imba-gun (Lat. 35°44′19″N, Long. 140°16′46″E). Kiyokawa Formation (7.4 m+) and Kamiiwahashi Formation (11.6 m).
Sample 264-1, medium-grained sand, immediately above the unconformity between Kiyokawa Formation and Kamiiwahashi Formation, with *Panopea japonica* in living position and abundant *Solen krusensterni*, *Glycymeris yessoensis*, *Spisula sachalinensis*, and *Pecten tokyoensis*.

Loc. 234 (Uchino-minami): a cliff, 500 m W of Sogoreido temple, Narita-shi (Lat. 35°45′25″N, Long. 140°16′36″E). Kiyokawa Formation (5 m+), Kamiiwahashi Formation (7.5 m) and Kioroshi Member of Kioroshi Formation (4.6 m).
Sample 234-1, shell lens in grey medium-grained sand, immediately above the unconformity between Kiyokawa Formation and Kamiiwahashi Formation, with abundant *Glycymeris yessoensis* and *Solen krusensterni*.

Loc. 259 (Ipponmatsu): a cliff, 500 m S 10 °W of Imba-mura village office, Imba-mura, Imba-gun (Lat. 35°46′16″N, Long. 140°12′26″E). Kamiiwahashi Formation (9.5 m+) and Kioroshi Member of Kioroshi Formation (11.5 m).
Sample 259-1, muddly sand, 8.5 m below the unconformity between Kamiiwahashi Formation and Kioroshi Formation, with abundant *Glycymeris yessoensis*, *Solen krusensterni*, and and *Mactra chinensis*.

Loc. 256 (Seto): a small exposure, 850 m S 60° E of Imba-mura village office, Imba-mura, Imba-gun (Lat. 35°46′17″N, Long. 140°13′00″E). Kamiiwahashi Formation (1 m+).
Sample 256-1, light brown medium-grained sand, 80 cm above the base of the exposure (level of the road), with *Panopea japonica* in living position and abundant *Glycymeris yessoensis*, *Mactra chinensis*, and *Spisula sachalinensis*.

Loc. 253 (Egawa): a small exposure, 800 m S 40 °E of Futago bridge of Imba waterway, Imba-mura, Imba-gun (Lat. 35°45′50″N, Long. 140°13′16″E). Kiyokawa Formation (1 m+) and Kamiiwahashi Formation (7.5 m+).
Sample 253-1, light brown medium-grained, well-sorted sand, 2.5 m above the unconformity between Kiyokawa Formation and Kamiiwahashi Formation, with abundant *Solen krusensterni*, *Anadara broughtoni*, and *Mactra chinensis*.

Loc. 238 (Yamada): a cliff of abandoned sand pit, 350 m S of Yamada bridge of Imba waterway, Imba-mura, Imba-gun (Lat. 35°46′04″N, Long. 140°13′52″E). Kiyokawa Formation (1.6 m+), Kamiiwahashi Formation (11 m) and Kioroshi Member of Kioroshi Formation (16 m).
Sample 238-2, medium-grained sand, immediately above the unconformity between Kiyokawa Formation and Kamiiwahashi Formation, with abundant *Glycymeris yessoensis* and *Solen krusensterni*.

Loc. 240 (Iwaido): a small exposure, 200 m E of Futago bridge of Imba waterway, Imba-mura, Imba-gun (Lat. 35°45′37″N, Long. 140°13′52″E) (Loc. 121 of Kozima, 1959a, b). Kiyokawa Formation (3 m+) and Kamiiwahashi Formation (10.5 m+).
Sample 240-1, coarse-grained sand, 1.8 m above the unconformity between Kiyokawa Formation and Kamiiwahashi Formation, with abundant *Glycymeris yessoensis* and *Crassostrea gigas*.

Loc. 251 (Kidouchi): a cliff, 1.2 km N 65° E of Yamada bridge of Imba waterway, Imba-mura, Imba-gun (Lat. 35°46′30″N, Long. 140°14′34″E). Kiyokawa Formation (6 m+), Kamiiwahashi Formation (11.5 m) and Kioroshi Member of Kioroshi Formation (7 m).
Sample 251-4, medium-grained well-sorted sand, 2.5 m above the unconformity between Kiyokawa Formation, with abundant *Glycymeris yessoensis* in irregular lenses.

Loc. 279 (Nakamura-minami): a cliff, 750 m N 03° E of Yamada bridge of Imba waterway, Imba-mura, Imba-gun (Lat. 35°46′40″N, Long. 140°13′52″E). Kamiiwahashi Formation (8 m+) and Kioroshi Member of Kioroshi Formation (19.5 m).
Sample 279-4, grey medium-grained well-sorted sand, 6.8 m below the unconformity between Kamiiwahashi Formation and Kioroshi Formation, with abundant *Glycymeris yessoensis*, *Solen krusensterni*, and *Schizothaerus keenae*.
Sample 279-3, bluish grey clay, 1m below the unconformity, with *Raeta yokohamensis*, *Anadara broughtoni*, and *Batillaria zonalis* scattered.

Loc. 228 (Narita): a small exposure, along the highway route 51, 300 m E of Keisei-Narita railway station (Keisei Line), Narita-shi (Lat. 35°46′23″N, Long. 140°19′18″E). Kamiiwahashi Formation (1.6 m+).
Sample 228-3, sandy silt, 30 cm above the base of the exposure (level of the road).
Sample 228-2, silty sand, 80 cm above the base of the exposure, with abundant *Raeta yokohamensis*.
Sample 228-1, bluish grey sandy silt, 1.2 m above the base of the exposure, with *Mactra chinensis* forming layers.

Loc. 224 (Otake): an exposure, 500 m N 40° W of Shimosa-manzaki railway station (J.N.R.), Narita-shi (Lat. 35°48′31″N, Long. 140°16′37″E) (Loc. 164 of Kozima, 1959a, b). Kamiiwahashi Formation (3.4 m+) and Kioroshi Member of Kioroshi Formation (4.7 m).
Sample 224-2, silt, 2 m below the unconformity between Kamiiwahashi Formation and Kioroshi Formation, with abundant plant fragments, burrows, *Dosinia angulosa* in living position, and *Raeta yokohamensis*.

Kioroshi Formation: The Kioroshi Formation is divisible into two members—namely, the Kioroshi Member and the Toyonari Member—based mainly on the dif-

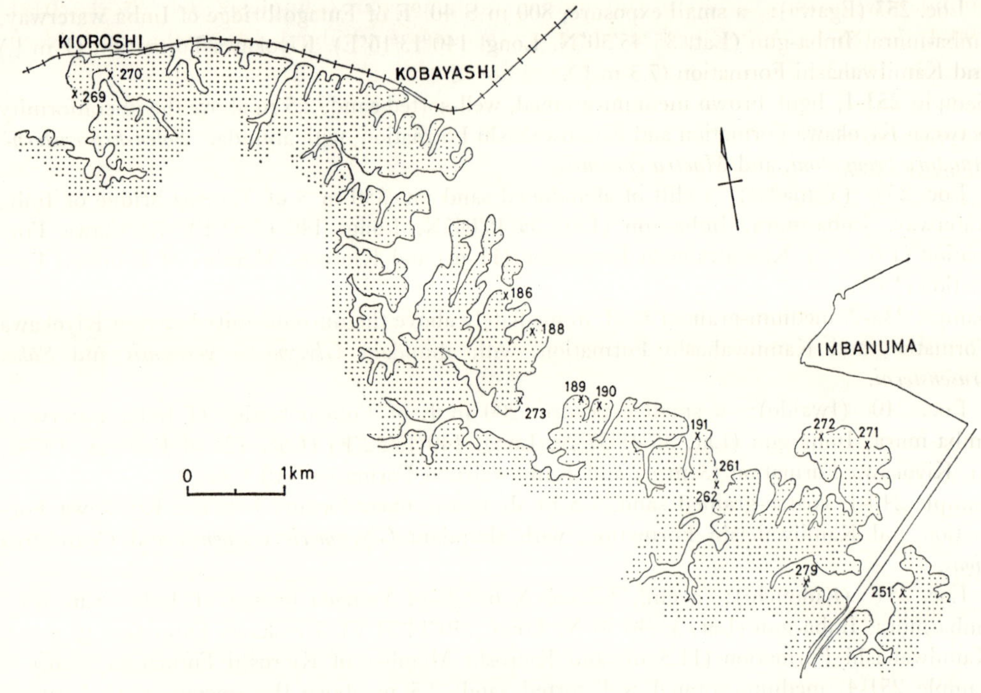

Text-fig. 8
Surveyed area of Kioroshi Member of Kioroshi Formation along the western coast of the northern half of Imbanuma Lake.

ference in area of distribution, sedimentary succession, and faunal assemblage. The Kioroshi Member crops out around Imbanuma Lake and extends southward to the area east to the Yoro River in the middle of the southern area. The Toyonari Member lies in a restricted area between the Yoro River and the Koito River of the Toyonari-Kisarazu area. The difference in sedimentary environment is seen in marine sand-nonmarine clay succession in the Kioroshi Member and gravel to silt to sand to silty-sand succession of the entire marine environment in the Toyonari Member. Molluscan assemblages suggest that the shallower environment dominated in the Kioroshi Member rather than in the Toyonari Member. Both members include K1P group of marker-tephras whose dates between 117,000 and 132,000 yr B.P. were obtained in the Oiso-Yokohama area (Sugihara et al., 1978). This formation probably represents the period of the Shimosueyoshi transgression.

I. Kioroshi Member of Kioroshi Formation. The Kioroshi Member of the Kioroshi Formation is composed mainly of sand about 15 m thick exposed along the western coast of the northern half of Imbanuma Lake (text-fig. 8). The deposition of this member begins with sand which fills burrows in the uppermost silt of the Kami-iwahashi Formation. *Panopea japonica* is found in a living position at the basal part (samples 279-2, 271-3, 272-1). The lower part consists of well-sorted, medium-grained sand with *Solen strictus*, *Tonna luteostoma*, *Rapana venosa*, *Tapes japonicus*, and abundant bryozoans (samples 271-2, 189-3, 2, 273-2, 1, 186-4, 3). *Mactra chinensis* occurs abundantly, forming layers in the middle and major part (samples 279-1, 262-1, 261-1, 190-4, 3, 188-1). A key bed of black scoriaceous sand about 10 cm thick is present (sample 190-2). Fine-grained sand containing type 2 burrows (Yajima, 1978) and beds of valves of *Mactra chinensis* overlies the scoriaceous sand (samples 251-1, 190-1). Aggregates of abundant fragments of *Mactra chinensis* with *Glycymeris yessoensis* and *Echinarachnius mirabilis* overlie the layers of *Mactra chinensis* (sample 189-1). The upper part consists of cross-laminated sand with layers of fragmented *Echinarachnius mirabilis*. It appears that the lower part of this member consists of shallow water deposits of an embayment; the middle part represents the deposits at the time of high sea-level; and the upper part serves as an example of retreating sea deposits. The fossiliferous sand is overlain by nonmarine clay, 1 to 3 m thick, containing plant fragments.

Near Kioroshi, the type locality of the Kioroshi Formation (locs. 269, 270), the most fossiliferous sand contains beds of *Echinarachnius mirabilis* (samples 269-3, 270-3) in its basal part. According to Sugihara (1970), a group of K1P marker-tephras were found at the uppermost horizon of the type locality (Loc. 269).

Around Narita (Locs. 232, 229, 227-224), the Kioroshi Member consists of 9 m thick, medium-grained sand, and the shell bed is reduced to 2 to 3 m. Fossils contained are abundant *Mactra chinensis*, *Tapes japonicus*, *Glycymeris vestita*, *Tellina venulosa*, and *Echinarachnius mirabilis* (samples 232-2, 1, 229-1, 227-1, 226-1, 225-1).

Near Hirayama and Okusa (Locs. 143, 145, 146, 150, 154, 156, 158–160, 280), one of the K1P marker-tephras, 50 cm thick, is observed clearly in 12 m+ thick light brown or grey medium-grained sand. Medium-grained sand with cross-laminations and trace fossils which are isopods of genus *Excilolana* are overlain by nonmarine clay. Sampling localities of this member are as follows.

Loc 232 (Ota): an exposure along the highway route 51, 500 m S 20° E of Narita railway station (J.N.R.) Narita-shi (Lat. 35°46′12″N, Long. 140°19′00″E). Kioroshi Member of Kioroshi Formation (9 m+).

Sample 232-2, coarse-grained sand, 1.4 m above the unconformity between Kamiiwahashi Formation and Kioroshi Formation, lowermost horizon of shell bed, with abundant *Mactra chinensis*.

Sample 232-1, coarse-grained sand, 2.9 m above the unconformity, with abundant *Echinarachnius mirabilis*.

Loc. 229 (Narita-nishi): an exposure along the Keisei Line, 450 m S 40° W of Keisei-Narita railway station (Keisei Line), 350 m E of Loc. 232, Narita-shi (Lat. 35°46′12″N, Long. 140°18′55″E). Kioroshi Member of Kioroshi Formation (8.2 m+).

Sample 229-1, brown medium-grained sand, with abundant *Mactra chinensis*, *Glycymeris vestita*, and *Anadara broughtoni*, 5.3 m below the 10 cm thick pumice tuff layer.

Loc. 227 (Nakadai): a small exposure along a small pond, 1.5 m N 50° W of Narita railway station (J.N.R.), Narita-shi (Lat. 35°47′00″N, Long. 140°18′14″ E). Kamiiwahashi Formation (1 m+) and Kioroshi Member of Kioroshi Formation (7 m+).

Sample 227-1, brown medium- to coarse-grained sand, 4.5 m above the unconformity between Kamiiwahashi Formation and Kioroshi Formation, with shell fragments of *Tapes japonicus* and *Mactra chinensis*.

Loc. 226 (Atago): an exposure, 2 km S 50° E of Shimosa-manzaki railway station (J.N.R.), Narita-shi (Lat. 35°47′33″N, Long. 140°17′50″E) (Loc. 157 of Kozima, 1959). Kioroshi Member of Kioroshi Formation (3 m+).

Sample 226-1, medium- to coarse-grained sand, 0.5 m above the base of the exposure, with abundant *Mactra chinensis*, *Tapes japonicus*, *Glycymeris vestita*, and *Echinarachnius mirabilis*.

Loc. 225 (Koshiro): an exposure, 1.7 km S 45° E of Simosa-manzaki railway station (J.N.R.), Narita-shi (Lat. 35°47′38″N, Long, 140°17′32″E) (Loc. 159 of Kozima, 1959). Kioroshi Member of Kioroshi Formation (3 m+).

Sample 225-1, medium- to coarse-grained sand, with abundant *Mactra chinensis* and *Echinarachnius mirabilis*, 1 m below the boundary between the 2 m thick shell bed and the 1 m thick nonfossiliferous medium-grained sand.

Loc. 251. See p. 161.

Sample 251-1, brown medium-grained well-sorted sand, 2 m above the unconformity between Kamiiwahashi Formation and Kioroshi Formation, with fragments of *Echinarachnius mirabilis* and abundant *Mactra chinensis* lying convex side upward.

Loc. 279. See p. 161.

Sample 279-2, grey fine- to medium-grained poorly-sorted sand, immediately above the unconformity between Kamiiwahashi Formation and Kioroshi Formation, with abundant *Anadara broughtoni*, *Pecten albicans*, and other shell fragments.

Sample 279-1, grey coarse-grained poorly-sorted sand, 2 m above the unconformity, with abundant *Mactra chinensis* lying convex side upward, *Tapes japonicus*, *Anadara broughtoni*, *Rapana venosa*, and *Tonna luteostoma*.

Loc. 271 (Nakamura-kita): a cliff, 1 km W of Jimbei bridge of Imbanuma Lake, Imba-mura, Imba-gun, (Lat. 35°48′20″N, Long. 140°14′28″E). Kiyokawa Formation (2.5 m+), Kamiiwahashi Formation (6.5 m) and Kioroshi Member of Kioroshi Formation (10.5 m).

Sample 271-3, medium-grained poorly-sorted sand, immediately above the unconformity between Kamiiwahashi Formation and Kioroshi Formation, with abundant *Panopea japonica* in living position, *Solen strictus*, and *Anadara broughtoni*.

Sample 271-2, medium-grained poorly-sorted sand, 1.3 m above the unconformity, with abundant *Tonna luteostoma*, *Glycymeris vestita*, and *Tapes japonicus*.

Loc. 272 (Nakamura-nishi): a cliff, 1.5 km W of Jimbei bridge of Imbanuma Lake, Imba-mura, Imba-gun (Lat. 35°47′26″N, Long. 140°14′10″E). Kiyokawa Formation (1.5 m+), Kamiiwahashi Formation (8.7 m), and Kioroshi Member of Kioroshi Formation (11.5 m).

Sample 272-1, fine-grained sand, immediately above the unconformity between Kamiiwahashi Formation and Kioroshi Formation, with abundant *Tonna luteostoma*, *Glycymeris vestita*, *Mactra chinensis*, and *Pecten albicans*.

Loc. 262 (Sakaida-minami): a small exposure, in front of Sakaida elementary school, Imba-mura, Imba-gun (Lat. 35°47′16″N, Long. 140°13′22″E) (Loc. 16 of Kozima, 1958a, b). Kioroshi Member of Kioroshi Formation (3.6 m+).

Sample 262-1, fine-grained sand with *Panopea japonica* in living position and abundant *Glycymeris vestita* and *Mactra chinensis*, about 20 cm above the base of the 60 cm thick shell layer in medium-grained sand.

Loc. 261 (Matsumushi): a small exposure, 500 m N 55° E of Matsumushi temple, Imba-mura, Imba-gun (Lat. 35°47′14″N, Long. 140°13′22″E). Kioroshi Member of Kioroshi Formation (3.8 m+).

Sample 261-1, coarse-grained sand, with abundant *Solen strictus*, 20 cm above the base of the 80 cm thick shell layer in medium-grained sand.

Loc. 190 (Sunaoshi): an exposure, 3.6 km S 42° E of Kobayashi railway station (J.N.R.), Imba-mura, Imba-gun (Lat. 35°47′50″N, Long. 140°12′46″E). Kioroshi Member of Kioroshi Formation (17.5 m+).

Sample 190-4, medium-grained poorly-sorted sand, with abundant *Mactra chinensis*, *Tonna luteostoma*, *Echinarachnius mirabilis*, and type 2 burrows (Yajima, 1978) with silt walls, 50 cm below the black scoriaceous sand layer.
Sample 190-3, medium-grained poorly-sorted sand, with abundant shells of the same fossil content as sample 190-4, 40 cm below the black scoriaceous sand layer.
Sample 190-2, the 20 cm thick black scoriaceous sand layer.
Sample 190-1, grey medium-grained well-sorted sand, with abundant *Mactra chinensis* lying convex side upward and sporadic *Glycymeris vestita*, 1.3 m above the black scoriaceous sand layer.

Loc. 189 (Wadadani): an exposure, 3.5 km S 20° E of Kobayashi railway station (J.N.R.), Imba-mura, Imba-gun (Lat. 35°47′52″N, Long. 140°12′38″E). Kioroshi Member of Kioroshi Formation (18 m+).
Sample 189-3, fine-grained sand, with *Solen strictus* and *Tonna luteostoma*, 5 m below the 20 cm thick black scoriaceous sand layer.
Sample 189-2, fine-grained sand, with layered *Solen strictus*, 2 m below the black scoriaceous sand layer.
Sample 189-1, red brown fine- to medium-grained sand, with abundant *Solen strictus* and *Mactra chinensis*, 3 m above the black scoriaceous sand layer.

Loc. 273 (Hetamai): an exposure, 3.3 km S 15° W of Kobayashi railway station (J.N.R.), Motono-mura, Imba-gun (Lat. 35°47′58″N, Long. 140°12′10″E). Kioroshi Member of Kioroshi Formation (16.5 m+).
Sample 273-2, light brown fine-grained sand, with abundant *Solen strictus*, 5 m below the 10 cm thick black scoriaceous sand layer.
Sample 273-1, light brown fine-grained sand, with abundant *Tonna luteostoma*, 4 m below the black scoriaceous sand layer.

Loc. 188 (Tozaki): an exposure, 200 m S of Motono-mura village office, Motono-mura, Imba-gun (Lat. 35°48′16″N, Long. 140°12′20″E) (Loc. 30 of Kozima, 1958a, b). Kioroshi Member of Kioroshi Formation (14.5 m+).
Sample 188-1, shell lens in coarse- to medium-grained sand, with abundant *Mactra chinensis* and *Glycymeris vestita*, 4.8 m below the base of nonfossiliferous, orange-brown, cross-laminated, coarse-grained sand.

Loc. 186 (Shimotani): an exposure, 250 m W of Motono-mura village office, 2 km S 10° E of Kobayashi railway station (J.N.R.), Motono-mura, Imba-gun (Lat. 35°48′30″N, Long. 140°12′10″E) (Loc. 33 of Kozima, 1958a, b). Kioroshi Member of Kioroshi Formation (13.5 m+).
Sample 186-4, grey coarse-grained sand, with abundant *Echinarachnius mirabilis* and *Mactra chinensis*, 7.5 m below the base of the shell lens of abundant fragments of *E. mirabilis* and *Glycymeris vestita* in orange-brown coarse-grained sand.
Sample 186-4, grey medium-grained sand, with *Solen strictus* and *Panopea japonica*, 6 m below the base of the shell lens.

Loc. 270 (Oshita): a cliff, 1 km S 83° E of Kobayashi railway station (J.N.R.), Inzei-machi, Imba-gun (Lat. 35°50′02″N, Long. 140°09′44″E). Kamiiwahashi Formation (3.8 m+) and Kioroshi Member of Kioroshi Formation (10 m).
Sample 270-3, medium- to coarse-grained sand, 80 cm above the unconformity between Kamiiwahashi Formation and Kioroshi Formation, with abundant layers of *Echinarachnius mirabilis*.

Loc. 269 (Kioroshi): type locality of the Kioroshi Formation, in front of Kioroshi high school, Inzei-machi, Imba-gun (Lat. 35°49′56″N, Long. 140°09′35″E) (Loc. 74 of Kozima,

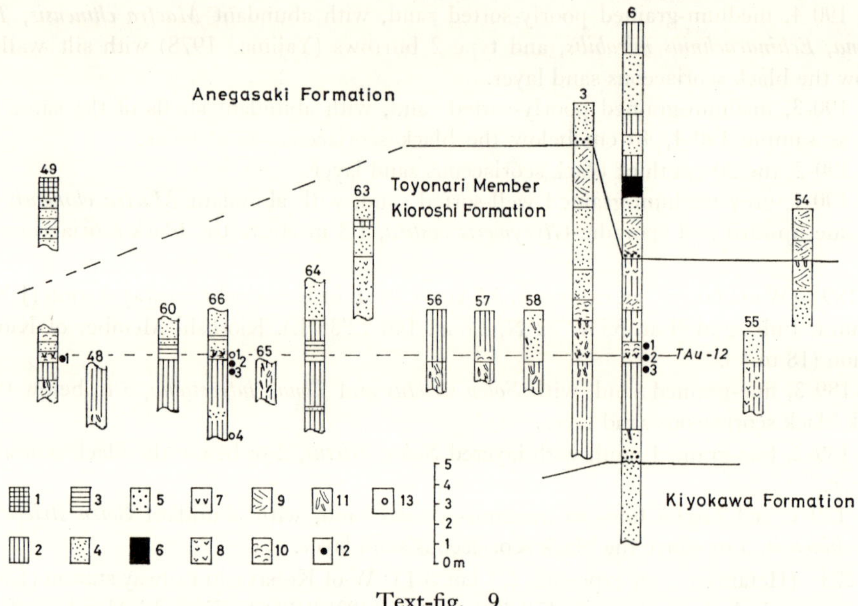

Text-fig. 9

Columnar sections of Toyonari Member of Kioroshi Formation. See explanation of legend of text-fig. 4.

1958a, b). Kamiiwahashi Formation (50 cm+) and Kioroshi Member of Kioroshi Formation (8 m).

Sample 269-3, coarse-grained sand, 1.4 m above the unconformity between Kamiiwahashi Formation and Kioroshi Formation, with abundant *Echinarachnius mirabilis* below shell bed of *Mactra chinensis*.

II. Toyonari Member of Kioroshi Formation: Basal gravel of the Toyonari Member, 1.6 m thick, unconformably overlies nonfossiliferous, well-sorted, coarse-grained sand of the Kiyokawa Formation (Loc. 6 of text-figs. 9, 10). The gravel is overlain by 3.3 m thick silt. The silt with plant fragments yields either *Dosinia angulosa* in living position (samples 6-3, 66-2, 3) or *Raeta yokohamensis* (Loc. 57) and is overlain by medium-grained sand. The medium-grained sand is 1 to 2 m thick, dominantly pumiceous (TAu-12 marker-tephra) and contains warm-temperate and shallow (20 to 50 m deep) water species of molluscs: *Pecten albicans*, *Limopsis forskalii*, *Paphia naganumana*, and *Siphonalia fusoides* (samples 49-1, 6-1, 2, 66-1). The date of TAu-12 marker-tephra of 147,000 yr B.P. was obtained in Kisarazu area (Sugihara et al., 1978). The upper silty sand which lies on fossiliferous medium-grained sand bears abundant pumice particles in which some of the K1P marker-tephra group may be present.

This sequence is similar to the succession of F Silt, G Sand, and lower part of H Sand in the Kisarazu area (Yajima, 1978). Molluscan fossils of this member suggest a more open sea environment than that of the Kioroshi Member. The embayment for the Kioroshi Formation may open toward south, similar to that of the Recent Tokyo Bay.

Sampling localities of this member are as follows.

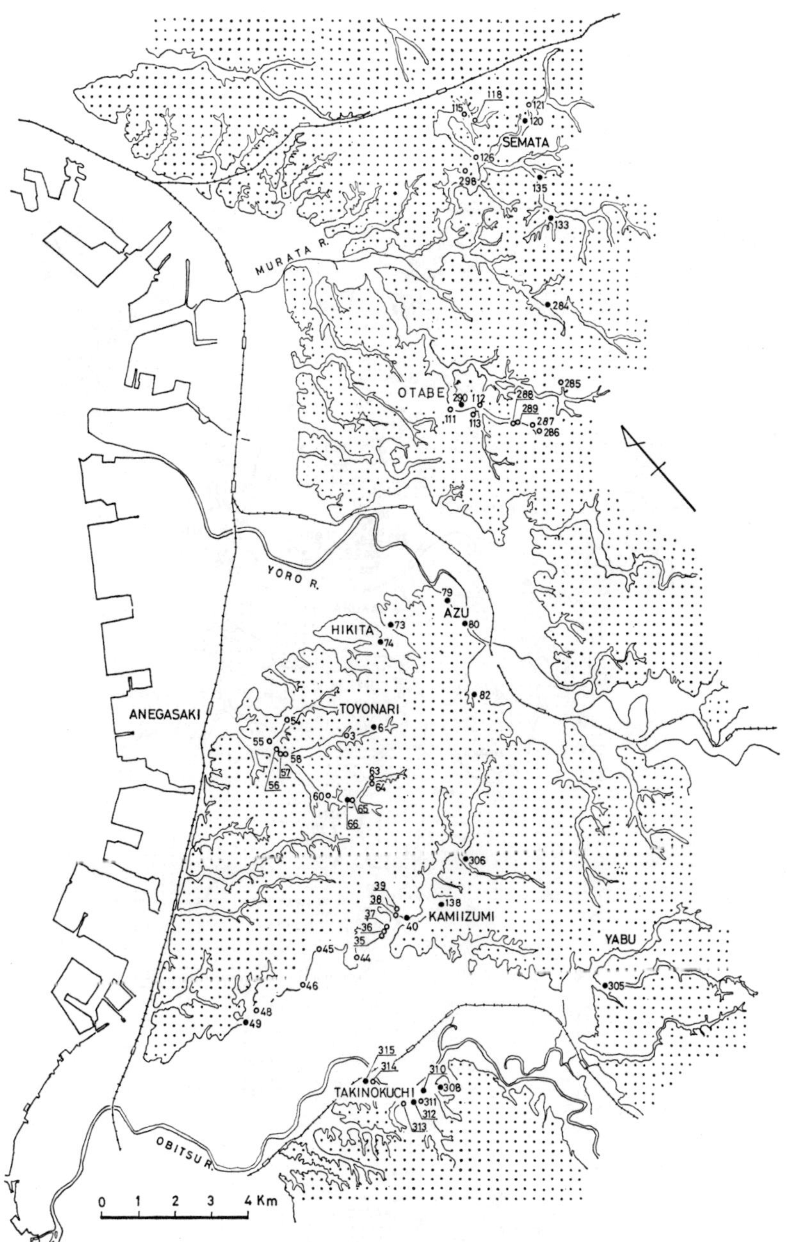

Text-fig. 10
Sampling localities in the southern half of surveyed area.

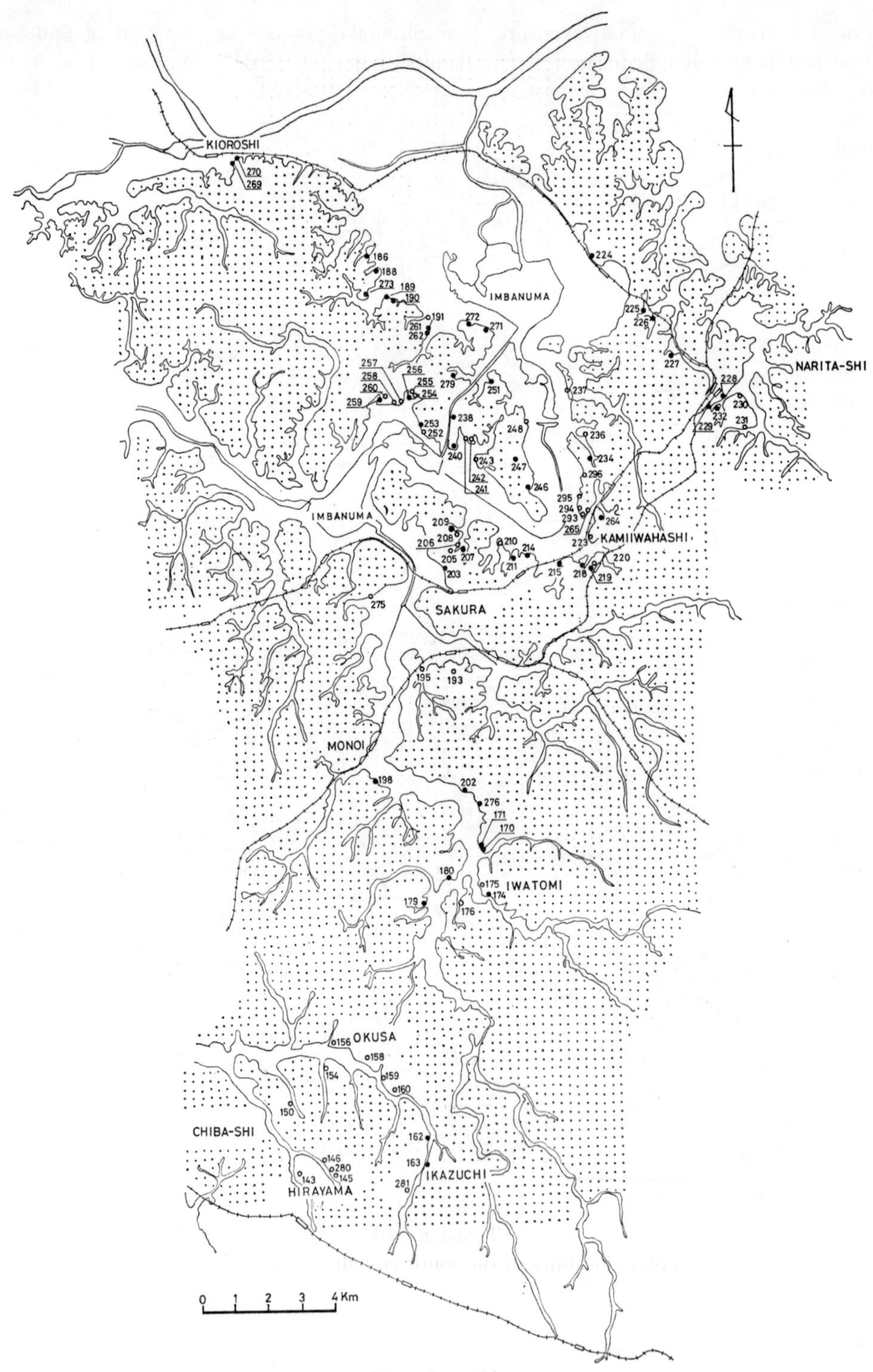

Text-fig. 11
Sampling localities in the northern half of surveyed area.

Loc. 49 (Iitomi): a small exposure, near Shimoike pond, 3 km S 60° E of Sodegaura railway station (J.N.R.), Sodegaura-machi, Kimitsu-gun (Lat. 35°24′57″N, Long. 139°59′30″E) (Loc. 30 of Kozima, 1966a, b; Iitomi of Aoki and Baba, 1971). Toyonari Member of Kioroshi Formation (3.2 m+) and Anegasaki Formation (3 m+).

Sample 49-1, medium-grined sand, with abundant *Pecten albicans* and *Limposis forskalii* and rare but characteristic *Cryptopecten vesiculosus*, 75 cm above mud layer of 15 cm thick, in the lower part of the exposure.

Loc. 66 (Fukashiro): a small exposure, 4.5 km S 15° E of Anegasaki railway station (J.N.R.), Ichihara-shi (Lat. 35°26′20″N, Long. 140°03′30″E) (Loc. 5 of Ikeba, 1936). Toyonari Member of Kioroshi Formation (6.5 m+).

Sample 66-3, pumiceous sandy mud, in *Dosinia angulosa* in living position obtained from pumiceous fine-grained sand, 40 cm below the base of fine-grained sand with abundant *Pecten albicans* and *Limopsis forskalii.*

Sample 66-2, pumiceous fine-grained sand, with abundant *Dosinia angulosa* in living position and with type 2 burrows (Yajima, 1978), 10 cm below the base of the fine-grained sand with *Pecten albicans* and *Limopsis forskalii.*

Loc. 6 (Toyonari): an exposure, near Toyonari shrine, 4.5 km S 40° E of Anegasaki railway station (J.N.R.), Ichihara-shi (Lat. 35°26′40″N, Long. 140°04′40″E) (Loc. 37 of Sugihara et al., 1978). Kiyokawa Formation (4 m+), Toyonari Member of Kioroshi Formation (8.5 m) and Anegasaki Formation (11.5 m+).

Sample 6-3, brown sandy silt, with abundant *Dosinia angulosa* in living position and pumice particles (TAu-12), 40 cm below the base of medium-grained coquina sand.

Sample 6-2, medium-grained sand, with abundant *Pecten albicans, Limopsis forskalii* and with burrows, 50 cm above TAu-12.

Sample 6-1, coarse-grained sand, with shell fragments of *Dosinia japonica* and *Chlamys farreri,* 80 cm above TAu-12.

Summary: The study deals with the sediment deposited in the Paleo-Tokyo Bay. The sediment consists of four formations—Yabu, Kiyokawa, Kamiiwahashi, and Kioroshi—of the Shimosa Group. Each formation shows a cyclic tendency of sedimentation reflecting a cycle of glacioeustatic sea level change. Because the bay was deeper and wider in the earlier deposition, marine components dominate in the lower part of each cycle. The cold climates are recorded by nonmarine facies of upper and lowermost horizon of each cycle and diastems between cycles (Sugihara et al., 1978, p. 595). The lower two, the Yabu and Kiyokawa Formations, consist mostly of sediments deposited in a cold-temperate shallow water (20–50 m deep) bay, which might have been opened toward the east and was under the influence of coastal water. During the maximum high sea level phase, however, warm water entered the bay (perhaps directly from the south), as suggested by molluscs from horizons of 120-3 (Yabu Member of Yabu Formation), 40-2, 1, 306-1 (Kamiizumi Member of Yabu Formation), and 82-1 (Kiyokawa Formation).

Sedimentation of the Kamiiwahashi Formation commenced with deposition of sand, which fills burrows of the underlying formation; the sand includes shallow water bivalves in living position in its lower part. The sand was succeeded by a shell bed characterized by occurrence of cold-temperate moluscs, which indicate a shallow water environment (20 to 50 m deep) of high sea level phase. The shell bed is covered by subtidal sand with abundant burrows, and sedimentary cycle terminates by deposition of brackish water clay. It is notable that there is no indication of influence of warm

Table 1
List of Upper Pleistocene ostracod species from the Boso Peninsula.

Table 1-1

Formation	Yabu																					
Member	Yabu																	Kamiizumi				
sample number / species	305-6	305-5	305-4	305-3	305-2	305-1	284-3	284-2	284-1	133-1	135-4	135-3	135-2	135-1	120-3	120-2	120-1	306-1	40-2	40-1	138-5	138-4
1 *Neonesidea oligodentata* (Kajiyama, 1913)		19	10	16	9	2	8	3	1		7	10			38	19	11		2		6	
2 *Propontocypris (Propontocypris)* sp.						1					1						1	1				
3 *Paracypris* sp.		1					1	1														
4 *Neocytherideis aoi* n. sp.					2	3						1					4					
5 *Pontocthere japonica* (Hanai, 1959)		23	4	6	8	21	1	1			1	3							1	1	2	
6 *P. kashiwarensis* (Hanai, 1959)																	3					
7 *P. miurensis* (Hanai, 1959)		11	4		6							2				4	6				14	3
8 *P. subjaponica* (Hanai, 1959)		13	17	14	13	7	17	25	16	2		8			3		15	16	15	7	11	
9 *P.* sp. 1																						
10 *P.* sp. 2								5	6								5					
11 *Parakrithella pseudadonta* (Hanai, 1959)								1											2	1	2	
12 *Pseudopsammocythere? tokyoensis* Yajima, 1978																1						
13 *Eucythere yugao* n. sp.				1		1										3	1					
14 *Museyella hatatatensis* Ishizaki, 1966								1	1						1		2				2	
15 *M. japonica* (Hanai, 1957)		3	1	2	2		1					1					3					
16 *M. oborozukiyo* n. sp.					2	1	1											8				
17 *M.* sp.		1	1																		1	
18 *Callistocythere alata* Hanai, 1957				1		2	1	1				1						1				
19 *C. hayamensis* Hanai, 1957					1													1	1			
20 *C. hotaru* n. sp.																						1
21 *C. nipponica* Hanai, 1957												1										
22 *C. reticulata* Hanai, 1957						3	2	1	2			3		1	1		5				1	1
23 *C. rugosa* Hanai, 1957			1	1	5	1	4	1		1		1					2		2		1	
24 *C. undata* Hanai, 1957		1		2		2	1	4				1							1		1	
25 *C. undulatifacialis* Hanai, 1957							1												2			
26 *C. japonica* Hanai, 1957	1	6	11	10	3	5	12	9	8	1	2	2	2	1			5		3			
27 *C. pumila* Hanai, 1957																						
28 *C. subjaponica* Hanai, 1957					1	4	3	6	7	2					2		3			1	1	
29 *C.* sp.		5	1									1				1	4	1				
30 *Cythere lutea omotenipponica* Hanai, 1959		1	1		1	5		4	2						1		2		3	4		1
31 *C. japonica* Hanai, 1959																						
32 *Loxocythere inflata* Hanai, 1959			1			1		3											1		1	
33 *Schizocythere asagao* n. sp.		12	4	19	3	33	11	15	20	1	3	12		1		1	26				2	
34 *S. kishinouyei* (Kajiyama, 1913)		33	45	31	46	6	32	34	18	3		15	1		5	11	37	4	27	10	22	5
35 *Hanaiborchella triangularis* (Hanai, 1970)		6	8	6	3	6	2	10	5	1	1	6			6	6	19	4	1	1		
36 *H. miurensis* (Hanai, 1970)		2	1	4	4	3	3	3	2			2					4				1	
37 *Neomonoceratina microreticulate* Kingma, 1948																			1	1	4	1
38 *Spinileberis quadriaculeata* (Brady, 1880)																			2	1	2	2
39 *S. furuyaensis* Ishizaki and Kato, 1976																						
40 *Trachyleberis scabrocuneata* (Brady, 1880)		3	1	5	11		5	5	6	8	4	18			6	13	4	1		2		
41 *T.? tosaensis* Ishizaki, 1968		5	1			3	4	1			1	4					5					
42 *T.* sp.		2	7	4	3			4				7			4	4			1		2	
43 *Acanthocythereis niitsumai* (Ishizaki, 1971)																						
44 *A.?* sp.												1										
45 *Actinocythereis kisarazuensis* Yajima, 1978																		1	3			
46 *Cletocythereis rastromarginata* (Brady, 1880)																						
47 *Stigmatocythere?* sp.									1								1					
48 *Rocaleberis?* sp.		1					2															
49 *Wichmannella bradyformis* (Ishizaki, 1968)																		1				
50 *W. bradyi* (Ishizaki, 1968)																		1				
51 *Carinovalva nipponica* (Yajima, 1978)				1								1										
52 *Bicornucythere bisanensis* (Okubo, 1975)							4					3							1	2	1	
53 *Buntonia hanaii* Yajima, 1978				4															2			
54 *Ambocythere japonica* Ishizaki, 1968				1																		
55 *Robertsonites? reticuliforma* (Ishizaki, 1966)												1								1		
56 *Australimoosella tomokoae* (Ishizaki, 1968)		8	2	1	3	3	6	6	5			1		1	1		1			2	6	1
57 *Campylocythereis? ukifune* n. sp.		3	2	2	2	1	7	5	4	2		1			4	6	1					
58 *Finmarchinella (Finmarchinella) uranipponica* Ishizaki, 1969			1																			
59 *F. (Barentsovia) japonica* (Ishizaki, 1966)		4	6	5	5		3	2	1			1				1			5	1	5	1
60 *Aurila cymba* (Brady, 1869)		1	5	5	4	8	9	8	4	1		5	2			1	15	7	9		12	

water even during the high sea-level phase of this stage.

By far the widest high sea-level phase, which is called the Shimosueyoshi transgression (140,000–130,000 yr B.P.), is represented by the Kioroshi Formation in the Boso Peninsula area. At the beginning of this phase, the bay was already filled in considerably. Cold climate is recorded in the sediment by plant remains found just above the unconformity at the base of the formation. However, the high sea-level phase of the Shimosueyoshi transgression is marked by abundant occurrence of warm water molluscs in the horizon G Member of the Kisarazu area (Yajima, 1978) and at horizon 269-3, 2, 1 in the Kioroshi area. It is highly probable that an extensive bay was formed,

Table 1-2

Formation	KIYOKAWA																			
Member																				
sample number / species	315-1	312-1	310-1	308-1	138-3	138-2	138-1	82-2	82-1	74-2	74-1	73-4	73-3	73-2	79-1	80-3	80-2	80-1	290-3	290-1
1	2	7	3	6		3		3		6		3				5	6	7		
2																				
3		1															1			
4	1						1					1					1			
5		5	2			1	8	25	18						2					
6		1					8			1		1								
7	4					13				3		4					2			
8		11	5	8	2		25	4	32	3		3				15	21	31		
9																				
10			11	4																
11		1		1		5				2		2						1		
12																				
13																				
14			1																	
15		1	2	1		10	3	1		7		12								
16					2				8											
17																				
18					1	10			4			1				1		1		
19				1					1	1										
20		4	1	1						3		1								
21																				
22				2												1	3	2		
23		1		4			2	1	1	4		6						2		
24	1		3	2				1												
25		2	2	3		1	4			12		9								
26		1	2				1	5				1			1	4	2	7		1
27																				
28		2		1			1									4		3		
29			1				1			2						1		2		
30		5		2		1	1	4	3	1		2				2	1	1		
31																				
32																				
33								22							1	22	12	20		
34	4	13	24	27			17	7		29		28		1	1	13	10	24	1	
35						5	2	3	16							3	2	4		
36				1			1					1				5	1	1		
37				1						2		1								
38		1	1	2			1		1	3										
39																				
40				1		4	1	5		1					2	2	1	7		
41								6								2	1			
42		1	1			21			5							3	2	4		
43																				
44																				
45																	1			
46										1										
47																				
48							1			1		1								
49							1											1		
50		1		1																
51																				
52				1		1						2			1	1	3			1
53		1		1								1								
54				1																
55			1							1										
56			1	1			1	3				3				4	3	3		
57																	2			
58																				
59	1		5	4			2			3		2				5	1	1		
60		4	3	8	1		5	5	9	8		3				2	4	8		

into which warm marine water flowed from the south at the time of the maximum phase of the Shimosueyoshi transgression and then retreated toward the east as well as toward the south. The molluscan assemblage of the Toyonari Member of the southern area suggests a more extensively offshore environment than that of the Kioroshi Member of the northern area. Thus, at the time of the deposition of the Kioroshi Formation, the area under discussion might have been already occupied by a bay which opened toward the south, similar to the present-day topography of Tokyo Bay.

Ostracod assemblage

Ostracods representing 121 species were found in 117 samples from 68 localities.

Table 1-3

sample number	KAMIIWAHASHI Formation																																							
species	163-1	162-2	179-2	180-1	174-3	174-2	170-3	170-1	171-2	171-1	276-1	202-1	198-2	198-1	203-3	203-2	203-1'	207-1	209-1	211-2	211-1	214-1	246-1	247-1	215-1	218-1	219-1	264-1	234-1	259-1	256-1	253-1	238-2	240-1	251-4	279-4	228-3	228-2	228-1	224-2	279-3
1	1										2	1																						1						
2																																									
3																																									
4									1			1				2	1																		1		3				1	
5																1									1													8	26	23	32	
6			1							1		1			1		2			1															2		1		1	1	
7			3	1	1	1									3	1		2	1				2		3		1	1				2	1		6	13	6	8	1	21
8			19								2	23			10	37	20				1		2	2												4	16	47	13	9	25	
9	1															7	3																			2					8
10					1				1			5								1																1	19				
11			1																																			3	2		
12																																									
13																																									
14																																									
15																																									
16																			2								1															
17																																									
18																			1													1										
19																																					4	3			
20			2																																						
21																																									
22					1							2				1	8				2			1											1	8	2	1		1	
23			5									1				1																						3	2	2	3	
24																																									
25																																									
26					1										6	2	1		1	2				1			1						1	1	5	7					
27																																									
28									2								11				1																19	7		1	2	
29			1													1																								
30			28									1																									11	10	5	9	
31																	1																							
32																																							1		
33																																									
34													2					1	1	2				1															12	2	5	9	
35					2				1		1	4			1	5	4		1					1		2	1					1	1		2	6	3	3	6		
36																																									
37			5									2		1						1															1		9	11	8	16	
38			11																																			14	8	11	6	23
39																																									51
40			5									1				4						1																2	3	4	9	
41					1						1	2			21	11				1				2		1	1	2				2	1		3	1						
42																1				1																				1		
43													2																												
44																																									
45																																									
46																																									
47																																									
48																																									
49			5					1				1					2																			2	5	1	1	7	
50																																									
51																	1																							
52			10								4	3				3	1	1	1							1				1					1		10	5	7	20	1
53																																							2		
54																																									
55																																					1					
56									2							3	7									1										8	2	2	1		
57																																									
58								2									23	38				2														2	1	43					
59																																					2		1		
60		1	16							1	1	8				6	2		2					2		3		1							4	1	26	18	24	23	

A complete list of ostracods from all samples is found in table 1.

Samples were collected carefully parallel to the bedding plane of fossiliferous sediments to get contemporaneous samples. Burrows filled with subsequent sediments were carefully exluded. Eighty grams dry weight of sediment from each sample was washed with water through a Tylor's 200-mesh sieve. The washed samples were divided by the quartering method into unit samples which contain approximately 200 ostracods. When the whole washed samples contained far less than 200 ostracods, a washed sample corresponding to only 5 g of sediment was picked because of economy of time. Knowledge on the ecology of modern Japanese species was summarized in the checklist (Hanai et al., 1977). (For additional information, refer to Ikeya and Hanai, 1980 MS.) Unpublished data on the ecology of ostracds of the Aburatsubo Cove were also useful

Table 1–4

Formation	KIOROSHI																																	
Member	Toyonari						Kioroshi																											
sample number / species	49-1	66-3	66-2	6-3	6-2	6-1	232-2	232-1	229-1	227-1	226-1	225-1	251-1	279-2	279-1	271-3	271-2	272-1	261-1	262-1	190-4	190-3	190-2	190-1	189-3	189-2	189-1	273-2	273-1	188-1	186-4	186-3	270-3	269-3
1						15																												
2	1									1										1								2			1			
3																								1										
4																	4				1		3	2	4	1	1	2	1	1				
5													3	8			5	1		2					1				1			5		
6																		9		3				1	4			2	2			4		
7			2	3		4							3	3	1		5						10	2	5				2			7		
8	9	4	11	26	10				2	1			24	7	18	1	18	7	1	7	11		8	18	23	1		8	16	1		6	1	1
9													2									1	8	2	1									
10													4	2			5	2					3	6	5	2		3						
11	2					8											1							1										
12	2																																	
13													2	3			1			1			2	2	6			2	1			2		
14																																		
15																																		
16	3		8	2	6																		1											
17																																		
18	15		3		3																													
19						6																												
20																																		
21																																		
22													2				1	10		5	1			4	6	1		1			1	3	1	
23																		1		1								1						
24													2										1			2								
25																																		
26																																		
27														1																				
28						1											2	2						2	2				2					
29																													1					
30			1	1										1				2		7	1			4	6				1	1				
31																																		
32																																		
33																																		
34										1			1	1			2	3		7	1			3	5			2	1					
35					6				1		1			11	3		8	5	1	3	1		5	8	12			6	5	1		7		
36																		2		1								2				2		
37			5	11									6	1			4	4	1	6	2			4	9			1	1					
38		2	26	18									4	4			1	4		6				5	5	1			1	1				
39														3						1														
40	25	10	46	39	53	8							3				3	3		4	4		1	13				2		1				
41													2	3											4									
42	3		2		1	4		1					5	7				4		6	1	2	5	2	10			6	7			4		
43														2				2							2				3					
44																																		
45																																		
46																																		
47																																		
48																																		
49	1	1	5	3	3	1							4			1	2			3			1	3	1			2				3		
50		1																1																
51																																		
52		1	18	18	1	1							22	4			8	10		6	2	1	13	24	14	1	1	4	1					
53	1		5																				1											
54																																		
55																																		
56									2					4				1							1			2	1			1		
57																																		
58																																		
59						25																												
60	5	1		2	1	34			5	4		1	3	20	1	1	2	23		39	3		4	5	15	3		31	24	2		30	1	2

for analytical interpretation of the environment represented by the fossil ostracod assemblages.

Yabu Member of Yabu Formation: Ostracods associated with warm-temperate water molluscs, which occur in coarse- to medium-grained sand just below the SY marker-tephra, are characterized by their rareness and high value of Simpson's index of species diversity (λ: $\lambda = \Sigma \frac{n_i(n_i-1)}{N(N-1)}$ where N is the number of specimens, and n_i is the number of specimens of i-th species). The index attains 0.103 in sample 120-3. The fact suggests that the sample is represented by a mixture of ostracods derived from various environments by postmortem transportation. Dominance of *Neonesidea oligodentata* and *Xestoleberis sagamiensis* suggests a warm more-or-less shallow coastal

Table 1-5

Formation	YABU																					
Member	Yabu																	Kamiizumi				
species / sample number	305-6	305-5	305-4	305-3	305-2	305-1	284-3	284-2	284-1	133-1	135-4	135-3	135-2	135-1	120-3	120-2	120-1	306-1	40-2	40-1	138-5	138-4
61 *A.? hataii* Ishizaki, 1968							1	1											1		1	
62 *A. kiritsubo* n. sp.		14	9	6	8	1	7	4	1		1	10					2	7			3	
63 *A. pseudoamygadala* Ishizaki, 1966		6	2	3	2	24	3		2		1	2				1	6	5			3	
64 *A. tosaensis* Ishizaki, 1968											1									7	2	
65 *A.* sp.							1					1					2					
66 *Pseudoaurila japonica* (Ishizaki, 1968)					1																	
67 *Robustaurila assimilis* (Kajiyama, 1913)					2	1	2	2				1			1		1		4		1	
68 *Urocythereis? gorokuensis* Ishizaki, 1966															1							
69 *Ambostracon ikeyai* Yajima, 1978				1	1		2		1			1					2		2			
70 *Caudites? posterocostatus* (Ishizaki, 1966)															2		1					
71 *Cornucoquimba tosaensis* (Ishizaki, 1968)			2	2	1	2	3		2			1			2	2	4		1			
72 *Coquimba ishizakii* Yajima, 1978		11	7	2	3	8	1	4	1	4		2					4	2	3	2	1	
73 *Bythocythere maisakensis* Ikeya and Hanai		1	2	2	5	7	1						1		2	1	16	2	3			
74 *Bythoceratina hanaii* Ishizaki, 1968		6			2		4					1				2	2		3		1	1
75 *B.* sp.			3																			
76 *Pseudocythere frydli* n. sp.		1				1									2							
77 *Eucytherura utsusemi* n. sp.			1		1										1	2	2					
78 *Hemicytherura cuneata* Hanai, 1957		33	25	20	33	3	3	5	2	1		6			6	23	15	26	28	3	19	2
79 *H. tricarinata* Hanai, 1957																						
80 *Howeina camptocytheroidea* Hanai, 1957								1	1													
81 *H. higashimeyaensis* Ishizaki, 1971																						
82 *Semicytherura? furuyaensis* (Ishizaki and Kato, 1976)																	4	2	1		2	
83 *S. henryhowei* Hanai and Ikeya, 1977		1				1											1	1				
84 *S. miurensis* (Hanai, 1957)		1			2	1									4	1	2	1	3	1		
85 *S. mukaishimensis* Okubo, 1980		2	1	2	4												1	2	1		6	
86 *S. skippa* (Hanai, 1957)																1					1	
87 *S. wakamurasaki* n. sp.		7		2	5											1	3	2	1		2	
88 *S.* sp.					1	1									1							
89 *Cytheropteron miurense* Hanai, 1957		7	9	5	7		2	6				2			2	2	9	2	2		1	
90 *C. uchioi* Hanai, 1957		6	4	6	1	2	3	1	4			2			5	8	15	1				
91 *Kangarina hayamii* n. sp.		2	1		2										1	1	1					
92 *Kabayashiina hyalinosa* Hanai, 1957																		1				
93 *Paracytheridea bosoensis* Yajima, 1978		10	16	11	9	7	3	8	5	4		4					8	3	7	3	7	2
94 *P. neolongicaudata* Ishizaki, 1966			1																			
95 *Loxoconcha hanachirusato* n. sp.		12	9	8	9	11	3	5	9								12		4			
96 *L. hattorii* Ishizaki, 1971																						
97 *L. japonica* Ishizaki, 1968																						
98 *L. kattoi* Ishizaki, 1968		4	1	2	4	4	6	2							2		3		3		1	
99 *L. laeta* Ishizaki, 1968															2	3		10	4	1	2	2
100 *L. optima* Ishizaki, 1968	2	28	17	11	8	41	19	4	3			16	3		1			16	9	7	6	3
101 *L. tamakazura* n. sp.		8	6	13	7	2	4	9	5	1	5	21			4	23	9	1	4		2	
102 *L. tosaensis* Ishizaki, 1968																		2		1		
103 *L. viva* Ishizaki, 1968																						
104 *Cytheromorpha acupunctata* (Brady, 1880)																	3	2	3	1	1	
105 *Nipponocythere bicarinata* (Brady, 1880)																		1	1			
106 *Phlyctocythere hamanensis* Ikeya and Hanai															1							
107 *Xestoleberis hanaii* Ishizaki, 1968		22	8	9	5	4	3	2	8			4				1	16	7	9		10	
108 *X. sagamiensis* Kajiyama, 1913		4	5	5	9	1	2	2	4		1	1			11	7	9	2		1		
109 *X. setouchiensis* Okubo, 1979		1																			2	
110 *X. suetsumuhana* n. sp.		2	1	1	2		2														3	1
111 *Paradoxostoma setosum* Okubo, 1977																					1	
112 *P. sohni* Okubo, 1980		1		1											2	2						
113 *Cytherois asamushiensis* Ishizaki, 1971								2	1													
114 *C.* sp.																	2					
115 *Paracytherois tosaensis* Ishizaki, 1968		3									1						1					
116 *Sclerochilus mukaishimensis* Okubo, 1977								1														
117 *S.* sp. 1																					1	
118 *S.* sp. 2																						
119 *Cytheroma? hanaii* Yajima, 1978																						
120 *Paracytheroma inflata* (Ikeya and Hanai)																			1		1	
121 *P.* sp.							1	1	1													
total number of individuals (a)	3	357	265	253	271	244	217	219	159	32	30	188	9	4	125	152	341	145	183	62	180	29
total number of species (b)	2	48	44	42	48	42	47	44	34	14	14	44	5	4	32	30	55	35	44	24	48	15
weight of sample (gramm) (c)	10	25	17	10	17	20	30	20	40	10	10	15	5	5	80	20	10	5	5	20	10	10

and open sea environment, and the presence of *Cytheropteron* may suggest a sand bottom more than 30 m deep. However, the abundant occurrence of Pteropoda gives the impression that the coarse-grained fossiliferous sand characterized by lack of fine matrix might have been deposited under the direct influence of warm Kuroshio current flowing into this area from the south.

In the coarse-grained sand above the SY marker-tephra, *Neonesidea oligodentata*, *Pontocythere subjaponica*, *Schizocythere kishinouyei*, *S. asgao*, *Loxoconcha optima*, *Hemicytherura cuneata*, and *Paracytheridea bosoensis* occur abundantly. Fifteen sand samples from five localities were analysed. Ostracods are generally abundant. The frequency is approximately 2,800 individuals per 80 g on the average. Three medium-

Table 1–6

Formation	KIYOKAWA																			
Member																				
sample number / species	315-1	312-1	310-1	308-1	138-3	138-2	138-1	82-2	82-1	74-2	74-1	73-4	73-3	73-2	79-1	80-3	80-2	80-1	290-3	290-1
61		5					2			8		18								
62		1	5	3		17	3	1	30			3				7		8		
63			2	1		1	5	12	1	5		2				1				
64		4	2	1		19	2			9		9								
65										1										
66		2		1						3		1		1		1	1	1		
67	1	6	1	1			3			11		6					3	2		
68																	1			
69												1				2		2		
70																				
71						2										3		1		
72		1	1				7	3		1						2	1	10		
73				1								1				2	1	7		
74		4		2						1						1				
75																				
76																				
77	1						2													
78	6		5	5	2	54	21		33	32		23	1		1	13	6	6		
79			1						1											
80				1																
81																				
82	1	1				1	1			2										
83						1	2		1									1		
84				1		3			1											
85							4			2						1				
86																				
87.			1	2		5	3		2	1						4	2	7		
88					1															
89			1	2			1									2	3	2		
90							1	1								3	1	3		
91																				
92																		1		
93		1	5	5		9	8	4		1						4	7	6		
94				2			1	1										1		
95	2			2			6	7										2		
96																				
97				3																
98		1	3	4		1	2	2		10		16			1	1	4	4		
99		1	4	1	3	37	2		7	2	1	1				2	2			
100		1	3	8		4	11	15	36	6		8				9	12	8	1	2
101							1	2				2				2	5	7		
102				1					5											
103																				
104			2	3			2		3			1				3	5	5		
105						2			1											
106																				
107	3	9	4	6	2	9	7	3		10	1	4				8	13	8		
108							2	5	2							2	2	5		
109	1						2													
110		1	1				1			3		2					1			
111						3														
112						1														
113								1								1	1			
114									1											
115																				
116		1														2		1		
117				1																
118																				
119																				
120																				
121																2		1		
(a)	28	102	110	142	14	244	189	152	222	202	2	190	1	2	10	173	151	229	2	4
(b)	13	34	34	48	8	29	47	28	25	39	2	38	1	2	8	43	40	44	2	3
(c)	20	20	25	20	5	20	15	20	20	5	5	5	5	5	5	5	5	5	5	20

to fine-grained sand samples (samples 305-5, 4, 2) show extreme abundance in the range of 11,000 to 13,000 individuals per 80 g. Species diversity is also very high ($\bar{\lambda}=$ 0.062). However, some of abundant species, for example *Paracytheridea bosoensis* which is associated quite often with *Pecten tokyoensis* fauna, may be inhabitants of the sand between SY and GoP marker-tephras. A fall in water temperature after the deposition of an SY marker-tephra may be recorded by the presence of a cold-water element, *Howeina camptocytheroidea*, which occurs in two samples (samples 284-2, 1) and another boreal species *Finmarchinella* (*Finmarchinella*) *uranipponica*, which occurs in one sample (sample 305-4). Another sample, 135-3, contains *Acanthocythereis*? sp. and

Table 1-7

	KAMIIHASHI Formation																																							
sample number / species	163-1	162-2	179-2	180-1	174-3	174-2	170-3	170-1	171-2	171-1	276-1	202-1	198-2	198-1	203-3	203-2	203-1	207-1	209-1	211-2	211-1	214-1	246-1	247-1	215-1	218-1	219-1	264-1	234-1	259-1	256-1	253-1	238-2	240-1	251-4	279-4	228-3	228-2	228-1	224-2	279-3
61																																				1			1	
62			6									2			2		10		1					2			1					2			4	1	17	7	1	6	
63			2									1																									2	3	2	2	
64			7																																						
65																																									
66																																									
67			2									1	1				1																				1	3	2		
68																																									
69												1																												2	
70																																									
71												1																									2		1	1	
72			10								3	19		1		1			1																1		13	10	5	16	
73												4																							1			2		3	
74																																									
75																																									
76																																									
77																1																									
78												1			1				1			1		10		2	2					1			2		4	1	3	4	
79																																								1	152
80																																									
81																1	4														1										
82																																									
83																																					1			1	
84			1																					2											1	1				3	
85																																						1			
86																																									
87												2																							1	7	2	7	3	2	
88																																									
89																			2			1		1			1														
90															1																						1				
91																																									
92																																									
93												17																													
94																																									
95												4																													
96			1																																						
97																																									
98																								1																	
99			7			1						4					1																				12	7	6	8	12
100			3				2					3			1		2																				3	3	11		
101																																									
102																																									
103																																									
104			17								1	2					9							2												4	24	14	21	24	4
105			1									1																									2	2	2	3	
106																																									
107			13									13				2	4																			1	4	3	3	2	
108																																							1		
109																																									
110																				1																					
111			1																																						
112												12			1																				1			1		1	
113																																					1				
114																																					4				
115																																									
116																																									
117																																									
118																																									
119																																									
120															1																						1	2	2	1	
121			1													2																					2	2		1	
(a)	2	1	184	1	7	2	4	1	7	2	13	149	2	2	48	117	135	2	18	8	6	3	3	29	1	13	8	4	1	1	1	7	5	6	36	137	298	185	186	247	272
(b)	2	1	28	1	6	2	2	1	5	2	7	34	2	2	12	21	24	2	13	7	4	3	2	13	1	7	7	3	1	1	1	5	4	5	17	19	40	33	35	35	8
(c)	5	5	15	5	5	5	5	5	5	5	10	40	5	5	80	40	80	5	5	5	10	5	5	10	5	10	5	5	5	5	5	5	5	5	80	80	5	5	15	25	10

Robertsonites? reticuliforma, which closely resemble boreal species of *Acanthocythereis dunelmensis* and *Robertsonites tuberculata*, respectively.

Kamiizumi Member of Yabu Formation: The lower bluish grey silty sand (sample 306-1) with warm sea molluscs living in a depth of 20 to 50 m—for example, *Pecten albicans* and *Chlamys farreri*—contains abundant ostracods. *Hemicytherura cuneata*, *Pontocythere subjaponica*, and *Loxoconcha optima* occur abundantly, suggesting that they might be *in situ* inhabitants or derived from the nearby environment. *Munseyella oborozukiyo* and *Loxoconcha laeta*, which are probably warm-water inhabitants, are characteristic.

Table 1–8

Formation	KIOROSHI																																	
Member	Toyonari						Kioroshi																											
Sample number / species	49-1	66-3	66-2	6-3	6-2	6-1	232-2	232-1	229-1	227-1	226-1	225-1	251-1	279-2	279-1	271-3	271-2	272-1	261-1	262-1	190-4	190-3	190-2	190-1	189-3	189-2	189-1	273-2	273-1	188-1	186-4	186-3	270-3	269-3
61																																		
62	34			20	8	2							14	4	1		7	6		4			3	11	8	1		3	1			6		
63					1	2																												
64																																		
65																																		
66																																		
67					1	3							1	1				1			1			1				4	2					
68																																		
69						1							1							1								1						
70																																		
71	1			2	1	18				1								2		5									1					
72						1							6	6	1		3	2		2	1		1	8	13	1		6	4	1		7		
73																				2					1	1		2	2			1		
74									2								2																	
75																									1									
76																				3														
77																																		
78	1				2	12		1	1				1	5	1	1	2	7		3				3	5			2	8			3		
79														3																				
80																																		
81																																		
82														1			1	1																
83						32																		2				1				1		
84						1							5				1	1		1				3	3				1			1		
85																	2							1				3	4			1		
86																																		
87						2			1	1			14	1			10	2			1		3	6	3			1	3	1				
88			1																															
89																	2								1									
90			1														1						1						1					
91																												1						
92																																		
93													2	3	1		3	3		2					1			7	5			7		
94																																		
95								1	7				3	21			4	33		3	1		3	10	33			30	29			40		
96													1				1																	
97																																		
98																																		
99				7	2	51			1	2			2	1			1							3	5			6				1		
100							1	1	4				1	3			3	10		10			2	2	10			6	6			5		
101																																		
102	1		1																															
103	12		51	6	7									3						2				2					1					
104			1	36		2							6				7	5	1	6	1		2	7	5			1	1			1		
105		1		1						1			4				1						1	1										
106																																		
107	1				1	2							1	3	1		18			5	3		3	3	7			6	10	1		1		
108	1																																	
109						1																												
110	1				1	2																												
111									1				9				3				2			18	10			1	2			3		
112									1				9				3				2			18	10			1	2			3		
113																				1														
114						3																		1										
115																	3	1		1			2						2					
116																																		
117																																		
118	1												1					1																
119						1														1														
120				1										1			2			1				2	1									
121				2													1									1		1	1					
(a)	121	29	187	198	108	247	1	4	27	12	1	1	164	141	30	4	150	171	4	162	38	4	87	197	248	16	2	161	155	11	2	154	3	3
(b)	21	9	17	18	18	29	1	4	11	8	1	1	34	32	10	4	39	34	4	37	18	3	25	39	38	12	2	36	37	10	2	26	3	2
(c)	40	5	20	10	80	5	5	5	20	5	5	5	80	80	80	5	20	40	5	20	10	5	80	40	15	5	5	10	10	5	5	20	20	10

The upper well-sorted, fine-grained sand with warm water molluscs, *Umbonium costatum*, also contains abundant ostracods. Species diversity being very hgih, the average of three samples (samples 40-2, 1, 138-5) is 0.059. Common sand flat dwellers of the present sea, *Schizocythere kishinouyei* and *Hemicytherura cuneata*, occur abundantly.

Kiyokawa Formation: Ten sand samples (samples 315-1, 312-1, 310-1, 308-1, 138-3, 2, 1, 82-2, 290-3, 1) were collected from the lower part of the Kiyokawa Formation (lower than TB-7 marker-tephra). Sand just above the basal gravel in the area near Kamiizumi contains an appreciable amount of sand bottom dwellers, *Loxoconcha laeta* and *Hemicytherura cuneata* (samples 138-3, 2). The upper horizon of the lowest sand (sample 138-1) contains abundant ostracods with high species diversity (1,000 individuals per 80 g and λ of 0.051). Sand flat dwellers, *Pontocythere subjaponica*, *Hemicytherura cuneata*, and *Schizocythere kishinouyei*, become dominant. In two samples (samples 290-3, 1) from lower black coarse-grained sand near Otabe, ostracods are rarely found. The grain size of the sand is larger than that of ostracod carapaces, and it is possible that their absence resulted from removal by sorting. The fine-grained sand (samples 315-1, 312-1, 310-1, 308-1) just above the silt, which may be correlated with the black clay, has abundant *Mactra chinensis* lying convex side upward and includes ostracods with very high species diversity near Takinokuchi ($\bar{\lambda}$=0.065). A sand flat dweller, *Schizocythere kishinouyei*, dominates, and a cold water inhabitant *of Howeina camptocytheroidea* occurs in one sample (sample 308-1). *Pontocythere subjaponica* and *Schizocythere asagao* are abundant and associated with cold-temperate water molluscs *Glycymeris yessoensis* in grey medium-grained well-sorted sand (sample 82-2), which is correlated with lower sand distributed around Takinokuchi.

Nine samples (samples 74-2, 1, 73-4, 3, 2, 79-1, 80-3, 2, 1) and one silty sand sample (sample 82-1) were examined from the upper part of this formation.

In the silty sand associated with warm-temperate molluscan fauna (sample 82-1) from a bed immediately above TB-7 marker-tephra, ostracods occur commonly (about 900 individuals per 80 g) and species diversity is relatively low (λ=0.122). *Loxonconcha optima* is dominant. Sand flat dwellers of *Hemicytherura cuneata* and *Pontocythere subjaponica* seem less dominant. *Aurila kiritsubo* and *Munseyela oborozukiyo* are characteristic.

Five sand samples (samples 74-2, 73-4, 80-3, 2, 1) include abundant ostracods with very high species diversity. They attain from 2,400 to 3,700 individuals per 80 g and $\bar{\lambda}$ is 0.054. Samples from the localities near Hikita (samples 74-2, 73-4) are represented by dominance of sand flat dwellers of *Schizocythere kishinouyei* and *Hemicytherura cuneata*, associated with *Aurila hataii*, *Loxoconcha kattoi*, *Callistocythere undulatifacialis*, and *Robustaurila assimilis* which may have been derived from a relatively shallow area. Samples from Azu (samples 80-3, 2, 1, 79-1) are characterized by dominance of *Schizocythere asagao*, and the fauna is similar to the assemblage of the Yabu Member of the Yabu Formation.

In general, no essential difference can be ascertained between ostracod assemblages of sand samples of the Yabu Formation and those of the Kiyokawa Formation. *Schizocythere asagao*, *Loxoconcha hanachirusato*, *L. tamakazura*, *Cytheropteron uchioi*, *C. miurense*, and *Xestoleberis sagamiensis* decrease, suggesting the dominance of a deeper environment in the Yabu Formation. *Loxoconcha kattoi*, *Aurila hataii*, and *Callistocythere undulatifacialis* increase in their number from the Yabu Formation to the Kiyokawa Formation.

Kamiiwahashi Formation: In the lower sand of the Kamiiwahashi Formation, 36 sand samples from 29 localities were obtained. Ostracods are rare in sand samples, ranging from 16 to 300 individuals per 80 g except for sample 179-2 (1,000 individuals per 80 g). Three horizons of molluscan fauna are recognized, and ostracods commonly occur in these three horizons. *Pontocythere subjaponica* occurs throughout the three horizons. In the lower horizon (samples 203-3, 211-2, 246-1, 247-1, 218-1, 264-1), associated with *Panopea japonica* in living position, *Trachyleberis*? *tosaensis* (sample

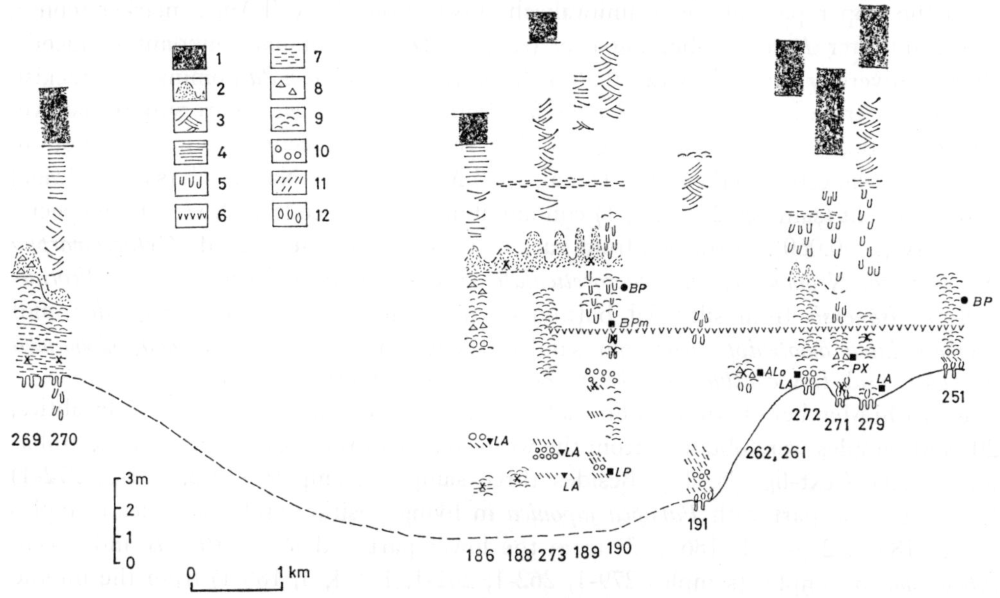

Text-fig. 12

Schematic stratigraphic succession of Kioroshi Member of Kioroshi Formation in the area of text-fig. 8, showing dominant species of ostracod assemblage and ratio of mud dwellers to total individuals.

(1) clay, (2) fragments of shells, (3) cross-lamination, (4) parallel lamination, (5) burrows, (6) black scoriaceous sand layer, (7) *Echinarachnius mirabilis*, (8) *Glycymeris vestita*, (9) *Mactra chinensis*, (10) *Tonna luteostoma*, (11) *Solen strictus*, (12) *Panopea japonica*. Dominant ostracode species: *B Bicornucythere bisanensis*, *P Pontocythere subjaponica*, *Pm P. miurensis*, *X Xestoleberis hanaii*, *L Loxoconcha hanachirusato*, *Lo L. optima*, *A Aurila cymba*. Ratio of mud dwellers to total individuals: ▼ 0–5%, ■ 12–16%, ● 20–40%. x: very rare occurrence of ostracods.

203-3) and *Hemicytherura cuneata* (sample 247-1) occurs characteristically. In the middle horizon (samples 174-3, 170-3, 203-2, 1, 207-1, 209-1, 211-1, 214-1, 215-1, 219-1, 234-1, 259-1, 256-1, 253-1, 238-2, 240-1, 251-4, 279-4) the cold water ostracod *Finmarchinella* (*Finmarchinella*) *uranipponica* is dominant and is associated with allochthonous molluscan fossils of cold-temperate water inhabitants. *Howeina higashimeyaensis*, also a cold-water inhabitant, occurs only in this horizon. *Callistocythere subjaponica* and *Aurila cymba* are characteristic in the middle horizon. In the upper horizon (samples 163-1, 162-1, 179-2, 180-1, 174-2, 170-1, 171-2, 1, 276-1, 202-1, 198-2, 1), species diversity increases ($\bar{\lambda}=0.069$ in samples 202-1 and 179-2). Species from various environments are mixed, yet the abundant occurrence of *Paracytheridea bosoensis* and *Coquimba ishizakii* suggests relatively deep water at the time of deposition of this horizon. Although the associated molluscan assemblage of the upper horizon is, in general, similar to that of the middle horizon, no cold-water ostracods occur in the upper horizon.

In the upper part of the Kamiiwahashi Formation above TAm-5 marker-tephra, brackish water *Batillaria*-bearing clay (sample 279-3) contains abundant ostracods. Species diversity is very low (λ=0.361). *Hemicytherura tricarinata* dominates. Brackish shallow water crawlers *Spinileberis quadriaculeata* and *S. furuyaensis* are subordinate yet characteristic. A similar assemblage is observed in the Recent shallow mud bottom of Hamana-ko estuary (Ikeya and Hanai, 1980 MS.). Around Narita, silt (sample 224-2) and sandy silt (samples 228-3, 2, 1) contain abundant ostracods with very high species diversity ($\bar{\lambda}$=0.060). Ostracods from various environments are mixed: *Cytheromorpha acupunctata*, *Neomonoceratina microreticulate*, *Spinileberis quadriaculeata*, and *Bicornucythere bisanensis* from silty flat bottom environment; *Pontocythere subjaponica*, and *Cythere lutea omotenipponica* from sand flat environment; *Aurila cymba*, *Coquimba ishizakii*, and *Loxoconcha laeta* from *Zostera* sand and tide pools.

Kioroshi Member of Kioroshi Formation: Among 28 sand samples from 18 localities, 20 sand samples were obtained from the western area of the northern half of the Imbanuma Lake (text-figs. 8, 12). Besides three samples (samples 279-2, 271-3, 272-1) from the basal part with *Panopea japonica* in living position and 7 samples (samples 271-2, 189-3, 273-2, 1, 186-4, 3) from the lower part with *Solen strictus* and *Tonna luteostoma*, 6 samples (samples 279-1, 262-1, 261-1, 190-4, 3, 188-1) from the middle part lower than black scoriaceous sand layer are characterized by dominance of possible *Zostera* sand dwellers, *Loxoconcha hanachirusato* and *Aurila cymba*, and a sand flat dweller, *Pontocythere subjaponica*. Species diversity is very high in the basal part ($\bar{\lambda}$= 0.060), in the lower part (λ=0.048 in sample 189-3 and 0.083 in samples 273-2), and in the lower half of the middle part (λ=0.076 in sample 262-1).

Ostracods from four sand samples (samples 251-1, 190-2, 1, 189-1) from the upper half of the middle part, within and above the black scoriaceous sand layer, are different from those of sand samples lower than the black scoriaceous sand layer in having a relatively abundant mud dweller, *Bicornucythere bisanensis*, a tide pool dweller, *Paradoxostoma sohni*, and a *Zostera* sand dweller, *Aurila kiritsubo*. Species diversity is very high ($\bar{\lambda}$=0.058 in samples 251-1, 190-2, 1). Although the middle part is characterized by abundant occurrence of *Mactra chinensis* convex side upward, the difference in ostracod assemblage seems to indicate that the bay becomes shallower from the lower toward the upper sand, with appreciable faunal change around the black scoriaceous sand layer.

Two samples near the type locality of the Kioroshi Formation (samples 270-3, 269-3) and six samples (samples 232-2, 1, 229-1, 227-1, 226-1, 225-1) around Narita, located in the eatern area of the northern half of the Imbanuma Lake, contain ostracods but not in abundance. This is perhaps because they were deposited in a relatively high energy environment. Ostracod assemblage is similar to that of the lower part in the western area of the northern half of Imbanuma Lake.

Toyonari Member of Kioroshi Formation: In two samples (samples 66-2, 6-3) from the lower sandy silt with *Dosinia angulosa* in living position, an inhabitant of the sandy silt bottom, *Trachyleberis scabrocuneata*, and shallow mud bottom crawlers *Cytheromorpha acupunctata*, *Neomonoceratina microretculate*, *Spinileberis quadriaculeata*, and *Bicornucythere bisanensis* occur abundantly. Species diversity is relatively low ($\bar{\lambda}$=0.143). This assemblage has close affinity with that of the upper part of the

F Silt in the Kisarazu area (Yajima, 1978). The lower sandy silt of the Toyonari area and the F Silt of the Kisarazu area might represent the same horizon. One sand sample (sample 6-3) from the closed valves of *Dosinia angulosa* contained partly dissolved carapaces of ostracods. The species contained are different from those of the matrix, yet they are still shallow mud bottom crawlers.

In three sand samples (samples 49-1, 6-2, 1) from the upper sand associated with shallow (20 to 50 m deep), warm-temperate molluscan fossils, *Zostera* sand dwellers, *Loxoconcha laeta*, *Aurila cymba*, and *A. kiritsubo*, and deeper water species, *Cornucoquimba tosaensis* and *Finmarchinella* (*Barentsovia*) *japonica*, are characteristic. *Trachyleberis scabrocuneata* is also abundant. Relatively low ($\bar{\lambda}$=0.170) species diversity suggests that the abundant species of this assemblage may be *in situ* inhabitants.

Summary: A cyclic tendency characterizes the upper Pleistocene sedimentation of the surveyed area. Ostracod assemblages vary in accordance with the sequence in the cycle of sedimentation. Most of the samples show high species diversity and do not warrant identification of even dominant autochthonous species or species composition. Yet the following four ostracod assemblages are discernible when one pays special attention to their occurrence in a cyclic sequence.

The following four assemblages of ostracods were recognized: subtidal sand, warm water sand, shallow water mud, and brackish water clay.

(1) Subtidal sand assemblage: The assemblage is recognizable in the sediments at the maximum transgressive phase. In general, the level subtidal environment was dominant throughout the deposition of the Paleo-Tokyo Bay. The subtidal sand assemblage is found in well-sorted sand containing abundant detached valves of *Mactra chinensis*. Ostracods are abundant and species diversity is high. This is perhaps due to the addition of allochthonous species from various shallow water environments. The main components of this assemblage are the shallow subtidal sand inhabitants *Schizocythere kishinouyei*, *Hemicytherura cuneata* and *Pontocythere subjaponica*, suggesting the existence of these species *in situ* or their derivation from nearby environments. This assemblage can be subdivided into two assemblages, based on the abundance of subordinate species, which is perhaps related to water depth.

1(a). The deeper water assemblage with which *Pecten tokyoensis* is associated is characterized by abundant *Paracytheridea bosoensis*, *Cytheropteron miurense*, and *C. uchioi*. Examples are seen in the middle part (high sea level) of the cycle in the middle horizon of the Yabu Member of the Yabu Formation (samples 305-6, 5, 4, 3, 2, 1, 284-3, 2, 1, 133-1, 135-4, 3, 2, 1, 120-3, 2, 1).

1(b). The shallower water assemblage is characterized by *Loxoconcha optima*, *L. laeta*, *L. kattoi*, *Aurila cymba*, *A. kiritsubo*, and *Callistocythere undulatifacialis*. Examples are seen in the lower middle and middle parts of cycles, as exemplified by the Kamiizumi Member of the Yabu Formation (samples 40-2, 1, 138-5, 4), the Kiyokawa Formation (samples 138-3, 2, 1, 290-3, 1, 315-1, 312-1, 310-1, 208-1, 82-2, 74-2, 1, 73-4, 3, 2, 79-1, 80-3, 2, 1) and the lower sand of the Kamiiwahashi Formation (samples 163-1, 162-2, 179-2, 180-1, 174-3, 2, 170-3, 1, 171-2, 1, 276-1, 202-1, 198-2, 1, 203-3, 2, 1, 207-1, 209-1, 211-2, 1, 214-1, 246-1, 247-1, 215-1, 218-1, 219-1, 264-1, 234-1, 259-1, 256-1, 253-1, 238-2, 240-1, 251-4, 279-4).

(2) Warm water sand assemblage: In the lower formations (Yabu and Kiyokawa

Formations) the assemblage is found inserted occasionally in the maximum transgressive sediments with clearly warm-water planktons and molluscs. Therefore, the criterion for classification of this assemblage is different from that of the rest, simply because this assemblage is primarily controlled not by the development of substrata due to eustatic sea level change but by the episodic insertion of warm current inflow at the time of the high sea level. However, in the upper formation (Kioroshi Formation) the assemblage dominates the entire formation. This assemblage is found either in the middle horizon of a cycle corresponding to high sea level phase offshore or nearshore at the base of a cycle, representing the extensively transgressive phase during which the underlying sediments were eroded away. Species diversity is high. The assemblage is subdivisible into s silty sand (20 to 50 m deep) assemblage and a nearshore sand assemblage.

2 (a). A silty sand (20 to 50 m deep) assemblage is found in sand or silty sand associated with warm-temperate water molluscan assemblages. The main components of this assemblage are *Loxoconcha laeta*, *L. optima*, *Aurila kiritsubo*, and deeper water inhabitants of *Munseyella oborozukiyo*. This assemblage occurs at the middle part (high sea level phase) of a cycle, in the silty sand of the Kamiizumi Member of the Yabu Formation (sample 306-1) and the silty sand of the middle part of the Kiyokawa Formation (sample 82-1). A similar in detail but somewhat modified assemblage characterized by subordinate species, *Cornucoquimba tosaensis* and *Finmarchinella* (*Barentsovia*) *japonica*, in addition to *Munseyella oborozukiyo*, is found in the upper sand of the Toyonari Member of the Kioroshi Formation (samples 49-1, 6-2, 1), suggesting a little deeper environment.

2 (b). The shallow nearshore sand assemblage is composed mainly of *Loxoconcha hanachirusato*, *Aurila cymba*, and *Pontocythere subjaponica*. It occurs in the transgressive basal sand of the Kioroshi Member of the Kioroshi Formation associated with *Panopea japonica* in living position at the base and with *Solen strictus* and *Tonna luteostoma* at the lower horizon (samples 279-2, 1, 271-3, 2, 272-1, 189-3, 2, 273-2, 1, 186-4, 3, 262-1, 190-4, 3, 188-1, 270-3, 269-3, 232-2, 1, 229-1, 227-1, 226-1, 225-1).

(3) Shallow water mud assemblage: This assemblage is found in the inner bay environment. Its main components are shallow mud bottom crawlers such as *Cytheromorpha acupunctata*, *Neomonoceratina microreticulate*, *Spinileberis quadriaculeata*, and *Bicornucythere bisanensis*. Associated subordinate species are *Trachyleberis scabrocuneata* or *Paradoxostoma sohni*. This assemblage occurs at the horizon just below the high sea-level phase of a cycle of the Toyonari Member of the Kioroshi Formation, in lower sandy silt with *Dosinia angulosa* in living position (samples 66-3, 2, 6-3). The species diversity is low, suggesting autochthonous occurrence of mud dwellers. The assemblage also occurs in the horizon of a regressive sand of a cycle of Kioroshi Member of Kioroshi Formation in fine-grained sand with *Mactra chinensis* (samples 251-1, 190-2, 1, 189-1). The species diversity is high, suggesting the derivation of mud dwellers from nearby environments. The assemblage is also found in the uppermost horizon of a cycle of Kamiiwahashi Formation, in silt with *Dosinia angulosa* in living position or sandy silt with *Raeta yokohamensis* (samples 224-2, 228-3, 2, 1). The relatively high species diversity suggests allochthonous occurrence of sand dwellers.

(4) Brackish water clay assemblage: This assemblage occurs in clay with *Batil-*

laria zonalis and is exemplified by the uppermost regressive part of a cycle of the Kamiiwahashi Formation (sample 279-3). The species diversity is very low, and ostracod abundance is high. This assemblage is composed of dwellers of brackish water clay of inner bay, represented by *Hemicytherura tricarinata*, *Spinileberis furuyaensis*, and *S. quadriaculeata*.

The subtidal sand, warm water sand, and shallow water mud assemblages can be correlated roughly with the sand flat, silty sand, and shallow mud assemblages distinguished by Yajima (1978) in the Kisarazu area. A brackish water clay assemblage is a new addition to the above three assemblages, showing the presence of an area with salinity lower than that of a shallow mud assemblage.

Most ostracod species preserved in the sediments of the interglacial period in this area are commonly shallow temperate water inhabitants under the influence of warm water. However, a few characteristic cold-water species, *Finmarchinella (Finmarchnella) uranipponica*, *Howeina camptocytheroidea*, *H. higashimeyaensis*, and *Robertsonites? reticuliforma* may be remnants of the cold-water fauna which may have been dominant in the glacial period, whose sediments are not preserved in this area (Hanai, 1957c; Ishizaki, 1971; Neale, 1974; Okada, 1979). The occurrence of these three species suggests that cold-water influence temporarily remained during the deposition of the Kamiiwahashi Formation and probably of the Yabu and Kiyokawa Formations.

Systematic description

Subclass Ostracoda Latreille, 1806
Order Podocopida Sars, 1866
Superfamily Cypridacea Baird, 1845
Family Pondtocyprididae G. W. Müller, 1894
Genus *Propontocypris* Sylvester-Bradley, 1947
Subgenus *Propontocypris* Sylvester-Bradley, 1947
Propontocypris (*Propontocypris*) sp.
Pl. 15, figs. 1, 2, 5.

Illustrated specimens.—A carapace, UMUT-CA 9801 (Pl. 15, fig. 5. L, 0.42; H, 0.21; W, 0.16); a left immature valve, UMUT-CA 9802 (Pl. 15, figs. 1, 2. L, 0.40; H, 0.20), sample 273-2, Kioroshi Member of Kioroshi Formation.

Remarks.—Surface is smooth. Outline resembles that of *Pontocypris*? sp. described by Ishizaki (1969, p. 214) from Recent mud to sand bottom of Nakanoumi Estuary, Shimane Prefecture, and later called *Propontocypris* (*Propontocypris*?) sp. by Hanai, Ishizaki and Ikeya (Hanai et al., 1977, p. 20). This species is, however, smaller in size and dorsal margin is more steeply arched.

Occurrence.—Rare in Yabu Member and Kamiizumi Member of Yabu Formation; also rare in Toyonari Member and Kioroshi Member of Kioroshi Formation.

Family Candonidae Kaufmann, 1900
Subfamily Paracypridinae Sars, 1923
Genus *Paracypris* Sars, 1866
Paracypris sp.
Pl. 15, figs. 6, 9; text-figs. 13-1, 2.

Illustrated specimens.—A right valve, UMUT-CA 9803 (Pl. 15, fig. 9; text-fig. 13-2. L, 1.03;

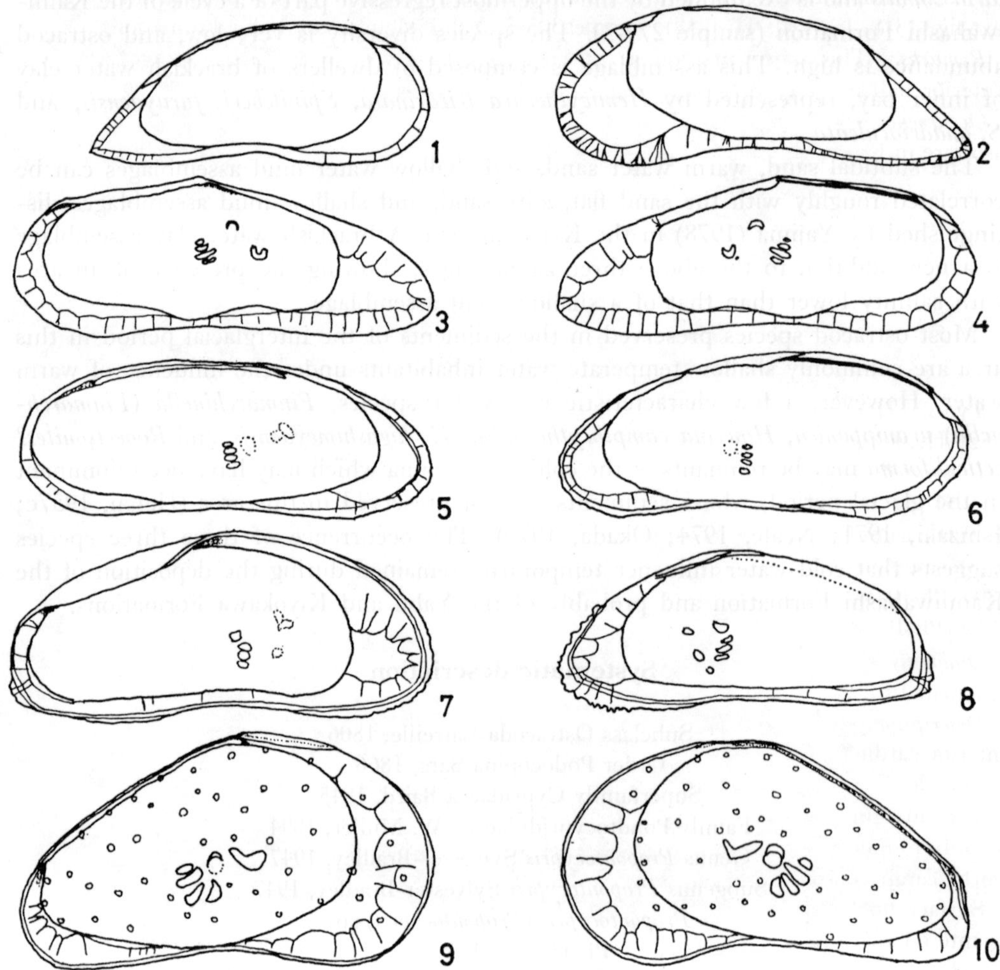

Text-fig. 13

Internal views. left row (odd number); left valves; right row (even number): right valves.

Figs. 1, 2. *Paracypris* sp. 1, CA 9804 (sample 305-5, Yabu Member of Yabu Formation). x 55. 2, CA 9803 (sample 80-2, Kiyokawa Formation). x 55.

Figs. 3, 4. *Neocytherideis aoi* n. sp. 3, holotype, female, CA 9805 (sample 189-3, Kioroshi Member of Kioroshi Formation). x 112. 4, female, CA 9806 (sample 189-3). x 112.

Figs. 5, 6. *Pontocythere* sp. 1. 5, CA 9812 (sample 190-2, Kioroshi Member of Kioroshi Formation). x 130. 6, CA 9811 (sample 190-2). x 130.

Figs. 7, 8. *Pontocythere* sp. 2. 7, CA 9814 (sample 228-3, Kamiiwahashi Formation). x 113. 8, CA 9813 (sample 228-3). x 101.

Figs. 9, 10. *Eucythere yugao* n. sp. 9, holotype, CA 9815, (sample 189-3, Kioroshi Member of Kioroshi Formation). x 98. 10, CA 9816 (sample 189-3). x 98.

H, 0.39), sample 80-2, Kiyokawa Formation; a left valve, UMUT-CA 9804 (Pl. 15, fig. 6; text-fig. 13-1. L, 0.90; H, 0.38), sample 305-5, Yabu Member of Yabu Formation).

Remarks.—This is the first report of the genus *Paracypris* from Japan. Hinge is adont. Anterior half of hinge margin of right valve slightly projects in dorsal view.

Occurrence.—Rare in Yabu Member of Yabu Formation; rare in Kiyokawa Formation; also rare in Kioroshi Member of Kioroshi Formation.

Superfamily Cytheracea Baird, 1850

Family Cytherideidae Sars, 1925

Subfamily Neocytherideinae Puri, 1957

Genus *Neocytherideis* Puri, 1952

Neocytherideis aoi n. sp.

Pl. 10, figs. 1–6; text-figs. 13-3, 4.

Types.—Holotype, a left female valve, UMUT-CA 9805 (Pl. 10, fig. 4; text-fig. 13-3. L, 0.52; H, 0.20), sample 189-3, Loc. 189, Kioroshi Member of Kioroshi Formation.

Illustrated specimens.—A right female valve, UMUT-CA 9806 (Pl. 10, fig. 2; text-fig. 13-4. L, 0.51; H, 0.19), sample 189-3; a right male valve, UMUT-CA 9807 (Pl. 10, fig. 1. L, 0.51; H, 0.18), sample 189-3; a male carapace, UMUT-CA 9808 (Pl. 10, fig. 5. L, 0.52; H, 0.24; W, 0.18), sample 273-1, Kioroshi Member of Kioroshi Formation; a female carapace, UMUT-CA 9809 (Pl. 10, fig. 6. L, 0.54; H, 0.24; W, 0.19), sample 120-1; a right female valve, UMUT-CA 9810 (Pl. 10, fig. 3. L, 0.55; H, 0.20), sample 120-1, Yabu Member of Yabu Formation.

Diagnosis.—A small species of genus *Neocytherideis* characterized by finely pitted and weakly reticulated surface ornamentation.

Description.—Viewed laterally, carapace elongate oblong, streamline in shape, highest at anterior cardinal angle. Anterior margin very narrowly and obliquely rounded; dorsoanterior margin long and straight, gently inclined anteriorly; dorsal margin straight and parallel to ventral margin. Ventral margin nearly straight and sinuated at middle. Posterior margin broadly rounded. Viewed dorsally, carapace lanceolate, widest at anterior fifth. Viewed anteriorly, carapace round.

Surface finely pitted and slightly reticulate in posterior half. Undulations running parallel to anterodorsal and anteroventral margins. Faint sulcus at middorsal area of carapace. Carapace translucent, except for anterior and posterior areas of marginal infold. Eye tubercle absent. Normal pore canals, sieve type, moderate in number, scattered. Radial pore canals simple and straight, short, few in number, and widely spaced.

Marginal infold broad anteriorly and narrow posteriorly. Line of concrescence running parallel to anterior, ventral and posterior margins. Anterior vestibule deep and sub-triangular; posteroventral vestibule shallow. Hingement weak, lophodont, and located posteriorly. Anterior hinge element smooth and long, having half length of median element, median element smooth and long, and posterior element short and weakly denticulated. Muscle scars located at middle of carapace, consisting of anteriorly inclined row of four adductor scars, a large V-shaped frontal scar, with an inverted U-shaped fulcral point. Dorsal two of adductor muscle scars very small, ventral two large and ovate. A frontal scar located anteroventrally at a distance from adductor muscle scars.

Sexual dimorphism weak. Male somewhat slender and ventral margin of male more strongly sinuate than that of female in lateral view.

Remarks.—This species resembles *Neocytherideis subulata* (Brady, 1868) illustrated by Wagner (1957) from Holocene sediment of the Netherlands, in general outline, but differs in shape of anterior and posterior vestibules. This species also resembles *Neocytherideis elongatus* described by Puri (1952) from Recent shore sand of South England in general outline and shape of vestibules, but differs in roundness of posterior margin in lateral view. It resembles *Neocytherideis punctata* Ikeya and Hanai, from Recent sediment of Hamana-ko Estuary, Japan, general outline and punctation of surface, but differs in the absence of vertical ridges and finer surface ornamentation.

Dimensions.—Measurements of specimens from samples of 189-3 and 271-2 of Kioroshi Member of Kioroshi Formation are given below.

Sp	Sa	Ho	N	Me	OR
LV ♀	189–3	Kioroshi M., Kioroshi F.	1	L	0.52
				H	0.20
RV ♀	189–3		1	L	0.51
				H	0.19
RV ♂	189–3		2	L	0.51–0.56
				H	0.18–0.19
C ♀	271–2	Kioroshi M., Kioroshi F.	1	L	0.52
				H	0.20
				W	0.20
C ♂	271–2		2	L	0.50–0.51
				H	0.19–0.20
				W	0.18–0.19
RV A-1	271–2		1	L	0.44
				H	0.18

Occurrence.—Rare in Yabu Member of Yabu Formation; rare in Kiyokawa Formation; rare in both sand and sandy silt facies of Kamiiwahashi Formation; common in Kioroshi Member of Kioroshi Formation.

Genus *Pontocythere* Dubowsky, 1939

Pontocythere sp. 1

Pl. 10, figs. 7, 8; text-figs. 13-5, 6.

Illustrated specimens.—A right valve, UMUT-CA 9811 (Pl. 10, fig. 7; text-fig. 13-6. L, 0.43; H, 0.16) and a left valve UMUT-CA 9812 (Pl. 10, fig. 8; text-fig. 13-5. L, 0.43; H, 0.16), sample 190-2, Kioroshi Member of Kioroshi Formation.

Remarks.—The species resembles *Pontocythere sekiguchii* Ikeya and Hanai, from Hamana-ko estuary, in general outline and carapace size, but differs in lateral outline of posteroventral margin and surface ornamentation. Entire surface of carapace is covered with weak undulation.

Occurrence.—Rare in both sand and clay facies of Kamiiwahashi Formation; also rare in Kioroshi Member of Kioroshi Formation.

Pontocythere sp. 2

Pl. 10, figs. 10, 11; text-figs. 13-7, 8.

Illustrated specimens.—A right valve, UMUT-CA 9813 (Pl. 10, fig. 10; text-fig. 13-8. L, 0.51; H, 0.20) and a left valve, UMUT-CA 9814 (Pl. 10, fig. 11; text-fig. 13-7. L, 0.51; H, 0.21), sample 228-3, Kamiiwahashi Formation.

Remarks.—The species resembles *Pontocythere subjaponica* (Hanai, 1959) in general outline, shape and broadness of marginal infold, and muscle scar pattern, but differs in having smaller carapace and in details of surface ornamentation. The species also resembles *Pontocythere minuta* Ikeya and Hanai, from Hamana-ko estuary in general outline and surface ornamentation, but differs in having crenulated anterior margin and in lack of ventrally protrudent flat posterior projection.

Occurrence.—Rare in Yabu Member of Yabu Formation; rare in Kiyokawa Formation; rare in sand facies of Kamiiwahashi Formation, except for one sample (sample 228-3) of sandy silt facies where the species occurs abundantly; common in Kioroshi Member of Kioroshi Formation.

Subfamily Krithinae Mandelstam, 1958

Genus *Pseudopsammocythere* Carbonnel, 1966

Pseudopsammocythere? *tokyoensis* Yajima, 1978

Pseudopsammocythere tokyoensis Yajima, 1978, p. 391–393, pl. 50, figs. 3a, b, text-figs. 7-1a, b.

Remarks.—Through courtesy of Dr. P. C. Carbonnel, the author had an opportunity of checking topotype specimens of *Pseudopsammocythere kollmanni* Carbonnel, 1966, (type-species of *Pseudopsammocythere*). *P. tokyoensis* differs from *P. kollmanni* in several characteristics including shape of posteroventral margin, broadness of posterior and anterior vestibules, and size of muscle scars. The difference between the two species seems to be enough to question assignment of this species to *Pseudopasmmocythere.*

Family Eucytheridae Puri, 1954

Subfamily Eucytherinae Puri, 1954

Genus *Eucythere* Brady, 1868

Eucythere yugao n. sp.

Pl. 10, figs. 13, 14, 16–19; text-figs. 13-9, 10.

Types.—Holotype, a left valve, UMUT-CA 9815 (Pl. 10, figs. 14, 16, 19; text-fig. 13-9. L, 0.58; H, 0.31), sample 189-3.

Loc. 189, Kioroshi Member of Kioroshi Formation.

Illustrated specimen.—A right valve, UMUT-CA 9816 (Pl. 10, figs. 13, 17, 18; text-fig. 13-10. L, 0.58; H, 0.30), sample 189-3.

Diagnosis.—*Eucythere* of moderate size, subtriangular in lateral outline with straight dorsal margin steeply inclined toward posterior. Surface weakly reticulated. Adductor scars elongate horizontally.

Description.—Viewed laterally, carapace subtriangular, highest at anterior cardinal angle. Anterior margin obliquely and broadly rounded. Anteroventral area compressed laterally. Dorsal margin straight, steeply inclined posteriorly. Ventral contact margin sinuated at middle. Posterior margin narrowly rounded. Viewed dorsally, carapace gently inflated at posterior one-third. Viewed anteriorly, carapace oblong.

Surface covered with faint reticulation in its ventral area and punctated with normal pore openings. Granules in two rows parallel to anterior margin in immature form. Eye tubercle absent. Normal pore openings large, sieve type with a small pore out of center, about 50 in number, scattered. Radial pore canals straight, simple and evenly spaced, about ten along anterior and a few along posterior margins.

Anterior marginal infold broad, line of concrescence subparallel to anterior margin. Posterior marginal infold narrow. Vestibule deep along anterior margin and shallow along posterior margin. Posterior margin denticulate internally. Hinge lophodont. Muscle scar field ventrocentral. Scars composed of a large V-shaped frontal scar and an oblique row of four elongate adductor scars. Posterior half of V-shaped frontal scar elongate. Ventral adductor scar crescent.

Dimensions.—Measurements of specimens from sample 189-3, Kioroshi Member of Kioroshi Formation, are given below.

Sp	N	Me	X̄	OR
LV	3	L	0.553	0.53–0.58
		H	0.307	0.30–0.31
RV	2	L		0.55–0.58
		H		0.30–0.31
LV A-1	1	L		0.47
		H		0.23

Remarks.—This is the first report of occurrence of this genus from Japan. The species resembles *Eucythere declivis* (Norman, 1865) illustrated by Wagner (1957) from the Quaternary of Netherlands in general outline, surface ornamentation, and muscle scar pattern, but difference can be found in distribution pattern of normal pores and more elongate adductor muscle scars. Anterior cardinal angle of this species is located more anteriorly than that of *E. declivis*. The specific name, *E. declivis*, has been applied to various forms slightly deviating from the original British form.

Occurrence.—Rare in Yabu Member of Yabu Formation; common in Kioroshi Member of Kioroshi Formation.

Subfamily Pectocytherinae Hanai, 1957

Genus *Munseyella* van den Bold, 1957

Munseyella oborozukiyo n. sp.

Pl. 10, figs. 9, 12.

Types.—Holotype, a right valve, UMUT-CA 9819 (Pl. 10, fig. 9. L, 0.44; H, 0.22), sample 66-2, Loc. 66, Toyonari Member of Kioroshi Formation.

Illustrated specimen.—A left valve, UMUT-CA 9820 (Pl. 10, fig. 12. L, 0.46; H, 0.25), sample 66-2.

Diagnosis.—A species of genus *Munseyella* characterized by thick and small carapaces. Anteroventral margin with small holes. The following broad ridges distinct in lateral view: (1) anterior marginal ridge along anterior margin, (2) posterior marginal ridge constricted at middle, (3) ventral ridge rising from anterocentral area, running straight and obliquely in posteroventral direction and ending on ventral ridge, (4) small ridge connecting posterior ridge and central undulate area. Deep hollow developed at anteroventral and posteroventral areas.

Description.—Viewed laterally, carapace thick, subrhomboidal, and highest at anterior cardinal angle. Anterior margin broadly and obliquely rounded. Six small holes along anteroventral marginal area. Dorsal margin straight, inclined posteriorly, with slight concavity at posterior fourth. Ventral margin straight, slightly sinuated at middle. Posterior margin truncated, nearly vertical to ventral margin, with two distinct spines.

Surface covered with undulated ridges and deep hollows. Anterior and posterior ridges distinct. Posterior ridge constricted at middle. Dorsal ridge straight and weak. Ventral ridge distinct, starting from anterocentral area, running straight toward posterior third of ventral ridge. Central ridge somewhat slender but distinct. Posterodorsal ridges connecting posterodorsal corner of posterior ridge and central ridge. Anteroventral and posteroventral hollows distinct.

For inner view, characteristics of pore canals, hinge and muscle scars, refer to the original description of *Munseyella japonica* (Hanai, 1957b).

Dimensions.—Measurements of pooled specimens are given below.

Sp	Sa	Ho	N	Me	X̄	Sd	V	OR
LV	66-2	Toyonari M., Kioroshi F.	4	L	0.453	0.010	2.32	0.44–0.46
				H	0.235	0.006	2.46	0.23–0.24
RV	66-2		3	L	0.437			0.43–0.44
				H	0.213			0.20–0.22
LV	6-2	Toyonari M., Kioroshi F.	4	L	0.410	0.008	1.99	0.40–0.42
				H	0.233	0.005	2.15	0.23–0.24
RV	6-2		1	L				0.41
				H				0.21
LV	82-1	Kiyokawa F.	4	L	0.428	0.024	5.53	0.41–0.46
				H	0.223	0.005	2.25	0.22–0.23
RL	82-1		1	L				0.42
				H				0.21
LV	306-1	Kamiizumi M., Yabu F.	2	L	0.410			0.40–0.42
				H	0.220			0.22
RV	306-1		4	L	0.415	0.013	3.11	0.40–0.43
				H	0.220	0	0	0.22

Remarks.—The species resembles *Munseyella japonica* (Hanai, 1957) in surface ornamentation, but differs distinctively in details of disposition of ridges as mentioned in diagnosis.

Occurrence.—Rare in Yabu Member and abundant but found only in one sample (sample 306-1) of silty sand facies from Kamiizumi Member of Yabu Formation; common in Kiyokawa Formation; rare in sand facies of Kamiiwahashi Formation; and common in Toyonari Member of Kioroshi Formation.

The species seems to be associated with warm-temperate water molluscan fauna.

Munseyella sp.

Pl. 10, fig. 15.

Illustrated specimen.—A carapace, UMUT-CA 9821 (Pl. 10, fig. 15. L, 0.35; H, 0.18; W, 0.14), sample 305-4, Yabu Member of Yabu Formation.

Remarks.—The species is ovate with broadly rounded anterior margin in lateral view and is similar to female form of *Munseyella japonica* (Hanai, 1957) in surface ornamentation.

Occurrence.—Rare in Yabu Member and Kamiizumi Member of Yabu Formation. A total of three carapaces are found.

Family Leptocytheridae Hanai, 1957

Remarks.—Hanai (1957a) recognized three species groups of *Callistocythere*. Guillaume (1976) fully reviewed the taxonomic situation of *Leptocythere pellucida* (Baird, 1850), the type-species of the genus. Keiji (1979c) redescribed and illustrated *Tanella gracilis* Kingma, 1948, the type-species of the genus. In the above genera, V-sahped sulci at dorsal area of carapace are distinct. Normal pores are sieve type with a well-defined subcentral opening, and are located on the edges of flat-topped ridges. For the other diagnostic features refer to Hanai (1957a, p. 436).

Genus *Callistocythere* Ruggieri, 1953

Callistocythere littoralis group

Remarks.—This group is strictly called *Callistocythere*.

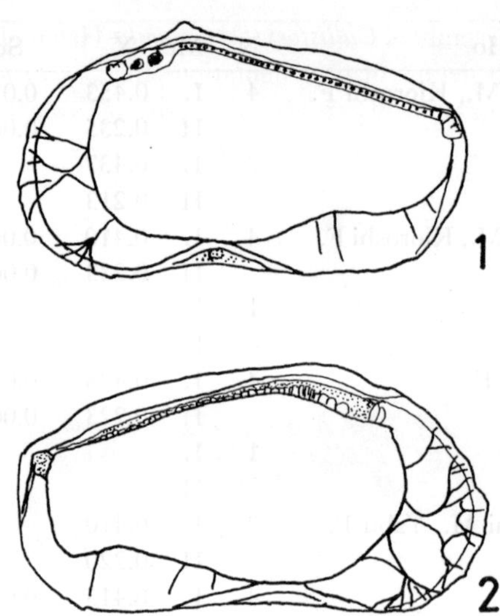

Text-fig. 14

Internal views of *Callistocythere hotaru* n. sp.

Fig. 1, right female valve, CA 9827 (sample 312-1, Kiyokawa Formation). x 100.

Fig. 2, left female valve, holotype, CA 9826 (sample 312-1). x 100.

Callistocythere hotaru n. sp.

Pl. 11, figs. 5, 6; text-figs. 14-1, 2.

Types.—Holotype, a left female valve, UMUT-CA 9826 (Pl. 11, fig. 5; text-fig. 14-2. L, 0.47; H, 0.24), sample 312-1, Loc. 312, Kiyokawa Formation.

Illustrated specimen.—A right female valve, UMUT-CA 9827 (Pl. 11, fig. 6; text-fig. 14-1. L, 0.45; H, 0.24), sample 312-1.

Diagnosis.—Small *Callistocythere*, subquadrate in lateral view, and sides parallel in dorsal view. Surface covered with large punctation. Puncta tending to connect to form grooves. Anterior marginal groove, anterodorsal groove just below eye spot, V-shaped groove in sulcus area and three posterodorsal obliquely parallel grooves distinct.

Description.—Carapace small, oblong in lateral view. In dorsal view, sides parallel, anteriorly pointed, and posterior bluntly truncated but with short posterior projection. Anterior margin obliquely rounded. Anterior cardinal angle slightly pronounced. Dorsal margin convex, ventral margin concave at middle. Posterodorsal margin nearly rectangular, posteroventral margin broadly rounded.

Surface ornamented with punctaation. Puncta tending to connect to form grooves. Marginal groove posterior to anterior marginal ridge, anterodorsal short groove just below eye spot, dorsocentral V-shaped groove in area of sulcus, three obliquely running posterodorsal grooves distinct. Irregularly shaped hollow at anterior of carapace distinct. Posteroventral marginal groove running parallel to ventral margin.

Eye spot distinct. Radial pore canals polyfurcated, being common to genus *Callistocythere*. Normal pore canals sunken sieve type.

Marginal infold moderately wide. Vestibule narrow anteriorly. Selvage distinct. Snap-knob indistinct. Hinge entomodont. Muscle scars consisting at least of a vertical row of four adductor scars.

Remarks.—The species resembles *Callistocythere pumila* Hanai (1957) in lateral outline and carapace size, but differs in details of ornamentation. In *C. hotaru*, carapace is covered with broad ridges and large punctation, where as in *C. pumila* carapace is ornamented with thin ridges and reticulation.

Dimensions.—Measurements of pooled specimens are as follows.

Sp		Sa	Ho	N	Me	OR
C	♀	179–2	Kamiiwahashi F.	1	L	0.45
					H	0.24
					W	0.24
C	♂	179–2		1	L	0.44
					H	0.21
					W	0.17
RV	♀	73–4	Kiyokawa F.	1	L	0.43
					H	0.22
C	♀	74–2	Kiyokawa F.	2	L	0.44
					H	0.22
					W	0.18–0.19
C	♂	74–2		1	L	0.43
					H	0.21
					W	0.16
C	♀	308–1	Kiyokawa F.	1	L	0.46
					H	0.23
					W	0.21
RV	♂	310–1	Kiyokawa F.	1	L	0.44
					H	0.23
LV	♀	312–1	Kiyokawa F.	2	L	0.45–0.47
					H	0.24
RV	♀	312–1		1	L	0.45
					H	0.24
C	♀	312–1		1	L	0.45
					H	0.23
					W	0.19

Occurrence.—Rare in Kiyokawa Formation; also rare in sand facies of Kamiiwahashi Formation.

Callistocythere rugosa Hanai, 1957
Pl. 11, fig. 7.

Callistocythere minaminipponica Ishizaki and Kato, 1976, p. 133. 134, pl. 2, figs. 6-10, pl. 3, fig. 1, text-fig. 1 [Types: Diluvium Furuya Mud, Shizuoka Pref.; IGPS-91737].

Remarks.—Specimens at hand are nearly identical with the type specimens of *Callistocythere rugosa* Hanai, 1957. Further reticulation pattern and carapace size of illustrated specimens are the same as those of *C. minaminipponica* Ishizaki and Kato, 1976. The species is distinguisable from *C. undulatifacialis* Hanai, 1957, by its smaller size, slight difference in reticulation pattern of the posterior half of carapace, and the presence of two horizontal ridges, of which the ventral one has two small nodes.

Callistocythere japonica type

Remarks.—This group closely resembles genus *Tanella* in lateral outline and surface ornamentation. However, hinge is similar to that of *C. littoralis* group in lacking antislip tooth in left valve.

Callistocythere sp.
Pl. 11, figs. 8, 9.

Illustrated specimens.—A right immature valve, UMUT-CA 9831 (Pl. 11, fig. 8. L, 0.53; H, 0.28) and a left immature valve, UMUT-CA 9832 (Pl. 11, fig. 9. L, 0.54; H, 0.28), sample 120-1, Yabu Member of Yabu Formation.

Remarks.—The species has a close affinity to *Callistocythere* sp. illustrated by Ikeya and Hanai (1980, MS) from Hamana-ko estuary.

Occurrence.—Rare in Yabu Member and Kamiizumi Member of Yabu Formation; rare in Kiyokawa Formation; rare in sand facies of Kamiiwahashi Formation; and also rare in Kioroshi Member of Kioroshi Formation.

Family Cytheridae Baird, 1850
Subfamily Schizocytherinae Mandelstam, 1960
Tribe Schizocytherini Mandelstam, 1960
Genus *Schizocythere* Triebel, 1950
Schizocythere asagao n. sp.
Pl. 11, figs. 14, 15.

Types.—Holotype, a left female valve, UMUT-CA 9833 (Pl. 11, fig. 15. L, 0.66; H, 0.35), sample 305-3, Loc. 305, Yabu Member of Yabu Formation.

Illustrated specimen.—A right female valve, UMUT-CA 9834 (Pl. 11, fig. 14, L, 0.64; H, 0.31), sample 135-3, Yabu Member of Yabu Formation.

Diagnosis.—A species of genus *Schizocythere* characterized by reticulation surface ornamented by three undulated ridges similar in thickness and strength.

Description.—Viewed laterally, general outline and carapace size similar to those of *Schizocythere kishinouyei* (Kajiyama, 1913). Surface reticulated by three undulated ridges similar in thickness and strength.

For inner view, feature of pore canals, hinge, and muscle scars, refer to the original descrip-

Sp	N	Me	$\bar{X}$	Sd	V	OR
RV ♀	2	L	0.615			0.61–0.62
		H	0.380			0.37–0.39
RV ♂	1	L				0.61
		H				0.33
RV A-1	4	L	0.530	0.016	3.08	0.51–0.55
		H	0.315	0.010	3.17	0.30–0.32
LV ♀	1	L				0.68
		H				0.39
LV ♂	1	L				0.60
		H				0.34
LV A-1	5	L	0.530	0.014	2.67	0.51–0.55
		H	0.318	0.016	5.17	0.29–0.33
LV A-2	1	L				0.42
		H				0.25

tion of *Schizocythere okhotskensis* (Hanai, 1970, p. 722).

Remarks.—This species is very close to *Schizocythere okhotskensis* Hanai, 1970, from Pliocene Sawane Formation, in posterior reticulation, but differs in anterior reticulation pattern and is smaller than *S. okhotskensis*. This species is similar to *Schizocythere kishinouyei* (Kajiyama, 1913) in size and general outline but differs in lateral ornamentation.

Dimensions.—Measurements of specimens of sample 284-1, Yabu Member of Yabu Formation are given below:

Occurrence.—Abundant in Yabu Member; also rare in Kiyokawa Formation.

Tribe Paijenborchellini Deroo, 1966

Genus *Neomonoceratina* Kingma, 1948

Neomonoceratina microreticulate Kingma, 1948

Pl. 11, figs. 10–13.

Neomonoceratina delicata Ishizaki and Kato, 1976, p. 136, 138, pl. 3, figs. 7–10, pl. 4, figs. 1–3, text-fig. 8 [Type: Diluvium Furuya Mud, Shizuoka Pref., IGPS-91733].

Remarks.—Surface ornamentation of *Neomonoceratina delicata* Ishizaki and Kato, 1976 agrees well with that of *N. microreticulate* Kingma, 1948. Sexual dimorphism is distinct. Female is wider than male in dorsal view.

Family Trachyleberididae Sylvester-Bradley, 1948

Diagnosis.—For diagnosis of Trachyleberididae, refer to Hazel (1967, p. 29). Additional diagnosis: normal pore canals with a sensory bristle, usually so-called simple type, with irregular sievelike openings of a deep-rooted apparatus around the basal part of the bristle.

Remarks.—The family include Trachyleberidinae, Rocaleberidinae, Pterygocythereidinae, and Buntoniinae.

Subfamily Trachyleberidinae Sylvester-Bradley, 1948

Diagnosis.—Carapace thick and large, elongate subtriangular to subrectangular with subtriangular posterior area in lateral view; oblong in dorsal view. Posterodorsal margin concave and posteroventral margin convex. Ventral contact margin sinuate at middle. Marginal infold narrow to moderate in breadth. Hinge holamphidont. A vertical row of four adductor scars with a V- or U-shaped scar in front. Dorsomedian adductor scar large and elongate. Sexual dimorphism distinct. Male very elongate.

Remarks.—The subfamily includes Trachyleberidini, Costini, Oertliellini, and a few species of uncertain tribe.

Tribe Trachyleberidini Sylvester-Bradley, 1948

Diagnosis.—Carapace elongate subtriangular in lateral view. Anterior margin with numerous short spines. Dorsal margin straight with approximately six regularly spaced spines. Two rows of small spines aligned along ventral to posteroventral margin. Posterior acutely pointed. Posterodorsal margin concave with a few spines. Carapace compressed along anterior, posterior, and posteroventral regions. Subcentral tubercle distinct. Surface varying from weakly reticulate with spines to smooth with tubercles, tubercles tending to align in three rows.

Remarks.—This tribe includes the following genera: *Trachyleberis* Brady, 1898, *Acanthocythereis* R. C. Howe, 1963, and *Actinocythereis* Puri, 1953. These three genera share the same lateral outline and the same tendency of surface ornamentation. Other genera such as *Hirsutocythere* Howe, 1951, and *Abyssocythere* Benson, 1971, have been referred to as Trachyleberidini by Hartmann and Puri (1974) and Liebau (1975).

Genus *Trachyleberis* Brady, 1898

Type-species.—Cythere scabrocuneata Brady, 1880.

Diagnosis.—Trachyleberidini with smooth surface having moderately numerous small tubercles. Anterodorsal short ridge often distinct.

Remarks.—Genus *Trachyleberis* was proposed by Brady (1898) on the basis of Japanese specimens in his study on a then imperfectly known group of Ostracoda. Sylvester-Bradley (1948) probably felt strongly the necessity of detailed study of both carapace and appendage structures of *Trachyleberis scabrocuneata* to clarify the content of his family Trachyleberididae. Thus, with Harding, he (1953) gave the detailed observation on appendage structure of the type specimens of *Trachyleberis scabrocuneata* kept in the British Museum.

Five species of *Trachyleberis* had been described from Japan before the publication of the checklist (Hanai et al., 1977). Two of them, belong actually to *Acanthocythereis* and *Actinocythereis*, and one species, which has been referred to *Trachyleberis* with a question mark, will very likely turn out to be a new genus.

Stratigraphic range.—Paleocene to Recent.

Trachyleberis scabrocuneata (Brady, 1880)

Cythere scabrocuneata: Puri and Hulings, 1976, p. 289, pl. 26, figs. 6, 8.

Trachyleberis scabrocuneata: Herrig, 1977, p. 1261, 1262, pl. 2, figs. 6, 7; Okubo, 1979b, p. 149-151, figs. 4, 7a–e; Okubo, 1980, p. 412.

Remarks.—Muscle scars are located on the wall of subcentral depression. Many authors mentioned the muscle scars of this species, because the species is the type-species of the genus and the genus is the type genus of the family Trachyleberididae. However, the Japanese specimens at hand show certain characters of muscle scars different from those which have hitherto been described. Among four adductor scars, the dorsal one is large, round, and separate from the lower three. A dorsomedian adductor scar is more elongate than other scars and inclined anteriorly. Ventromedian and ventral adductor scars are placed close to each other. This proves that *Trachyleberis scabrocuneata* is not an exception among those found in other trachyleberids.

Subcentral node has characteristic four tubercles. The specimens at hand are close to topotypes (BMNH 1974 342; Puri and Hulings, 1976, p. 26, figs. 6, 8; SEM photos) than paratypes (BM 1948. 3. 10. 1, 2, 5; Harding and Sylvester-Bradley, 1953, pl. 1, figs. 5, 6, 8; drawing sketches) in lateral outline, and the shape and arrangement of tubercles on carapace surface. The species may have considerable variation in the size and shape of tubercles.

Trachyleberis? *tosaensis* Ishizaki, 1968

Pl. 12, figs. 8, 9, 11; text-fig. 15-3.

Remarks.—Exterior shape suggests close relation of this species to Trachyleberidini. However, because of the unique surface ornamentation, it is different to assign the species to any known genus of the tribe.

The species closely resembles *Reymentia taiwanica* Hu, 1977 from the Pleistocene of Taiwan, described by Hu (1977b) and illustrated by Hu and Yeh (1978), in lateral outline and broad and fine ornamentation. Moreover, the strong and feeble ridges of *T.*? *tosaensis* correspond closely with those of *R. taiwanica.* These species are, therefore, quite likely to be conspecific. However, the generic assignment of them to *Reymentia* is doubtful, for the pattern of surface ornamentation, especially unique ridge arrangements and the absence of distinct anterior marginal furrows, is quite different from that of *Reymentia ijebuorum* Omatsola, 1970, type species of the genus. Muscle scars characterized by a long and anteroventrally inclined dorsomedian scar are of Trachyleberidini type.

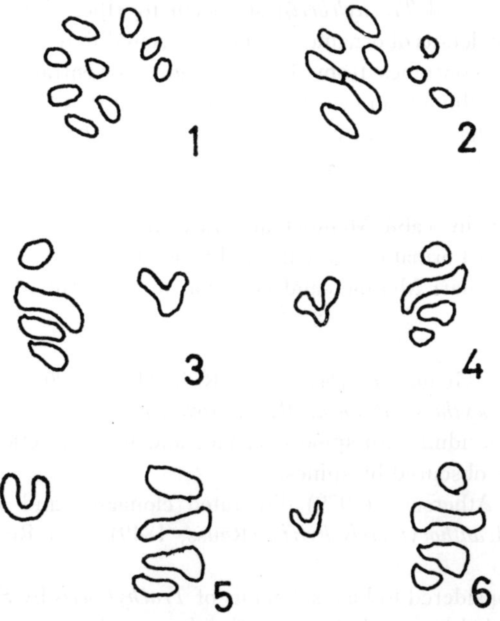

Text-fig. 15

Muscle scar patterns.

Fig. 1. *Aurila* sp., a left valve, CA 9866 (sample 120-1, Yabu Member of Yabu Formation). x 150.

Fig. 2. *Caudites*? *posterocostatus* (Ishizaki, 1966), a carapace, outer right valve view, CA 9871 (sample 120-1, Yabu Member of Yabu Formation). x 215.

Fig. 3. *Trachyleberis*? *tosaensis* Ishizaki, 1968, a left female valve, CA 9842 (sample 135-3, Yabu Member of Yabu Formation). x 320.

Fig. 4. *Australimoosella tomokoae* (Ishizaki, 1968), a right valve, CA 9855 (sample 73-4, Kiyokawa Formation). x 310.

Fig. 5. *Rocaleberis*? sp., a left valve, CA 9846 (sample 74-2, Kiyokawa Formation). x 215.

Fig. 6. *Wichmannella bradyformis* (Ishizaki, 1968), a right partly broken valve, CA 9851 (sample 251-5, Kioroshi Member of Kioroshi Formation). x 390.

Sexual dimorphism is distinct. Male form is longer and slenderer than female form. Maximum width is at the posterior third in male, and at the middle in female. Immature form is characterized by two parallel and obliquely running ridges, and by ornamentation of round to gourd-shaped pits arranged in three rows along anterior margin and in one row on both sides of each oblique ridge.

The Pleistocene specimens at hand are slightly larger than Ishizaki's specimens.

Trachyleberis? sp.

Trachyleberis sp. Yajima, 1978, p. 398, 399, pl. 49, figs. 1a, b.

Remarks.—The species is known only in immature form. As suggested by Ishizaki (1971, p. 93), mature form of this species may be *Acanthocythereis niitsumai* (Ishizaki, 1971). In fact,

Acanthocythereis niitsumai and *Trachyleberis*? sp. occur together in Kioroshi Member of Kioroshi Formation, but the details of ornament structure seem different. In the present specimens, anterior marginal ridge continues from dorsal margin to ventral margin and two distinct longitudinal ridges run obliquely in central area, suggesting that this species and *T.*? *tosaensis* may be congeneric and have certain characteristics close to tribe Costini.

A fragment of A-1 instar of *A. niitsumai* occurs in the sample of 189-3 (Kioroshi Member of Kioroshi Formation).

Occurrence.—Common in Yabu Member and rare in Kamiizumi Member of Yabu Formation; rare in Kiyokawa Formation; rare in sand facies and sandy silt facies of Kamiiwahashi Formation; rare of Toyonari Member and common in Kioroshi Member of Kioroshi Formation.

Genus *Acanthocythereis* R. C. Howe, 1963

Type-species.—Acanthocythereis araneosa R. C. Howe, 1963.

Diagnosis.—Trachyleberidini with spinose surface and weblike reticulation between spines. Subcentral muscle node obscured by spines.

Remarks.—Recently, Athersuch (1979) illustrated elongated and anteroventrally inclined dorsomedian scar of *Acanthocythereis hytrix* (Reuss, 1849) from Recent sediments, off the north coast of Cyprus.

Acanthocythereis is considered to be a subgenus of *Trachyleberis* by Siddiqui (1971), because difference between *Trachyleberis* and *Acanthocythereis* seems less prominent than that between *Trachyleberis* and *Actinocythereis* in the Eocene of West Pakistan. In this paper, however, *Trachyleberis*, *Acanthocythereis*, and *Actinocythereis* are all treated as being of the same rank.

Stratigraphic range.—Paleocene to Recent.

Acanthocythereis niitsumai (Ishizaki, 1971)

Remarks.—In the present species, eye tubercle is not so clear as in the case of Gulf Coast *Acanthocythereis*, but is continuous with anterior marginal ridge, which is broken into spines in ventral half. The species resembles *Trachyleberis scabrocuneata* in general outline, but is different in ornamental structure and smaller in size. A feeble short horizontal ridge runs across the area of indistinct subcentral tubercle. Surface of posterodorsal area is covered with fine reticulation, which is strengthened at each intersection of reticulation by a small tubercle. Small tubercles in posterocentral area have similar distribution pattern to those of *Trachyleberis nodosa* Bassiouni, 1969, illustrated by Cronin and Khalifa (1979) from the Eocene of Egypt.

Acanthocytereis? sp.

Pl. 12, fig. 5.

Illustrated specimen.—A partly broken left valve, UMUT-CA 9843 (Pl. 12, fig. 5. L, 0.96), sample 135-3, Yabu Member of Yabu Formation.

Remarks.—The species resembles *Acanthocythereis dunelmensis* (Norman, 1865) from Russian Harbor, Novaya Zemlya illustrated by Neale and Howe, 1975 in the arrangement of tubercles along anterior margin of carapace. Development of a few large round tubercles in posteroventral area found both in this species and in *Rocaleberis*? sp. (Pl. 12, figs. 1, 2) seems to be one of the characteristic tendencies among certain Trachyleberididae.

Liebau (1975) is of the opinion that *Acanthocythereis dunelmensis* group might be separated from Gulf Coast *Acanthocythereis* based on the difference of reticulation pattern.

Occurrence.—Only the illustrated specimen occurs in Yabu Member of Yabu Formation.

Genus *Actinocythereis* Puri, 1953

Type-species.—Cythere exanthemata Ulrich and Bassler, 1904.

Diagnosis.—Trachyleberidini with two or three longitudinal rows of spines.

Remarks.—As suggested by Howe and Howe (1975) and Liebau (1975), *Archicythereis* Howe, 1936, is certainly a valid genus, but because the type species of *Archyicythereis*, *Cythereis yazooensis* Howe and Chambers, 1935, is based on a late larval stage of one of the Eocene *Actinocythereis* species, the generic name *Actinocythereis* is temporally used.

Hazel (1967) is of the opinion that spine alignment of lateral surface is gradational from very strong alignment of *Actinocythereis* to no or weak alignment of *Trachyleberis*. As a result of his classification of North American and Caribbean species of *Trachyleberis* and *Actinocythereis* based on carapace shape and ornamentation, *Trachyleberis* includes certain species with weak alignment of spines.

In Japan, a species doubtfully assignable to *Actinocythereis* is characterized by three longitudinal rows of spines.

Stratigraphic range.—Eocene to Recent.

Tribe Oertliellini Liebau, 1975

Diagnosis.—Carapace large, and subtriangular in lateral view. Anterior margin broadly and obliquely rounded. Dorsal and ventral margins tapering posteriorly. Posterior margin concave dorsally, convex ventrally, making acute angle at posterior junction. Surface strongly reticulate with undercut ridges. Dorsal ridge pronounced. Ventrolateral ridge straight and strong, giving arrowhead shape in ventral view. Venter flat and reticulate. Eye tubercle usually distinct but sometimes absent.

Remarks.—This tribe includes the following genera: *Oertliella* Pokorný, 1964, *Cletocythereis* Swain, 1963, and *Agrenocythere* Benson, 1972. In Oertliellini, reticulation pattern is very conservative.

Benson (1972) described appendages of female specimens of *Agrenocythere radula* (Brady, 1880) and classified *Agrenocythere* in Trachyleberididae-Trachyleberidinae based on six-segmented first antenna, second antenna with abbreviated exopodite, and mandibule with "five-fingered" exopodite, despite "knee apparatus" on thoracic legs. Liebau (1975) proposed Oertliellini as ancestral of Hemicytherinae emphasizing the presence of "knee apparatus" on thoracic legs and its close resemblance to *Hermanites*-like advanced hemicytherine. *Agrenocythere* shows hemicytherine characters including "knee apparatus," but all the carapace characters seem to suggest that it is best included in Trachyleberidinae.

Genus *Cletocythereis* Swain, 1963

Type-species.—Cythere rastromarginata Brady, 1880.

Diagnosis.—Oertliellini with strong lateral outgrowth of tegumen to form celation. Surface regularly reticulated without spines. Eye tubercle distinct.

Remarks.—Cletocythereis of Swain's sense is no more acceptable, and the genus covers a small group closely related to the type species *C. rastromarginata*. Hazel (1967) indicated that *C. rastromarginata* has close relationship with the type-speiies of *Hermanites*, but later Liebau (1975) pointed out the difference in ventral ornamentation between the two genera.

Stratigraphic range.—Neogene to Recent.

Cletocythereis rastromarginata (Brady, 1880)

Pl. 12, figs. 3, 4.

Cythere rastromarginata Brady, 1880, [partim], p. 83, pl. 16, figs. 1a–d, 2a–d; Puri and Hulings, 1976, p. 286, pl. 9, figs. 9–14.

Cletocythereis bradyi Holden, 1967, p. 40–42, figs. 9-14; Okubo, 1979b, p. 152, figs. 1e-g; Okubo, 1980, p. 412.

Cletocythereis rastromarginata: Benson, 1972 p. 22, 26, pl. 1, figs. 1-4.

Remarks.—Holden (1967) proposed a new species *Cletocythereis bradyi* based on specimens from Easter Island (late Cenozoic) and included it an inflated form of *Cythere rastromarginata* (Brady, 1880, pl. 16, fig. 2). Benson (1972) designated a less inflated specimens from the reefs off Honolulu as lectotype (BMNH 80. 38. 105) of *C. rastromarginata*. Puri and Hulings (1976, pl. 9, fig. 9) considered the "lectotype" to be a probable male, and an inflated paralectotype (BMNH 80. 38. 104) a probable female. A specimen at hand (UMUT-CA 9844) is inflated in dorsal view and covered with reticulations. Recticulum is large and deep, has flat floor, and cloverleaf-shaped tegument. Frontal muscle scar is V-shaped as Benson (1972, p. 22) observed. Japanese specimen is slightly larger than the lectoholotype, but it resembles those illustrated by Holden and Benson suggesting conspecific relation of all of those specimens.

The specimen has a row of six strong spinelike projection along posteroventral margin and a series of short, blunt, and subequal projections along anterior margin. Ishizaki (1968, p. 40, pl. 8, fig. 9) reported *Cletocythereis bradyi* from Recent coarse sand of Uranouchi Bay, Kochi Prefecture. Ishizaki's specimen has only four projections along posteroventral margin and frill-like projection along anterior margin.

Tribe uncertain

Genus *Stigmatocythere* Siddiqui, 1971

Type-species.—*Stigmatocythere obliqua* Siddiqui, 1971.

Diagnosis.—Moderate in size, subrectangular in lateral view, characteristically compressed anteriorly and posteriorly. Anterior margin armored with a row of short spines. Posteroventral margin nearly straight and posterodistal corner higher than the middle of the posterior margin. Anterior marginal rim very high with depressed furrow just behind it. Surface ornamented with three longitudinal ridges and reticulation between ridges, or with blunt tubercles on smooth undulate surface, Two short ridges starting from the eye tubercle, anterior one forming anterior marginal rim, posterior one running downward forming arch anteriorly and connecting with subcentral muscle node. Muscle scars consist of an oval frontal scar and a vertical row of four adductor scars.

Remarks.—Among five species of the genus *Stigmatocythere* described by Siddiqui (1971), two species including type species of the genus have three longitudinal costate ridges. Thus, Hartmann and Puri (1974) included the genus in tribe Costini. However, the unique character of very high anterior marginal rim common to Siddiqui's five species seems to separate them from the genera of tribe Costini, which have a low anterior marginal ridge.

Stratigraphic range.—Middle and Upper Eocene in West Pakistan.

Stigmatocythere? sp.

Pl. 12, fig. 12.

Illustrated specimen.—A right valve, UMUT-CA 9845 (Pl. 12, fig. 12. L, 0.75; H, 0.44), sample 120-1, Yabu Member of Yabu Formation.

Remarks.—Flat and compressed carapace, very high anterior marginal rim and furrow behind it, posterior marginal rim and four dorsal blunt tubercles, including tubercle at distinct anterior cardinal angle, suggest that this species belongs to the genus *Stigmatocythere*. However, the tapered posterior, which has posterodistal corner lower than middle of the posterior margin, makes generic identification doubtful. This species is somewhat larger than the five species of *Stigmatocythere* described by Siddiqui (1971).

Species is closely similar to and perhaps congeneric with *Cythereis keyi* Kingma, 1948, from the Pleistocene Poetjangan beds and Pliocene lower Kaling beds of Bodjonegoro, eastern Java. The species resembles each other in carapace size, general outline, tapered posterior, three distinct tubercles along dorsal margin, and reticulation in central area. In the present specimens, eye tubercle is distinct at anterior cardinal angle, and posterodorsal tubercle extends into a low ridge toward posterocentral area. Ventral short ridge is distinct and somewhat alate. Surface is reticulated without "second swelling."

A new genus may be needed to accomodate the above two species.

Occurrence.—Rare in Yabu Member of Yabu Formation. Only two specimens are found in all samples.

Subfamily Rocaleberidinae Bertels, 1969

Diagnosis.—Carapace large, subrectangular to subtrapezoidal in lateral view, ovate in dorsal view. Posterior margin broadly rounded or truncated obliquely in its lower half. Anterior and posterior margins armored with small spines. Surface reticulate or tuberculate with longitudinal low and broad undulations. Tubercles on reticulation ridges, especially at the junctions of muri. Subcentral tubercle distinct. Eye tubercle and postero-ocular depression distinct. Radial pore canals moderate in number, simple, straight, and sometimes bifurcated. Normal pore canals of open type. Marginal infold moderately broad. Vestibule narrow but distinct. Hinge holamphidont. Frontal scar V-shaped, among vertical row of four adductor scars, dorsal scar large, separated from the other three, dorsomedian scar longer and inclined anteriorly, ventral two scars closely placed. Sexual dimorphism distinct, male elongate.

Remarks.—Bertels (1969) proposed the subfamily Rocaleberidinae to accomodate *Rocaleberis*, *Wichmannella*, and *Neoveenia*. In study on the evolution of the Argentine Rocaleberidinae, Bertels (1976) thought that the distribution of Rocaleberidinae was restricted to the South Atlantic basins. Recently, however, species from many parts of the world, especially from the deep sea sediment turned out to belong to this subfamily.

The Rocaleberidinae seems close to Echinocythereidinae in lateral outline and surface ornamentation. In fact, Bertels (1969) compared the two subfamilies at the time of her proposal of a new subfamily Rocaleberidinae. However, both subfamilies differ distinctly in the number of frontal scars.

There is some evidence that suggests a close relationship between Trachyleberidinae and Rocaleberidinae. Japanese trachyleberids show a similar pattern of muscle scars as rocaleberids. The genus *Henryhowella*, usually included in Trachyleberidinae-Trachyleberidini, resembles *Rocaleberis* as pointed out by Bertels (1976).

Genus *Rocaleberis* Bertels, 1969

Type-species.—*Rocaleberis nascens* Bertels, 1969.

Diagnosis.—Rocaleberidinae with three longitudinal undulations on carapace surface. Surface ornamented with reticulations superimposed often by small tubercles or spines.

Remarks.—Difference of surface ornamentation between *Rocaleberis* and *Wichmannella* is gradational. *Rocaleberis araucana* (Bertels, 1969) was originally described as a species of genus *Wichmannella* in having surface ornamentation intermediate between the two genera.

Rocaleberis? sp.

Pl. 12, figs. 1, 2; text-fig. 15-5.

Illustrated specimen.—A right valve, UMUT-CA 9846 (Pl. 12, figs. 1, 2; text-fig. 15-5. L, 1.14; H, 0.65), sample 74-2, Kiyokawa Formation.

Remarks.—A large and flat species tentatively assigned to the genus *Rocaleberis* characterized by subtrapezoidal lateral outline, cluster of large spines at posteroventral marginal corner, large and small round tubercles superimposed on regular reticulation, pore openings on the top of large tubercle, indistinct eye tubercle, bundles of two or three radial pore canals along anteromedian and anteroventral margins, and a V-shaped frontal scar and a long and anteroventrally inclined dorsomedian adductor scar. All of these characteristics clearly indicate that the specimens at hand belong to a new species. However, since only a right valve is available for study, the species will be left in open nomenclature and only a tentative description of right valve will be given. Viewed laterally carapace large, subtrapezoidal, highest at anterior cardinal angle. Anterior margin broadly rounded with 25 to 26 small and round spines. Dorsal margin straight with six small tubercles, gently sloping posteriorly. Ventral margin slightly convex, sinuated at posterior third, ornamented with numerous small spines. Posterior margin nearly straight, making nearly vertical angle with ventral margin and with numerous small spines. In dorsal view, carapace compressed laterally. Surface covered with reticulation. Reticulum subquadrate, subpentagonal, or subhexagonal, having small and occasionally large tubercles at each intersection. Large tubercles with pore openings on their top. Small tubercles of posteroventral area often with cluster of very small radiating spines on their top. Anterior marginal rim distinct, with numerous large and small tubercles. Posteroventral corner with cluster of large tubercles. Eye tubercle distinct but obscured by covering of tubercles. Radial pore canals straight and numerous, in bundles of two or three along anteromedian and anteroventral margins, and evenly spaced along ventral and posterior margins. Normal pores of sieve type, situated at the top of large round tubercles. Marginal infold broad along anterior margin and narrow along posterior and posteroventral margins. Posteroventral outer margin slightly undulated. Two striations distinct along anterior margin. Snap-knob at middle of ventral margin in right valve. Anterior vestibule weak. Hinge holamphidont, median groove smooth and narrow. Muscle scars in subcentral depression consisting of a U-shaped frontal scar and a vertical row of four adductor scars, of which dorsomedian scar long and inclined anteroventrally.

The species resembles *Hirsutocythere*? *nozokiensis* (Ishizaki, 1963) in lateral outline and in having posteroventral cluster of tubercles. Surface is, however, reticulated clearly in the present species. Eye tubercle is not so distinct as in *H. nozokiensis*.

Occurrence.—Rare in Yabu Member of Yabu Formation; also rare in Kiyokawa Formation. Altogether six right valves found.

Genus *Wichmannella* Bertels, 1969

Type-species.—Wichmannella meridionalis Bertels, 1969.

Diagnosis.—Rocaleberidinae without distinct rim or furrow in anterior marginal area. Eye tubercle distinct. Surface covered with regular reticulation without undulations. Reticulation of fossae of the same size, arranged parallel to carapace margin and occasionally without superimposed small tubercles at junction of muri.

Remarks.—Bertels (1969, 1975) pointed out that *Wichmannella* differs from *Echinocythereis* in the number of frontal scars and presence of vestibule. Japanese species hitherto placed in *Echinocythereis* has only one V-shaped frontal scar and seems to belong to *Wichmannella*.

Two species groups may exist in the genus *Wichmannella*: one consists of five species from the Lower Maastrichtian to Oligocene of Argentina, and the other includes Recent psychrospheric species such as *W. circumdentata* (Brady, 1880), *W. digitalis* (Levinson, 1974), and *W.*? *dasyderma* (Brady, 1880). These three species have no distinct eye, as Benson (1975) pointed out.

Stratigraphic range.—Upper Cretaceous to Recent.

Wichmannella bradyformis (Ishizaki, 1968)
Pl. 12, figs. 14–17; text-fig. 15-6.

Echinocythereis? bradyformis: Hanai et al., 1977, p. 51.

Echinocythereis bradyi Ishizaki 1968, Okubo, 1979b, p. 152–155, figs. 5, 6, 7f-h; Okubo, 1980, p. 412, 413.

Remarks.—Muscle scars consist of a V-shaped frontal scar and a vertical row of four adductor scars, of which the dorsal scar is round and separated from three other lower scars; dorsomedian scar is horizontally elongated and not inclined in anteroventral direction; ventromedian and ventral scars are crescent shaped. Surface is covered with reticulation having a small tubercle at intersection of the reticulation, which is distinct especially in the area posterior to the subcentral node. Vestibule is not distinct.

Sexual dimorphism is not distinct; the male form is elongate and steeply tapering toward posterior. Immature form A-1 is cylindrical, coarsely reticulated with a marginal ridge running along anterior and ventral margins. Forms younger than A-1 are subtriangular in lateral view. Surface is smooth or sometimes faintly reticulated, with distinct marginal ridges along anterior and ventral margins, and with a spine in the posteroventral area of both valves.

Echinocythereis miaoliensis (Hu and Yang, 1975) from the Plio-Pleistocene of Taiwan resembles this species in general features and therefore needs comparison of details of ornamentation.

Wichmannella bradyi (Ishizaki, 1968)

Echinocythereis? bradyi: Hanai et al., 1977, p. 51, 52.

Remarks.—Distinct ridges starting from the eye spot and running parallel to anterior and ventral margins, characteristic of *W. bradyformis*, are not present in this species. Surface is reticulated.

Echinocythereis arachis (Hu and Cheng, 1977) from the Late Pleistocene of Taiwan has two frontal scars yet resembles this species in lateral outline and surface ornamentation.

Subfamily Pterygocythereidinae Puri, 1957

Diagnosis.—Carapace subrectangular, sometimes oblong in lateral view, ovate to arrowhead-shaped in dorsal view with sharp pointed or inflated alae or with long spines. Surface smooth and pellucid, sometimes punctate or reticulate. Marginal infold moderate to narrow, vestibule narrow to none, except for *Incongruellina*. Vestibule deep along anteroventral and posteroventral margins in *Incongruellina*. Eye tubercle not distinct. Radial pore canals straight and feeble, moderately numerous. Hinge holamphidont with the exception of hemiamphidont in *Alatacythere* and *Pterygocythere*. Muscle scars consisting of a V-shaped or U-shaped frontal scar and a vertical row of four adductor scars. Dorsal adductor scar sometimes separated from the lower three; dorsomedian scar elongate; ventromedian and ventral scars placed closely to each other. Sexual dimorphism not very distinct.

Remarks.—Hazel (1967, p. 39) observed muscle scar pattern of Pterygocythereidinae as having "four adductor scars; the upper scar is usually heart-shaped and sometimes divided." Most muscle scar patterns of Pterygocythereidinae illustrated or described in generic diagnoses seem close to those of Trachyleberidinae.

Pterygocythereidinae includes the following three groups of genera which are different in lateral and dorsal views and surface ornamentation: Pterygocythereidini Puri, 1957; *Incongruellina* group; and *Bicornucythere* group.

Incongruellina group

Diagnosis.—Carapace subtrapezoidal in lateral view, with well-developed and inflated ventrolateral alae with or without convex in dorsal view, not strongly carinate, ponticulate, or broken up into row of spines. Inflated subtriangular in anterior view. Surface smooth and pellucid. Hinge holamphidont.

Remarks.—This genus group includes the following genera: *Incongruellina* Ruggieri, 1968; *Lixouria* Uliczny, 1969; and *Carinovalva* Sissingh, 1979.

Incongruellina and its allied genera have been placed in the tribe Pterygocythereidini (Hartmann and Puri, 1974). Liebau (1975) questioned the systematic position of this new genus group in Pterygocythereidini, a tribe recognized by him as a synonym of Brachycytherini.

Genus *Carinovalva* Sissingh, 1973

Type-species.—Incongruellina (*Lixouria*) *keiji* Sissingh, 1972.

Diagnosis.—Incongruellina group with obliquely and broadly rounded anterior margin furnished with several short spines. Ventrolateral keel nearly parallel, or sometimes overhanging ventral margin. Marginal pore canals moderately numerous. Vestibule absent. Eye tubercle small or indistinct.

Remarks.—Sissingh (1973) clarified the status of genera *Incongruellina* and *Lixouria* and proposed a new genus *Carinovalva* to cover lixourid species found in Oligocene to Recent sediments, restricting genus *Lixouria* to a rare Oligocene species, *Cythereis unicostulata* Kuiper, 1918. Some species, "*Cythere*" *tetsudo* Namias, 1900, illustrated by Uliczny (1971), *Incongruellina* (*Lixouria*) *unicostulata* illustrated by Uliczny (1969), *Ruggieria* (*Keija*) [*sic*] *carinata carinata* and *R*. (*K*.) [*sic*] *carinata fongolinii* illustrated by Carbonnel (1969) may well be included in the genus *Carinovalva*.

Stratigraphic range.—Oligocene to Recent.

Carinovalva nipponica (Yajima, 1978)

Lixouria nipponica Yajima, 1978, p. 400, 401, pl. 50, figs. 7a-c, text-figs. 9-2a, b.

Remarks.—McKenzie (1979, pers. comm.) suggested that this species belongs to the genus *Carinovalva* rather than to *Lixouria*. Japanese species resembles *Lixouria unicostulata* from the Oligocene of Netherlands, illustrated by Uliczny (1971) and Sissingh (1973), in ventrolateral alae without posterolaterally directed spines and subovate dorsal outline; species differs in subtrapezoidal lateral outline, position of the maximum width, and presence of spines along anterior margin. General outline, which reflects the shape of the soft body and some detailed characters, such as spines along anterior and posterior margins and the number of radial pore canals along posterior margin, seem to suggest a close relation of this species to genus *Carinovalva*. *C. nipponica* resembles *C. marginata* (Terquem, 1878) from Late Cenozoic of the Rhodos, illustrated by Sissingh (1972, pl. 5, fig. 12) in lateral subtrapezoidal outline, shape of ventrolateral alae and spinose anterior and posterior margins. *C. nipponica* is, however, smaller than *C. marginata*, and details of posteroventral spines and shape of ventrolateral alae of *C. marginata* are different from those of *C. nipponica*.

Some specimens at hand have a shallow groove and a weak ridge running just above and along ventrolateral edge of alae.

Bicornucythere group

Diagnosis.—Carapace subrectangular to subovate in lateral view, ovate in dorsal view with a posteriorly pointed spine. Anterior margin armored with a row of strong quadrate spines. Surface smooth with broad, low and longitudinal ridges and large pits between ridges, or regularly reticulate and without ridges.

Remarks.—This genus group includes the following genera: *Ruggieria* Keij, 1957; *Keijella* Ruggieri, 1967; and *Bicornucythere* Schornikov and Schaitarov, 1979.

Genus *Bicornucythere* Schornikov and Schaitarov, 1979

Type-species.—Leguminocythereis bisanensis Okubo, 1975.

Diagnosis.—Bicornucythere group with ovate carapace in lateral and dorsal views with a distinct long posterolaterally directed spine. Surface reticulate and characterized by more or less longitudinal arrangement of fossae.

Remarks.—In the checklist of the Southeast Asian Ostracoda (Hanai et al., 1980), 12 species are included in the genus *Bicornucythere. Bicornucythere* has been reported only from the Western Pacific. The center of the distribution seems to be in Southeast Asia and the distribution extends as far north as Amur Bay (Schornikov and Schaitarov, 1979).

The genus has no distinct lateral alae but is considered to belong to the Pterygocythereidinae. Ala in this genus may be represented by a distinct posteriorly pointed spine.

Stratigraphic range.—Pliocene to Recent.

Bicornucythere bisanensis (Okubo, 1975)

Ruggieria (*Keijella*) *bisanensis*: Hanai et al., 1978, p. 42.

Bicornucythere bisanensis: Schornikov and Schaitarov, 1979, p. 45–47, figs. 1–3; Okubo, 1980, p. 413.

Remarks.—Muscle scars are similar in their shape and distribution to those of Pterygocythereidinae.

Immature carapaces in one sample vary in surface ornamentation from the common form with reticulation over the entire surface, through a form with reticulation in the posterior half of the carapace only, to an entirely smooth form.

Hu and Cheng (1977) and Hu (1977a, b) reported three species of the genus, *B. elongata* (Hu, 1977), *B. ovalis* (Hu and Cheng, 1977), and *B. taiwanensis* (Hu, 1977), from the Plio-Pleistocene of Taiwan. These three species resemble *B. bisanensis* in lateral outline and surface ornamentation but differ in details of ornamentation.

Subfamily Buntoniinae Apostolescu, 1961

Diagnosis.—Carapace pyriform to subtriangular with characteristic convergence toward posterior, or subrectangular with nearly parallel dorsal and ventral margins and truncated posterior margin narrowly rounded, posteriorly projected area in its lower half in lateral view, in dorsal view subovate with sides mostly parallel. Anterior margin broadly rounded, sometimes with marginal rim. Subcentral tubercle indistinct. Surface smooth with pits, reticulate, or with longitudinal ridges. Anterior marginal infold broad with numerous, usually unbranched, straight or curved marginal pore canals. Inner margin nearly parallel to outer margin. Hinge holamphidont. Muscle scars on flat wall of central part of inner carapace consisting of a J-shaped or V-shaped or subcrescent frontal scar and a vertical row of four round to ovate adductor scars.

Remarks.—This subfamily includes Buntoniini and Phacorhabdotini.

Tribe Buntoniini Apostolescu, 1961

Diagnosis.—Carapace pyriform with characteristic convergence toward posterior in lateral view. Anterior marginal rim present. Posterior cardinal angle distinct in left valve, overlapping posterior cardinal tooth of right valve. Surface ornamented with posteriorly converging longitudinal ridges and reticulation between ridges. Ornamentation often restricted to posterior half of carapace.

Remarks.—The tribe includes the following genera: *Buntonia* H. V. Howe, 1935; *Protobuntonia* Grekoff, 1954; *Quasibuntonia* Ruggieri, 1958; *Isobuntonia* Apostolescu, 1961; *Soudanella* Apostolescu, 1961; *Togoina* Apostolescu, 1961; *Huantraiconella* Bertels, 1968; *Argenti-*

cytheretta Rossi de Garcia, 1969; *Bensonia* Rossi de Garcia, 1969; *Grekoffiana* Rossi de Garcia, 1969; *Asymmetricythere* Bassiouni, 1971; *Rectobuntonia* Sissingh, 1972; *Harringtonia* Bertels, 1975; *Chiliella* Rose, 1975; and *Magallanella* Rose, 1975.

Genus *Buntonia* H. V. Howe, 1935

Tyep-species.—Buntonia shubutensis Howe, 1935.

Diagnosis.—Buntoniini with smooth or reticulate surface at posterior half. Hinge holamphidont, located posteriorly.

Stratigraphic range.—Upper Cretaceous?, Eocene to Recent.

Tribe Phacorhabdotini Gründel, 1969

Diagnosis.—Carapace subrectangular in lateral view, compressed with subparallel sides in dorsal view. Anterior marginal rim reduced. Dorsal and ventral margins nearly straight and subparallel. Posterior margin truncated and with posteroventrally projecting denticulate flange. Surface smooth to reticulate. Two short longitudinal ridges confined to posterior half. Posterodorsal short ridge present. Marginal infold extremely broad with numerous curved marginal pore canals, a few of them bifurcate. Eye spot indistinct. Hinge holamphidont, short, located posteriorly.

Remarks.—This tribe includes the following three genera: *Phacorhabdotus* Howe and Laurencich, 1958; *Imhotepia* Gründel, 1969; and *Ambocythere* van den Bold, 1957.

Gründel (1969) proposed Phacorhabdotini including *Phacorhabdotus*, *Ambocythere*, *Atjehella*, and *Imhotepia*. *Atjehella* is, however, different in course of inner margin and number of radial pore canals. Undulating inner margin is found in several unrelated genera (Keij, 1979 b; Liebau, 1975). *Atjehella* may form a triad of genera together with *Hemikrithe* and *Kalingella* as discussed by Keij (1979b).

Genus *Ambocythere* van den Bold, 1957

Type-species.—Ambocythere keiji van den Bold, 1957.

Diagnosis.—Phacorhabdotini with distinct posteroventral projection of flange in lateral view. Anterior margin with distinct marginal rim and marginal furrow just behind marginal rim. Anterior vestibule narrow. Frontal scar large and hook-shaped.

Remarks.—Japanese *Ambocythere* deviates from Gulf Coast species. Distribution of this form of *Ambocythere* seems endemic, being restricted to Japan and adjacent seas.

Stratigraphic range.—Miocene to Recent.

Ambocythere japonica Ishizaki, 1968

Remarks.—A short vertical sulcus in dorsomedian area is characteristic of this species. The species resembles *Atjehella tricarinata* Keij, 1979, from the Holocene of Phillipines in lateral outline and surface ornamentation, especially in distinct anterior marginal rim, truncated posterior margin, and posteroventral longitudinal ridges. However, the inner margin running nearly parallel to outer margin and details of hinge structure seem to warrant assignment of the species to *Ambocythere*.

Tribe uncertain

Genus *Robertsonites* Swain, 1963

Type-species.—Robertsonites gubikensis Swain, 1963 [=*Cythereis tuberculata* Sars, 1865].

Diagnosis.—Carapace large, subquadrate in lateral view, sides flat and subparallel in dorsal view. Surface reticulate regularly, characterized by two posterior nodes and a muscle node surrounded by shallow and narrow groove.

Remarks.—Swain (1963) established the genus designating *Robertsonites gubikensis* Swain, 1963, as the type species. Hazel (1967), however, found that both "*Robertsonites gubikensis* Swain, 1963, and *R. tuberculatina* Swain, 1963, are junior synonyms of *Robertsonites tuberculata* (Sars, 1865)." Carapace outline and surface ornamentation of *R. tuberculata* vary greatly, as seen in description by Brady (1868).

Muscle scars consist of a V-shaped frontal scar and a vertical row of four adductor scars. Dorsomedian adductor scar is elongate as illustrated by Sylvester-Bradley and Benson (1971).

Hazel (1967) tentatively included the genus in Trachyleberidinae and called for further study. Hartmann and Puri (1974) classified the genus into Costini with a question mark, and Liebau (1975) included it in Trachyleberidini with a question mark. The genus is different from Trachyleberidini in lateral outline, especially in shape of posterior margin, and from Costini and Hiltermannicytherini in having no distinct longitudinal ridges but distinct posterior nodes.

Stratigraphic range.—Pleistocene to Recent.

Robertsonites? *reticuliforma* (Ishizaki, 1966)
Pl. 12, fig. 13.

Buntonia? *reticuliforma*: Hanai et al., 1977, p. 53.
Buntonia? *japonica* Ishizaki, 1966, Yajima, 1978, p. 387, table 5.

Remarks.—Carapace is subquadrate with broadly rounded anterior margin. Posterior margin is bluntly and obliquely rounded. Lateral surface is covered with large semicircular puncta of similar size. Posterodorsal and posteroventral nodes and ventral ridge are distinct. Surface ornamentation does not quite fit with that of genus *Buntonia.* Although subcentral tubercle and compressed area around subcentral tubercle are not developed, it may be possible that the species is close to genus *Robertsonites.* The species resembles *R. tuberculata* (Sars, 1865), type-species of the genus, from North Atlantic Sea (Hazel, 1967; Neale and Howe, 1975; Siddiqui and Grigg, 1975) in lateral outline, surface ornamentation, and broadness of marginal infold, but it differs from *R. tuberculata* in detailed surface ornamentation. Therefore, generic assignment is still questionable.

Subfamily uncertain
Tribe Moosellini Hartmann, 1978

Diagnosis.—Carapace moderate to large in size, elongate elliptical in lateral view, and fusiform in dorsal view. Dorsal and ventral margins nearly straight and subparallel. Posterior margin often with a few spines. Surface with three marginal ridges along anterior margin, several longitudinal ridges running from anteroventral area obliquely to posterodorsal area, and with reticulation posteriorly. Vestibule narrow to almost absent. Appendages similar to those of trachyleberids, except for second antenna having a long exopodite similar to hemicytherids.

Remarks.—This tribe includes the following genera: *Moosella* Hartmann, 1964; *Australimoosella* Hartmann, 1978; and *Doratocythere* McKenzie, 1967.

Based on description of *Moosella* in 1964 and *Australimoosella* in 1978, Hartmann (1978) proposed tribe Moosellini, emphasizing the presence of a long exopodite on second antenna, characteristic lateral outline, longitudinal ornamentation, and other carapace structure as diagnostic features. Tribe Moosellini is close to subfamily Campylocytherinae of family Hemicytheridae. In fact, *Doratocythere* was classified by McKenzie (1967) in Campylocytherinae based on the thickened shell wall behind the anterior hinge elements. The appendage characters of *Doratocythere,* however, show close relationship to Trachyleberididae except for a well-developed exopodite of second antenna. Hartmann (1978) classified tribe Moosellini in

Phacorhabdotinae of Trachyleberididae "with some reservation," but when Phacorhabdotinae is restricted to a group of Ostracoda developed from *Phacorhabdotus* to *Ambocythere*, Moosellini seems to be better classified in Trachyleberidinae. It has the same tendency of surface ornamentation with Hiltermannicytherini.

Genus *Australimoosella* Hartmann, 1978

Type-species.—*Australimoosella liebaui* Hartmann, 1978.

Diagnosis.—Moosellini with distinct distal-dorsal angulation. Posterior margin smooth without spines. Lateral surface with strong and numerous longitudinal ridges and grooves. Hinge teeth not crenulated. Second antenna with two terminal claws. Maxilla with small subquadrate terminal segment.

Remarks.—The writer has had no opportunity to examine the appendage structure of the genus.

Stratigraphic range.—Pleistocene to Recent.

Australimoosella tomokoae (Ishizaki, 1968)
Pl. 12, figs. 18, 19; text-fig. 15-4.

Proteoconcha tomokoae: Hanai et al., 1977, p. 49.

Remarks.—Muscle scars consist of a large V-shaped frontal scar and a vertical row of four adductor scars, of which the dorsal one is large and separated from the lower three, and dorsomedian one is elongate and inclined anteroventrally.

In the genus *Proteoconcha*, anterior radial and false radial pore canals are "usually unevenly spaced, the false radial pore canals indent the line of concrescence distally" and anterior vestibule is short and shallow (Plusquellec and Sandberg, 1969, p. 448). In the present species, however, radial pore canals are numerous and evenly spaced and anterior vestibule is not developed.

The species resembles *Australimoosella liebaui* Hartmann, 1978, type-species of the genus *Australimoosella*, in general outline and surface ornamentation, but differs in details of ornamentation, especially in the number of longitudinal ridges and grooves. *A. tomokoae* is larger than *A. liebaui*.

Moosella anterosulcata Herrig, 1976, from the Plio-Pleistocene of the coast of the Tonquin Bay, is similar in its general appearance but is smaller than *A. tomokoae*. Poor preservation of the Indochinese specimens does not warrant further discussion on the relation between the two species.

Similarity of ***Basslerites taiwanensis*** Hu and Yeh, 1978, from the Pleistocene Liushuang Formation in Taiwan, to this species in details of outline, broad and fine surface ornamentation, as well as several characters of inner valve surface suggests a close relation between the two species.

Genus *Campylocythereis* Omatsola, 1971

Type-species.—*Campylocythereis sandbergi* Omatsola, 1971.

Diagnosis.—Moosellini with five concentric ridges convex upward at midlength. Frontal muscle scars either one (V-shaped), two, or three. Adductor muscle scars a vertical row of four; dorsomedian scar longest and inclined anteriorly.

Remarks.—The genus is very similar to *Australimoosella* but lateromedian ridges are convex upward at midlength in *Campylocythereis* and at posterodorsal area in *Australimoosella*.

Campylocythereis was classified in Campylocytherinae of Hemicytheridae in original description (Omatsola, 1971) and in Campylocytherinae of Trachyleberididae by Hartmann and Puri (1974). There has been no description of appendage structure of *Campylocythereis*, yet single V-shaped frontal scar and very similar ornamental structure of the carapace of *Australimoosella* and *Campylocythereis* suggest the affinity of this genus to Moosellini.

Stratigraphic range.—Pleistocene? to Recent.

Campylocythereis? *ukifune* n. sp.

Pl. 12, figs. 6, 7, 10.

Types.—Holotype, a left female valve, UMUT-CA 9856 (Pl. 12, fig. 10. L, 0.88; H, 0.45), sample 284-3, Loc. 284, Yabu Member of Yabu Formation.

Illustrated specimens.—A left male valve, UMUT-CA 9857 (Pl. 12, fig. 7. L, 0.84; H, 0.46), sample 284-3; a right male valve, UMUT-CA 9858 (Pl. 12, fig. 6. L, 0.89; H, 0.46), sample 120-2, Yabu Member of Yabu Formation.

Diagnosis.—A large species tentatively assigned to the genus *Campylocythereis* characterized by less elongate carapace than the type-species with broadly rounded anterior margin in lateral view. Surface ornamented with concentric ridges in central area. Puncta semicircular, large, and aligned between ridges.

Description.—Viewed laterally, carapace thick, ovate, highest at anterior cardinal angle. Anterior margin broadly rounded with 10 to 13 small denticles along its lower half. Dorsal margin straight, nearly parallel to ventral margin. Posterior cardinal angle distinct. Ventral margin slightly sinaute at middle. Posterodorsal margin slightly compressed. Posterior margin truncated obliquely at middle. Viewed dorsally, carapace elongate ovate, widest a little posterior to middle.

Surface coarsely punctated. Ridges mostly concentric and parallel to outer margins. Each fossa semicircular and similar in size. Eye tubercle present but not distinct. Normal pores sunken sieve type, located on muri. Radial pore canals numerous, straight, with bulbous enlargement.

Marginal infold narrow. Inner margin coinciding with line of concrescence and running parallel to outer margin. Selvage distinct with a few striations running parallel to outer margin. Hinge holamphidont with finely denticulate median element. Muscle scars consisting of a heart-shaped frontal scar, two dorsal scars, and a vertical row of four adductor scars, of which ventral one smallest, and dorsocentral one longest, inclined obliquely toward anterior.

Sexual dimorphism distinct. Female form elongate box-shaped and male form triangular in lateral view.

Dimensions.—Measurements of specimens from sample 284-3, Yabu Member of Yabu Formation are given below:

Sp	N	Me	$\overline{X}$	OR
LV ♀	1	L		0.88
		H		0.45
LV ♂	2	L	0.850	0.84–0.86
		H	0.465	0.46–0.47
RV ♀	1	L		0.89
		H		0.45
RV ♂	1	L		0.86
		H		0.44

Remarks.—The species is close to *Leguminocythereis dumonti* Keij, 1958, from the Eocene of Belgium in general outline, ornamentation, and hinge structure, but differs in having a heart-shaped frontal scar. Ornamental structure of *Campylocythereis sandbergi* Omatsola,

1971, from Recent Niger Delta is similar in pattern to those of this species. This species is, however, easily distinguishable from the African species in having large punctae between main ridges and in being shorter in lateral view.

Occurrence.—Common in Yabu Member of Yabu Formation; also rare in Kiyokawa Formation.

Family Hemicytheridae Puri, 1953

Diagnosis.—For diagnosis of Hemicytheridae refer to Hazel (1967, p. 13).

Remarks.—The family includes Hemicytherinae, Thaerocytherinae, and Campylocytherinae. Siddiqui (1971) retained the trachyleberids and hemicytherids in the family Trachyleberididae. In this paper, however, Trachyleberididae and Hemicytheridae are treated as being of the same rank. Taken into consideration here, besides characters of carapace, are all characters of appendages, including number of podomeres in the antennule, because any particular character is not expected to be unique to one group.

Subfamily Hemicytherinae Puri, 1953

Diagnosis.—Carapace thick and large, auriform to subquadrate sometimes produced posteroventrally in lateral view, usually elliptic in dorsal view. Anterior margin without spines. Surface smooth, pitted to reticulate. Ventral contact margin sinuate at middle. Marginal infold moderately wide with numerous marginal pore canals. Normal pore canals sieve-type. Vestibule usually present. Hinge holamphidont, modified amphidont, or merodont. Frontal muscle scars two or three. A vertical row of adductor scars of which dorsal one single, dorsomedian mostly divided, ventromedian single or divided, and ventral one again single.

Remarks.—Hanai (1961) studied hinge-type in Hemicytherinae. Included are merodont *Finmarchinella* type, amphidont with crenulated posterior protogeneric tooth of *Urocytheris* type, and strong holamphidont of *Hemicythere-Mutilus* type. In *Mutilus*, posterior tooth of the right valve has an incision in its middle and corresponding tooth in the middle of posterior socket in the left valve.

Hemicytherinae includes the following tribes and genus group: *Finmarchinella* group, Aurilini, Urocythereidini, and Orioninini.

Finmarchinella group

Diagnosis.—Carapace subquadrate in lateral view, elliptic to hexagonal in dorsal view. Anterior margin with distinct selvage. Surface pitted to rugose. Marginal infold moderate to narrow with numerous radial pore canals with bulbous enlargement. Vestibules at anterior and posterior. Hinge merodont. Three frontal scars. Adductor scars inclined anteriorly, median two adductor scars clearly divided. Sexual dimorphism strong, male elongate.

Remarks.—*Finmarchinella* group is based mainly on its unique hinge structure. The group is so far monogeneric.

Genus *Finmarchinella* Swain, 1963

Types-species.—*Cythereis finmarchica* Sars, 1865.

Diagnosis.—For diagnosis, refer to that of genus group.

Remarks.—A detailed information on the species of this genus was given by Neale (1974). Japanese *Finmarchinella* were described by Ishizaki (1966, 1969) and Okada (1979).

Finmarchinella includes subgenera *Finmarchinella* and *Barentsovia* Neale, 1974.

Stratigraphic range.—Miocene to Recent.

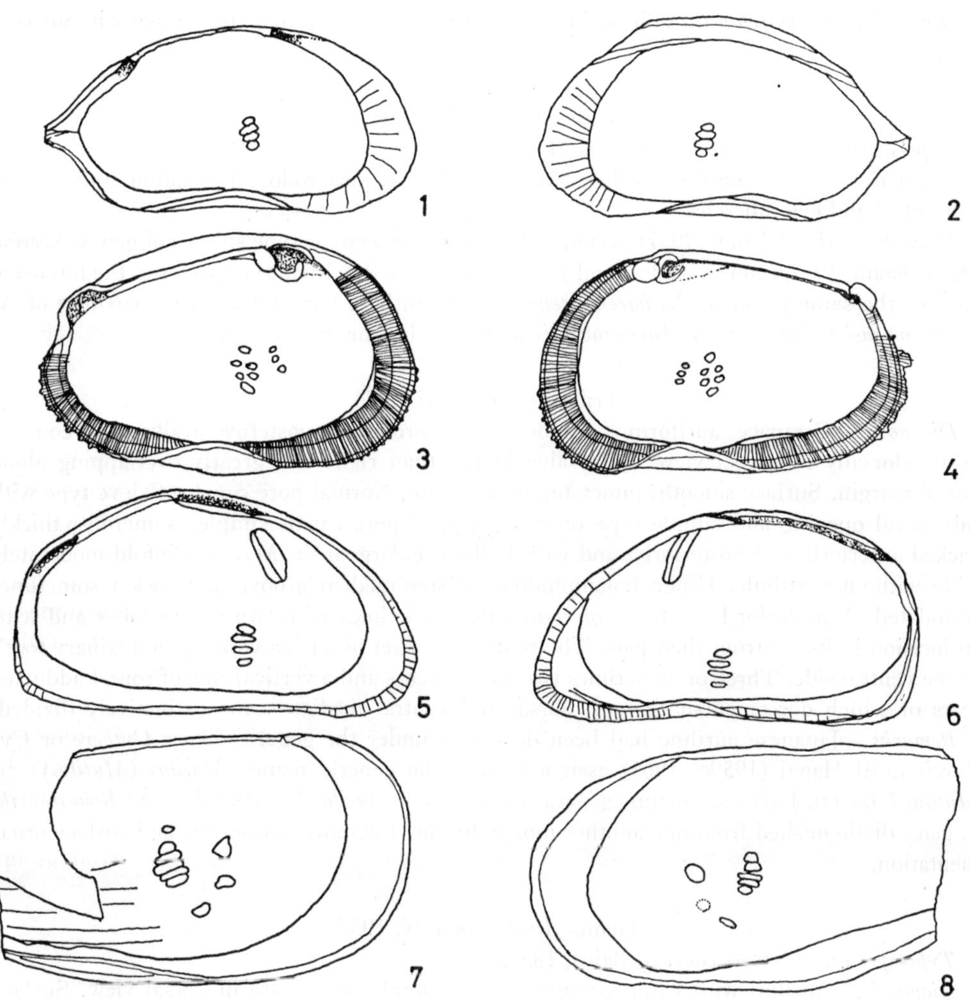

Text-fig. 16

Internal views. left row (odd number): left valves; right row (even number): right valves.

Figs. 1, 2. *Semicytherura wakamurasaki* n. sp. 1, female, CA 9885 (sample 190-1, Kioroshi Member of Kioroshi Formation). x 150. 2, female, CA 9890 (sample 189-3, Kioroshi Member of Kioroshi Formation). x 152.

Figs. 3, 4. *Aurila kiritsubo* n. sp. 3, holotype, CA 9863 (sample 49-1, Toyonari Member of Kioroshi Formation). x 69. 4, CA 9864 (sample 49-1). x 69.

Figs. 5, 6. *Xestoleberis suetsumuhana* n. sp. 5, holotype, CA 9913 (sample 138-5, Kamiizumi Member of Yabu Formation). x 86. 6, CA 9914 (sample 284-3, Yabu Member of Yabu Formation). x 79.

Figs. 7, 8. *Pseudocythere frydli* n. sp. 7, CA 9877 (sample 120-3, Yabu Member of Yabu Formation). x 98. 8, CA 9878 (sample 305-5, Yabu Member of Yabu Formation). x 98.

Subgenus *Finmarchinella* Swain, 1963

Diagnosis.—*Finmarchinella* elliptic in dorsal view, pitted or reticulate surface without costation.

Subgenus *Barentsovia* Neale, 1974

Type-species.—*Nereina barenzovensis* Mandelstam, 1957.

Diagnosis.—*Finmarchinella* with hexagonal dorsal view, posterodorsal costation, and distinct subcentral and eye tubercles.

Remarks.—Mandelstam (1957) designated *N. barenzovoensis* as type species of genus *Nereina* Mandelstam, 1957, (non Cristofori and Jan, 1832), but he descirbed the species in the following page of the same paper as *N. barenzovensis*. Other authors have followed his spelling of *N. barenzovoensis*; however, *N. barenzovensis* seems to be the correct spelling.

Tribe Aurilini Puri, 1974

Diagnosis.—Carapace auriform to subquadrate produced posteroventrally and concave posterodorsally in lateral view. Left valve higher than right and greatly overlapping along dorsal margin. Surface smooth, punctate, or reticulate. Normal pore canals of sieve type with subcentral opening and simple type on muri. Radial pore canals simple, numerous, thickly packed anteriorly and posteriorly, and with bulbous enlargement. Marginal infold moderately wide without vestibule. Hinge holamphidont. Posteromedian groove and socket sometimes crenulated. A posterior large tooth of right valve sometimes with two to five lobes and with an incision in its ventromedian part. The posterior socket of left valve with an auxiliary tooth in the ventral side. Three or sometimes two frontal scars and a vertical row of round adductor scars of which dorsomedian or both dorsal and ventral median scars distinctively divided.

Remarks.—Japanese auriline had been described under the generic names *Cythere* or *Cythereis* until Hanai (1959c, 1961) assigned them the generic names *Mutilus* (*Mutilus*) and *Mutilus* (*Aurila*). Japanese auriline genera so far include *Aurila*, *Pseudoaurila*, and *Robustaurila* n. gen., distinguished from one another mostly by the differences of outline and surface ornamentation.

Genus *Aurila* Pokorný, 1955

Type-species.—*Cythere convexa* Baird, 1850.

Diagnosis.—Aurilini with carapace auriform in lateral view, ovate in dorsal view. Surface smooth, pitted, or finely reticulated. Muscle scars consisting of three frontal scars, always divided dorsomedian, and sometimes divided ventromedian adductor scars.

Remarks.—The ventromedian adductor scar in Japanese *Aurila* is often divided into two, whereas those of European and American species are mostly single and elongate. Three frontal scars and adductor scars with divided median scars are all round and nearly equal in size.

Stratigraphic range.—Oligocene to Recent.

Aurila kiritsubo n. sp.

Pl. 13, figs. 9–11; Pl. 15, figs. 14, 16; text-figs. 16-3, 4.

Types.—Holotype, a left valve, UMUT-CA 9863 (Pl. 13, fig. 10; pl. 15, figs. 14, 16; text-fig. 16-3. L, 0.75; H, 0.47), sample 49-1, Loc. 49, Toyonari Member of Kioroshi Formation.

Illustrated specimens.—A right valve, UMUT-CA 9864 (Pl. 13, fig. 9; text-fig. 16-4. L, 0.75; H, 0.45, sample 49-1; a carapace, UMUT-CA 9865 (Pl. 13, fig. 11. L, 0.89; H, 0.53; W, 0.47 sample 49-1.

Diagnosis.—A species of genus *Aurila* characterized by anterior marginal ridges ending at anteroventral margin of carapace, not running parallel to anteroventral margin. Ventral ridge strong and convex, ending at posterocentral caudal area. Dorsal ridge gently arched continuing obliquely to posterodorsal short ridge. Other ridges running more or less radially from center toward anterior and posterior marginal area.

Description.—Carapace auriform and subquadrate, highest at anterior cardinal angle. Difference between left and right valves large in lateral outline. Left valve higher than right. Anterior margin broadly and obliquely rounded, with some small denticulation in its lower half. Dorsal margin slightly arched. Ventral margin covex in left valve, sinuated at anterior fourth in right valve. Posterior cardinal angle distinct. Upper half of posterior margin steeply and concave, with a small spine near caudal projected area. Lower half of posterior margin broadly rounded with four denticulations and continuing to ventral margin smoothly.

Surface covered with strong ridges and reticulation. Anterior marginal ridge beginning at anterior part of dorsal margin, passing through eye tubercle, wavy in its middle, ending at anterior part of ventral margin. Ventral marginal ridge distinct, beginning at a little posterior to anterior marginal ridge in anteroventral area, slightly expanding at middle, and ending at posterior caudal area. Dorsal ridge gently arched. Posterodorsal ridge short and weak, parallel to posterodorsal margin. Other ridges running more or less radially from center to anterior and posterior marginal areas, three or four ridges toward anterior, and seven or eight toward posterior. Size of fossae small in anterocentral area, large in posterocentral area, and smallest in central area.

Eye tubercle distinct. Radial pore canals numerous, thickly spaced, straight, and inflated near middle Normal pore openings of sieve type with subcentral opening for sensory bristle. Some simple type pores opening on muri. Marginal infold moderate in width. Inner margin parallel to outer margin. Selvage distinct along free margin. List distinct along posterior margin. Three or four striae running parallel to inner margin. Anterior vestibule very shallow. Hinge holamphidont of *Aurila* type with crenulated median element. Muscle scars located a little anterior to middle, consisting of an inclined row of three frontal scars and a posteriorly inclined row of round adductor scars, of which dorsal and ventral scars single, dorsomedian and ventromedian scars distinctly divided.

Dimensions.—Measurements of specimens from sample 49-1, Toyonari Member of Kioroshi Formation are given below.

Sp	N	Me	$\bar{X}$	Sd	V	OR
LV	2	L	0.755			0.75–0.76
		H	0.480			0.47–0.49
RV	5	L	0.770	0.026	3.67	0.75–0.81
		H	0.462	0.016	3.56	0.45–0.49
C	4	L	0.823	0.060	7.38	0.75–0.89
		H	0.493	0.033	6.70	0.45–0.53
		W	0.433	0.050	11.53	0.36–0.47

Remarks.—This new species resembles *Aurila* (*Cimbaurila*) *cimbaeformis* (Sequenza, 1884), type species of subgenus *Cimbaurila* illustrated by Ruggieri, 1975, in general outline, reticulation pattern, and course of marginal ridges, but the reticulation is coarser in this species than *A.* (*C.*) *cimbaeformis*.

The new species is distinguishable from species of *Radimella* and *Malzella* by its divided dorsomedian and ventromedian adductor scars and absence of auxilliary denticles in the median hinge element and of posterior hinge element.

Occurence.—Common throughout all formations.

Aurila sp.

Pl. 13, fig. 5; text-fig. 15-1.

Illustrated specimen.—A left valve, UMUT-CA 9866 (Pl. 13, fig. 15; text-fig. 15-1. L, 0.68; H, 0.44), sample 120-1, Yabu Member of Yabu Formation.

Remarks.—The species is characterized by round auriform lateral outline, ventral inflation, and carapace surface covered entirely with large round punctae. *Aurila uranouchiensis* Ishizaki, 1968 from Recent mud and coarse-grained sand, Uranouchi, Kochi Prefecture, resembles this species in lateral outline but differs in the much smaller size of puncta.

Occurrence.—Rare in Yabu Member of Yabu Formation; also rare in Kiyokawa Formation. Altogether five left valves found.

Genus *Pseudoaurila* Ishizaki and Kato, 1976

Type-species.—Pokornyella japonica Ishizaki, 1968.

Diagnosis.—Aurilini with carapace auriform in lateral view. Surface finely reticulated, and reticulation arranged longitudinally. Radial pore canals numerous anteriorly, moderately spaced posteriorly, consisting of two frontal scars and adductor scars of which dorsal and dorsomedian scars divided and ventral scar undivided.

Remarks.—Ishizaki and Kato (1976) described the dorsal adductor scar as divided, but it seems to vary from undivided to clearly divided, showing several phases of subdivision. In Aurilini, *Pseudoaurila* is unique in the number of frontal scars.

Stratigraphic range.—Pleistocene to Recent.

Genus *Robustaurila* n. gen.

Type-species.—Cythereis assimilis Kajiyama, 1913.

Etymology.—From *robutus*+*auris* [Latin: "robust," "ear"].

Diagnosis.—Aurilini with carapace subquadrate in lateral view. Center flattened. Posterior caudal process distinct. Surface very coarsely costate with blunt ridges. Solid large subcentral node and winglike angulation of ventrolateral ridge distinct. Hinge without auxiliary denticles of median hinge element. Sieve-type pores on the floor of reticulation. Simple-type pores on muri.

Remarks.—The genus resembles *Mutilus* Neviani, 1928, in reticulate surface and angular posterodorsal margin, but differs in presence of subcentral node and radial ridges on reticulate surface. Reticulate ridges are not thin, sharp, and high but thick, blunt, and low.

Stratigraphic range.—Pleistocene to Recent.

Robustaurila assimilis (Kajiyama, 1913)

Pl. 13, figs. 6–8.

Mutilus assimilis: Hanai et al., 1977, p. 45; Okubo, 1980, p. 403–405, figs. 5, 7a, b, 11a-d.

Illustrated specimens.—A male carapace, UMUT-CA 9867 (Pl. 13, fig. 8. L, 0.62; H, 0.41; W, 0.32), sample 74-2, Kiyokawa Formation; a right male valve, UMUT-CA 9868 (Pl. 13, fig. 6. L, 0.59; H, 0.35), sample 74–2, and a left male valve, UMUT-CA 9869 (Pl. 13, fig. 7. L, 0.63; H, 0.41), sample 74-2.

Remarks.—The spines deviate considerably from the type species of *Mutilus*. However, type-species resembles *M. keiji* Ruggieri, 1962, illustrated by Doruk (1973) from the Tortonian of Italy but differs in having very thick ridges, posterodorsal earlike projection, and solid large subcentral node. Divided ventromedian adductor scar of Japanese species is different from that of *M. keiji*.

Tribe Urocythereidini Hartmann and Puri, 1974

Diagnosis.—Carapace elongate-subrectangular in lateral view. Surface coarsely pitted, reticulate, extension of distal surface of muri often creating small enclosure of fossae. Subcentral node present. Marginal infold moderately wide with numerous radial pore canals and without vestibule. Normal pore canals of sieve type with subcentral hair opening. Hinge holamphidont sometimes with lobate posterior tooth in right valve. Posterior element of left valve often having an auxiliary tooth in the ventral side of socket. Two or three frontal scars rarely making a row, median scar usually located anteriorly. Dorsomedian adductor scar always divided, ventromedian scar often divided, and ventral scar occasionally divided.

Remarks.—Urocythereidini differs from Aurilini in general outline and to some extent in arrangement of frontal scars. Urocythereidini resembles Orioninini also in general outline and certain aspects of ornamentation but differs by lack of pillar structure—that is, pillar-shaped fusion between inner carapace surface and free part of inner lamella and in disposition of frontal scars.

The tribe includes two groups of genera as follows: Group 1—*Urocythereis* Ruggieri, 1950, and *Muellerina* Bassiouni, 1965; Group 2—*Ambostracon* Hazel, 1962, *Australicythere* Benson, 1964, and *Patagonacythere* Hartmann, 1962.

Genus *Urocythereis* Ruggieri, 1950

Type-species.—*Cytherina favosa* Roemer, 1838.

Diagnosis.—Urocythereidini with coarsely reticulate surface and extension of distal surface of muri often creating small enclosure of fossae. Anterior marginal rim distinct. Anterior marginal groove, postjacent to anterior marginal rim, narrow but conspicuous. Two or three frontal scars and dorsomedian adductor scar usually divided, and ventromedian scar often divided. Ventral adductor scar usually undivided.

Remarks.—Pokorný (1955) described the hinge type of *Urocythereis* as merodont with a trend toward amphidont which may be seen in the enlargement of the anteromedian hinge element. Van Morkhoven (1963) discussed the hinge of *Urocythereis* starting that "the development of the anteromedian tooth in the left valve and the shape of the terminal teeth in the right valve (especially the anterior one) are shown by different authors to vary considerably." Recently, Doruk (1974) and Athersuch and Ruggieri (1975) showed that hinge varies even within one species: posterior tooth of right valve with or without lobation; posterior socket of left valve with or without an auxiliary tooth.

According to Howe (1961), adductor muscle scars vary within the genus from ventromedian paired to both median paired. Doruk (1974) and Athersuch and Ruggieri (1975) also showed a variation in one species in the number of frontal scars from two to three and in division of four adductor scars.

Stratigraphic range.—Miocene to Recent.

Urocythereis? *gorokuensis* Ishizaki, 1966

Pl. 13, fig. 14.

Remarks.—Hinge of this species is holamphidont without lobation in posterior tooth in the right valve. Frontal scars of specimens at hand are not observable, but those appearing in original description are two. The present specimen and paratype of *U.*? *gorokuensis* (Ishizaki, 1966, pl. 19, fig. 10, IGPS coll. cat. no. 87060, Pliocene Tatsunokuchi Formation) have corresponding characteristics: immature specimen with similar general outline, carapace size, and surface ornamentation. The specimen differs from the Recent specimens from Aomori Bay, northern Honshu, Japan, illustrated by Ishizaki (1971, p. 83, pl. 3, figs. 4, 5) in surface ornamentation.

The species is tentatively assigned to genus *Urocythereis* with a question mark, since it has ornamentation similar to the *Ambostracon-Patagonacythere* group classified into four groups based on the difference of surface ornamentation. The Japanese species has reticulate surface between the main ridges and is included in *A.* (*P.*) *tricostata* group of Valicenti (1977).

Stratigraphic range.—Eocene to Recent.

Tribe Orioninini Puri, 1974

Diagnosis.—Carapace elongate to subquadrate often with acute posterior margin in lateral view. Surface with two distinct horizontal ridges, a little anteriorly inclined, the rest of surface mostly smooth, sometimes reticulate. Pillar structure between inner carapace surface and free part of inner lamella distinct. Muscle scars consisting of three frontal scars and single dorsal, paired dorsomedian, paired ventromedian, and single ventral adductor scars.

Remarks.—Pillar structure is known only in this tribe. The tribe includes the following genera: *Orionina* Puri, 1954; *Caudites* Coryell and Fields, 1937; *Palaciosa* Hartmann, 1959; and *Anterocythere* McKenzie and Swain, 1967.

Genus *Caudites* Coryell and Fields, 1937

Type-species.—Caudites medialis Coryell and Fields, 1937.

Diagnosis.—Orioninini with subquadrate to subtriangular carapace in lateral view. Posterior caudal process acutely protruding. Sides compressed in dorsal view. Surface with strong ridges. Anterior marginal and ventral ridges strong. A strong ridge running from anteroventral corner to posterodorsal corner. Lateral surface between ridges mostly smooth, sometimes reticulate. Muscle scars consisting of three or four frontal scars and often divided dorsal, divided dorsomedian and ventromedian, and a single ventral adductor scar.

Remarks.—Pokorný (1971) studied nine species of *Caudites* from Recent of the Galapagos Islands in detail. He illustrated intraspecific and interspecific variation in number and pattern of muscle scars of genus *Caudites*. He explained the surface ornamentation of *Caudites* by a uniform descriptive terminology.

Stratigraphic range.—Miocene to Recent.

Caudites? posterocostatus (Ishizaki, 1966)

Pl. 13, figs. 12, 13; text-fig. 15-2.

Hermanites? posterocostatus: Hanai et al., 1977, p. 48.

Remarks.—Posteroventral margin of the present specimen is broadly rounded with two blunt spines. Marginal pore canals are numerous, straight, with a bulbous enlargement. Muscle scars are located in subcentral tubercle and consist of three frontal scars and a vertical row of four adductor scars, of which dorsomedian and ventromedian and ventral scars are elongate and obliquely extended. Characteristic posterodorsal sinuation present in lateral view. Carapace is compressed laterally in dorsal view. Skeletal structure species to genus *Caudites*. Single complete, tightly closed carapace does not permit first-hand observation of internal features of carapace.

The species resembles *Caudites javanus* Kingma, 1948, from Recent sand of Manila, illustrated by Key [Keij] (1954), in reticulate laternal surface and disposition of minor ridges. *C.? posterocostatus* is larger than *C. javanus* and subcentral tubercle is more distinct in *C. javanus*. Species of *Caudites* described by Ohmert, 1971, from the Upper Pliocene of Chile contain a few forms resembling this species.

Subfamily Thaerocytherinae Hazel, 1967

Diagnosis.—Carapace subtriangular to subquadrate in lateral view, subrhomboidal or subovate in dorsal view. Surface heavily ornamented by reticulation, costation, and several nodes. Subcentral node distinct. Marginal infold moderately broad with moderate to numerous number of marginal pore canals. Vestibule usually absent. Normal pore canals of sieve type with subcentral opening on sola and of simple type on muri. Hinge strong, amphidont. Muscle scars consisting of two frontal scars and a nearly vertical row of four adductor scars of which dorsomedian and ventromedian scars elongated. Dorsal and dorsomedian scars sometimes divided.

Remarks.—Benson (1972) elevated subfamily Thaerocytherinae to the family level, because the soft parts are only in part similar to those of the Hemicytheridae and because numerous genera have quite long geologic ranges indicating separate evolution. However, the soft parts of species of *Bradleya* and *Poseidonamicus* seem to be close to those of hemicytherids. Hartmann and Puri (1974) mentioned in the description of Bradleyini that "only the character 'respiratory of maxilulla' with special process for first 'Strahl' may be trachyleberidid."

The subfamily includes Thaerocytherini, Bradleyini, and Coquimbini.

Tribe Thaerocytherini Hazel, 1967

Diagnosis.—Carapace subquadrate or rectangular with caudate posterior in lateral view, subrhomboidal in dorsal view. Surface reticulate or coarsely ornamented. Posterodorsal and ventrolateral ridges distinct. Muscle scar node distinct. Lower frontal scar sometimes U-shaped.

Remarks.—This tribe includes the following genera: *Thaerocythere* Hazel, 1967; *Hornibrookella* Moos, 1952; *Quadracythere* Hornibrook, 1952; *Hermanites* Puri, 1955; *Jugosocythereis* Puri, 1957; *Limburgina* Deroo, 1966; and *Cornucoquimba* Ohmert, 1968.

Genus *Cornucoquimba* Ohmert, 1968

Types-species.—Cornucoquimba aligera Ohmert, 1968.

Diagnosis.—Thaerocytherini with distinct subcentral node, posterodorsal tubercle, and ventrolateral ala-like ridge terminating posteriorly with posteroventral tubercle. Posterodorsal tubercle located posterior to posteroventral tubercle. Anterior and posteroventral vestibules usually present. Characteristic muscle scars on inner wall of subcentral pit consisting of two frontal scars and anteriorly inclined row of four adductor scars, of which dorsomedian scar clearly divided and ventromedian scar sometimes divided.

Remarks.—Characteristic dorsal outline of carapace, distinct ventrolateral alae, presence of subcentral node, and of posterodorsal tubercle suggest that the genus is related to Thaerocytherini, in spite of the apparent difference in muscle scar pattern and in development of anterior and posteroventral vestibules.

Adductor muscle scars were described in original description (Ohmert, 1968) as consisting of two upper vertically set scars and three lower obliquely set scars. This adductor muscle scar pattern, however, seems to be explicable as an oblique row of four adductor scars seen in ordinary thaerocytherines.

Stratigraphic range.—Eocene to Oligocene, Pliocene to Recent.

Tribe Coquimbini Ohmert, 1968

Diagnosis.—Carapace moderate to small in size, subquadrate in lateral view and subovate in dorsal view. Posterior bluntly angular with several denticules along posteroventral margin. Anterior marginal rim, subcentral tubercle, and posterodorsal and posteroventral nodes present.

Anterior and posteroventral vestibules usually developed. Muscle scars on wall of subcentral pit, consisting of two frontal scars and anteriorly inclined adductor scars of which dorsomedian scar is clearly divided and ventromedian scar is obliquely elongate and sometimes divided.

Remarks.—This tribe includes *Coquimba* Ohmert, 1968; *Nanocoquimba* Ohmert, 1968; and questionably *Puriana* Coryell and Fields, 1953.

Genus *Coquimba* Ohmert, 1968

Type-species.—*Coquimba hermi* Ohmert, 1968.

Diagnosis.—Coquimbini with distinct anterior marginal rim, anterior marginal groove, ventral ridge terminating in posteroventral tubercle, subcentral node, posterodorsal tubercle, and posterior flat platform. Eye tubercle distinct.

Remarks.—Ornamentation of general surface varies among species from celation through punctation of reticulation. In some species, extension of distal surface of muri often creates small enclosure of fossae, giving appearance similar to that of certain species of *Urocythereis*.

Stratigraphic range.—Eocene to Oligocene, Pliocene to Recent.

Family Bythocytheridae Sars, 1926

Subfamily Bythocytherinae Sars, 1926

Genus *Bythoceratina* Hornibrook, 1952

Bythoceratina sp.

Pl. 13, fig. 16

Illustrated specimens.—A right valve, UMUT-CA 9875 (Pl. 13, fig. 16. L, 0.65; H, 0.35), sample 305-4, Yabu Member of Yabu Formation.

Remarks.—A species of the genus *Bythoceratina* characterized by a lateral ala starting from anteroventral area, running posteroventrally and then turning in posterodorsal direction, forming an angulate tip of wing and ending at caudal process. Wing slightly obscures posteroventral contact margin in lateral view. Dorsal ridge runs parallel to dorsal margin in lateral view, bordering wide flat dorsal area of carapace. Surface is evenly punctated.

The species resembles *Bythoceratina* sp. (originally *Monoceratina* sp. A) from Celebes Sea (depth 1,299 m) illustrated by Key [Keij], 1953, in lateral outline and shape of lateral alae, but differs in details of surface ornamentation.

Occurrence.—Rare in Yabu Member of Yabu Formation; also rare in Kioroshi Memberof Kioroshi Formation. Three right valves and an immature left valve found in all samples.

Genus *Pseudocythere* Sars, 1865

Pseudocythere frydli n. sp.

Pl. 13, fig. 15; text-figs. 16-7, 8.

Etymology.—Named in honor of Paul M. Frydl, University of Tokyo.

Types.—Holotype, a carapace, UMUT-CA 9876 (Pl. 13, fig. 15. L, 0.76; H, 0.38; W, 0.39), sample 262-1, Loc. 262, Kioroshi Member of Kioroshi Formation.

Illustrated specimens.—A broken left valve, UMUT-CA 9877 (text-fig. 16-7. H, 0.34), sample 120-3, Yabu Member of Yabu Formation; a broken right valve, UMUT-CA 9878 (text-fig. 16-8), sample 305-5, Yabu Member of Yabu Formation.

Diagnosis.—A large and elongate ovate species, tentatively assigned to the genus *Pseudocythere*, characterized by anterior smooth surface and approximately 16 ridges on posterior surface running nearly parallel to ventral margin and converging to the area just below the posterior caudal process.

Description.—Viewed laterally, carapace oblong to ovate, highest at anterior cardinal angle. Anterior margin broadly rounded. Dorsal margin straight. Posterodorsal caudal process distinct. ventral margin downwardly arched. Viewed dorsally, carapace subovate and slightly inflated in ventrocentral area. Surface smooth in anterior half. Approximately 16 weak ridges running nearly parallel to ventral margin and converging to area just below posterodorsal caudal process. Weak and shallow groove running along anterior margin. Marginal infold wide anteriorly and moderate ventrally and posteriorly. Inner margin nearly parallel to outer free margin. Vestibule moderately deep along anterior margin and shallow ventrally and posteriorly. Selvage weak, not sinuate at middle of ventral margin. Hinge straight lophodont, long median element smooth in left valve, and slightly crenulate in right valve. Muscle scars consisting of a vertical row of five adductor scars, two large frontal scars and one mandibular scar. Dorsal and ventral adductor scars crescent, median three elongated and set closely to each other.

Dimensions.—Measurements of two specimens are as follows.

Sp	Sa	Ho	N	Me	OR
C	262-1	Kioroshi M., Kioroshi F.	1	L	0.76
				H	0.38
				W	0.39
RV	120-3	Yabu M., Yabu F.	1	L	0.78
				H	0.34

Remarks.—All of the characters mentioned above suggest that the specimens clearly belong to a new species. This species resembles *Pseudocythere fuegiensis* Brady, 1980, from Challenger St. 311 (Lat. 52°50′S, Long. 73°53′W) of South Chile (depth 245 fathoms; mud bottom), in general outline and surface ornamentation, but this species is shorter and smaller than *P. fuegiensis*. From the Pacific species of *Baltrella* reported by Keij (1979a), the species is easily distinguished by lack of internal rib and sinuation at the middle of ventral margin as well as by surface ornamentation.

Occurrence.—Rare in Yabu Member of Yabu Formation; also rare in Kioroshi Member of Kioroshi Formation. Altogether seven specimens found in all samples.

Family Cytheruridae G. W. Müller, 1894

Subfamily Cytherurinae G. W. Müller, 1894

Genus *Eucytherura* G. W. Müller, 1894

Eucytherura utsusemi n. sp.

Pl. 14, figs. 12, 13, 15, 16.

Types.—Holotype, a right valve, UMUT-CA 9879 (Pl. 14, figs. 12, 15. L, 0.28; H, 0.15), sample 120-3, Loc. 120, Yabu Member of Yabu Formation.

Illustrated specimens.—A left valve, UMUT-CA 9880 (Pl. 14, fig. 16. L, 0.33; H, 0.18), sample 305-2, Yabu Member of Yabu Formation); a carapace, UMUT-CA 9881 (Pl. 14, fig. 13. L, 0.27; H, 0.16; W, 0.18), sample 120-2, Yabu Member of Yabu Formation.

Diagnosis.—*Eucytherura* with small carapace characterized by obliquely and broadly rounded anterior margin, straight and posteriorly inclined dorsal margin rectangular caudal margin, straight and slightly sinuated ventral margin. Lateral outline of posteroventral corner of wing smoothly rounded. Anterior marginal ridge distinct.

Description.—Carapace very small and robust, subrhomboidal in lateral outline. Highest at anterior cardinal angle. Anterior margin obliquely and broadly rounded. Dorsal margin straight and slightly sinuate at middle with its posterior half obscured by posteroventral winglike inflation. Posterior margin with a small rectangular caudal area in its upper half and obliquely truncate in its lower half. Viewed dorsally, carapace subtriangular, winglike inflation distinct, and dorsocentral area slightly sulcate.

Surface, including surface of winglike posteroventral projection, reticulated. Reticula deep, with sunken sieve plate in some fossae in posterodorsal area. Anterior marginal ridge distinct. Ventral weak ridges running from anteroventral area obliquely to posteroventral margin. Reticulum bifoliate, especially in dorsal area.

Eye tubercle distinct. Radial pore canals numerous along anterior margin, sparse along posterior margin. Normal pores located at crossings of reticulation ridges, of simple type with a very small pore opening beside normal pore opening. Marginal infold wide anteriorly and posteriorly. Inner margin parallel to outer margin. Line of concrescene coincides with inner margin. Selvage distinct. Hingement lophodont. Muscle scars located on the extension of weak median sulcus, consisting a backwardly inclined vertical row of four adductor scars and a quadrilobate frontal scar.

Dimensions.—Measurements of pooled specimens are below.

Sp	Sa	Ho	N	Me	OR
LV	203-2	Kamiiwahashi F.	1	L	0.29
				H	0.17
C	138–1	Kiyokawa F.	2	L	0.30
				H	0.16–0.18
				W	0.18–0.20
C	120-1	Yabu M., Yabu F.	2	L	0.31
				H	0.18
				W	0.18–0.19
C	120-2	Yabu M., Yabu F.	2	L	0.27–0.30
				H	0.15–0.16
				W	0.18–0.19
RV	120-3	Yabu M., Yabu F.	1	L	0.28
				H	0.15
C A-1	120-3	Yabu M., Yabu F.	1	L	0.16
				H	0.10
				W	0.11
LV	305-2	Yabu M., Yabu F.	1	L	0.33
				H	0.18
C	305-4	Yabu M., Yabu F.	1	L	0.29
				H	0.15
				W	0.17

Remarks.—This species resembles *Eucytherura neoalae* (Ishizaki, 1966) from Miocene Hatatate Formation of Sendai, Japan, in general outline but differs in its acute angulation of the tip of winglike posteroventral projection.

Occurrence.—Rare in Yabu Member of Yabu Formation; Rare in Kiyokawa Formation; also rare in Kamiiwahashi Formation.

Genus *Semicytherura* Wagner, 1957

Semicytherura wakamurasaki n. sp.

Pl. 14, figs. 1–8, 17; text-figs. 16-1, 2.

Types.—Holotype, a left female valve, UMUT-CA 9882 (Pl. 14, fig. 6. L, 0.36; H, 0.18), sample 190-1, Loc. 190, Kioroshi Member of Kioroshi Formation.

Illustrated specimens.—A right female valve, UMUT-CA 9883 (Pl. 14, figs. 3, 17. L, 0.36; H, 0.18), sample 190-1; a right male valve, UMUT-CA 9884 (Pl. 14, fig. 1. L, 0.33; H, 0.15), sample 190-1; a left female valve, UMUT-CA 9885 (Pl. 14, fig. 7; text-fig. 16-1. L. 0.34; H, 0.15), sample 190-1; a right female valve, UMUT-CA 9886 (Pl. 14, fig. 8. L, 0.36; H, 0.18), sample 190-1; a left male valve, UMUT-CA 9887 (Pl. 14, fig. 2. L, 0.32; H, 0.18), sample 271-2, Kioroshi Member of Kioroshi Formation; a male carapace, UMUT-CA 9888 (Pl. 14, fig. 4. L, 0.32; H, 0.15; W, 0.15), sample 271-2; a female carapace, UMUT-CA 9889 (Pl. 14, fig. 5. L, 0.37; H, 0.21; W, 0.16), sample 271-2; a right female valve, UMUT-CA 9890 (text-fig. 16-2. L, 0.34; H, 0.18), sample 189-3, Kioroshi Member of Kioroshi Formation.

Diagnosis.—A small species of genus *Semicytherura* characterized by elongate subovate lateral outline with protrudent caudal process, finely pitted surface, and by numerous fine longitudinal ridges parallel to dorsal and ventral margins.

Description.—Viewed laterally, carapace elongate subovate, highest at middle. Anterior margin obliquely and narrowly rounded. Dorsal margin gently arched. Ventral margin straight, very slightly sinuate at middle. Posterior caudal process developed in middle of posterior margin.

Surface finely pitted in posterior and central area,, and without pits in anterodorsal area. Fine longitudinal ridges running parallel to ventral and dorsal margins and converging toward anteroventral area. Weak ridges running obliquely between longitudinal ridges in anterodorsal area and vertically in central area. Area of posterior caudal process compressed, not pitted, but coarsely reticulate.

Eye tubercle obscure. Normal pore opening of simple type with lip. Radial pore canals about ten along anterior margin, wavy, and widely spaced. Marginal infold moderately wide anteriorly and narrow along posteroventral margin. Hinge of *Cytherura* type lophodont. Anterior and posterior elements of right valve represented by slight swelling at terminal selvage. Median element smooth. Hingement of left valve complementary. Muscle scars located in central area of carapace, consisting at least of a vertical row of four ovate adductor scars.

Sexual dimorphism distinct. Male form more slender in lateral view and more plump in dorsal view.

Dimensions.—Measurements of specimens from sample 251-1, Kioroshi Member of Kioroshi Formation are given below:

Sp	N	Me	$\bar{X}$	Sd	V	OR
RV ♂	2	L				0.31–0.33
		H				0.14–0.15
LV ♀	2	L				0.33–0.34
		H				0.15
RV ♂	2	L				0.35–0.36
		H				0.18
LV ♀	4	L	0.340	0.022	6.35	0.32–0.37
		H	0.175	0.006	3.30	0.17–0.18

Remarks.—Inner margin is not S-shaped in posterior area. Diagnostic character selected by the original author of genus *Semicytherura* seems to be no longer diagnostic in this species.

The species resembles *Semicytherura? miurensis* (Hanai, 1957) in general outline and surface ornamentation, but is smaller in carapace size and not translucent. Despite wide separation of sampling localities, the species is conspicuously similar to *Cytherura vestibulata* Hall, 1965, illustrated by Swain and Kraft, 1975, (Pl. 1, figs. 4a, b, 5a–c, pl. 2, figs. 1a–d) from tidal bay of southern Delaware. Similarities include general outline, carapace size, and surface ornamentation, but in detail, the species differs from *Cytherura vestibulata* in the higher position of posterior caudal process.

Occurrence.—Rare in Yabu Member and Kamiizumi Member of Yabu Formation; common in Kiyokawa Formation; common in sandy silt facies,and rare in sand facies of Kammiiwahashi Formation; rare in Toyonari Member; common in Kioroshi Member of Kioroshi Formation.

Semicytherura sp.
Pl. 14, fig. 9.

Illustrated specimen.—A left valve, UMUT-CA 9891 (Pl. 14, fig. 9. L, 0.35; H, 0.18), sample 305-1, Yabu Member of Yabu Formation.

Remarks.—The species resembles *Semicytherura skippa* (Hanai, 1957) in carapace size, general outline, and distribution of strong ridges but differs in having higher carapace and fine secondary reticulation between strong ridges.

Occurrence.—Rare in Yabu Member of Yabu Formation; rare in Kiyakawa Formation; also rare in Toyonari Member of Kioroshi Formation. Altogether five valves found.

Subfamily Cytheropterinae Hanai, 1957
Genus *Kangarina* Coryell and Fields, 1937
Kangarina hayamii n. sp.
Pl. 14, figs. 10, 11, 14.

Etymology.—Named in honor of Itaru Hayami, University Museum, University of Tokyo.

Types.—Holotype, a carapace, UMUT-CA 9892 (Pl. 14. fig. 10. L, 0.36; H, 0.23; W, 0.20), sample 120-3, Loc. 120, Yabu Member of Yabu Formation.

Illustrated specimens.—A left valve, UMUT-CA 9893 (Pl. 14, fig. 14. L, 0.35; H, 0.22), sample 120-1, Yabu Member of Yabu Formation; a right valve, UMUT-CA 9894 (Pl. 14, fig. 11. L, 0.34; H, 0.23), sample 273-2, Kioroshi Member of Kioroshi Formation.

Diagnosis.—A small species of the genus *Kangarina* characterized by subtrapezoidal dorsal outline, well-developed ventrolateral alae without posterior small spine.

Description.—Carapace small and thick; subrhomboidal in lateral view; highest at middle. Anterior margin obliquely and narrowly rounded. Dorsal margin strongly arched, slightly concave in posterior half. Ventral contact margin concave in anterior half and convex in posterior half. Posterior half of ventral contact margin obscured by strongly overhung lateral area. Posterior caudal process triangular, distinct and compressed. Viewed dorsally, carapace subtrapezoidal, widest at middle, tapering steeply toward anterior and posterior. Projection of lateral geniculate alae giving appearance of truncate posterior margin with projection of caudal process in dorsal view.

Dorsal marginal ridge strong, running from anterior cardinal angle to posterior cardinal angle, then turning downward and ending in poteroventral area. Ventrolateral ridge strong, starting from anteroventral area toward posterior forming edge of without distinct spine. From node, short ridges running radially; one running horizontally to anteroventral margin, one running to anterior cardinal angle, one running to posterior cardinal angle, and one running downward to middle of ventrolateral ridge. Space between ridges finely punctate.

Eye tubercle obscure. Type of normal pore opening unknown because of ill-preserved surface. Radial pore canals long, simple, moderate in number along anterior margin, long and sparse along posterior margin. The longest one bifurcated. Marginal infold wide anteriorly, narrow posteriorly. Line of concrescence coincides with inner margin along anterior and ventral margins but is separated along posteroventral margin. Posteroventral vestibule narrow.

List distinct. Hinge merodont, of *Cytherura* type. Muscle scars consisting at least of a vertical row of four adductor scars.

Dimensions.—Measurements of pooled specimens are given below.

Sp	Sa	Ho	N	Me	OR
RV	273-1	Kioroshi M., Kioroshi F.	1	L	0.34
				H	0.23
LV	120-1	Yabu M., Yabu F.	1	L	0.35
				H	0.22
RV	120-2	Yabu M., Yabu F.	1	L	0.40
				H	0.26
C	120-3	Yabu M., Yabu F.	1	L	0.36
				H	0.23
				W	0.20
LV	305-2	Yabu M., Yabu F.	2	L	0.38
				H	0.23–0.24
LV	305-4	Yabu M., Yabu F.	1	L	0.33
				H	0.21
LV	305-5	Yabu M., Yabu F.	1	L	0.34
				H	0.23
RV	305-5	Yabu M., Yabu F.	1	L	0.37
				H	0.23

Remarks.—This species is similar to "*Kangarina*" sp. A reported by Valentine (1976, Pl. 5, figs. 12, 13) from the Holocene shelf deposits of Santa Monica Bay, west coast of North America, in general outline and surface ornamentation, but differs in lack of fine ridges on ventrolateral ridge and on posterodorsal caudal process. This species also resembles *Kangarina abyssicola* G. W. Müller, 1894) illustrated by G. W. Müller (1894, p. 302, 303, Pl. 20, fig. 5) and later by Puri, Bonaduce and Malloy (1964, p. 163) from Recent sediment of Gulf of Naples in general outline, but ventrolateral ridge running on the edge of ala ends without leaving any distinct spine posteriorly in this species.

Occurrence.—Rare in Yabu Member of Yabu Formation; also rare in Kioroshi Member of Kioroshi Formation.

Family Loxoconchidae Sars, 1925

Genus *Loxoconcha* Sars, 1866

Loxoconcha hanachirusato n. sp.

Pl. 11, figs. 1–4.

Types.—Holotype, a male right valve, UMUT-CA 9895 (Pl. 11, fig. 2. L, 0.44; H, 0.24), sample 189-3, Loc. 189, Kioroshi Member of Kioroshi Formation.

Illustrated specimens.—A female right valve, UMUT-CA 9896 (Pl. 11, fig. 4. L, 0.44; H, 0.26); a male left valve, UMUT-CA 9897 (Pl. 11, fig. 1. L, 0.44; H, 0.24); a female left valve, UMUT-CA 9898 (Pl. 11, fig. 3. L, 0.45; H, 0.28), sample 189-3.

Diagnosis.—A small species of genus *Loxoconcha*, characterized by finely punctated lateral ornamentation. Punctae small and surface becoming smooth toward dorsal half of carapace.

Description.—Viewed laterally, carapace subrhomboidal, highest at anterior cardinal angle. Anterior margin broadly and slightly obliquely rounded. Dorsal margin nearly straight. Anterior cardinal angle not distinct. Posterior fourth of dorsal margin slightly sinuated in left valve. Ventral margin nearly straight and parallel with dorsal margin. Ventral contact margin slightly obscured by poteroventral inflation of carapace. Posterior margin narrowly rounded without distinct caudal process. Viewed dorsally, carapace subovate, sides subparallel with acute anterior and blunt posterior taperings. Posterior end protruding slightly.

Surface finely reticulate ventrally and finely punctate dorsally. Reticulation gradually changing upward to punctation and finally to smooth dorsal surface. Ventral marginal ridges weak.

Eye tubercle distinct. For radial pore canals, normal pores, marginal infold, hingement, and muscle scars, refer to description of genus.

Sexual dimorphism distinct. Male carapace slender and female higher than male.

Dimensions.—Measurements of specimens from sample 189-3 of Kioroshi Member of Kioroshi Formation are given below.

Sp	N	Me	$\bar{X}$	Sd	V	OR
LV ♂	4	L	0.428	0.010	2.34	0.42–0.44
		H	0.239	0.003	1.26	0.24
LV ♀	8	L	0.433	0.015	3.46	0.41–0.45
		H	0.262	0.014	5.29	0.24–0.28
RV ♂	2	L	0.425			0.41–0.44
		H	0.228			0.22–0.24
RV ♀	10	L	0.475	0.013	3.11	0.40–0.44
		H	0.247	0.008	3.24	0.24–0.25
C ♂	3	L	0.420			0.40–0.45
		H	0.237			0.23–0.24
		W	0.222			0.22
C ♀	1	L				0.41
		H				0.24
		W				0.24

Remarks.—The species resembles *Loxoconcha laeta* Ishizaki, 1968, in lateral outline but differs in surface ornamentation and smaller carapace.

Occurrence.—Common in Yabu Member and rare in Kamiizumi Member of Yabu Formation; rare in Kiyokawa Formation; rare in sand facies of Kamiiwahashi Formation; abundant in Kioroshi Member of Kioroshi Formation.

Loxoconcha tamakazura n. sp.

Pl. 11, figs. 16, 17.

Types.—Holotype, a left valve, UMUT-CA 9903 (Pl. 11, fig. 17. L, 0.59; H, 0.40), sample 120-1, from Loc. 120, Yabu Member of Yabu Formation.

Illustrated specimens.—A right valve, UMUT-CA 9904 (Pl. 11, fig. 16. L, 0.57; H, 0.40), sample 120-1.

Diagnosis.—A large and round species of genus *Loxoconcha* characterized by inflated carapace, coarse and uniform reticulation, and three posteroventral ridges.

Description.—Viewed laterally, carapace large, subrhomboidal. Anterior margin obliquely and broadly rounded. Dorsal margin arched and subparallel to ventral margin. Ventral margin convex. Posterior margin protrudent at slightly above middle, forming triangular caudal process. Ventral half of posterior margin broadly rounded. Viewed dorsally, carapace subovate with a narrow projection of frill-like rim along anterior margin.

Surface coarsely and evenly reticulated. Each fossa deep. Three weak posteroventral ridges running parallel to posteroventral margin. Short spine tending to develop at posteroventral corner.

Eye tubercle prominent at a little below anterior cardinal angle. Radial pore canals straight, widely spaced. Normal pores of sieve type, opening at intersections of partition walls of reticulation.

Marginal infold moderately wide. Inner margin parallel to outer margin, except for posteroventral margin where inner margin runs rather straight and obliquely from middle of posterior margin to middle of ventral margin. Selvage distinct. Two or three striations running parallel to anterior margin. Vestibule absent. Hinge gongylodont, dorsal margin hanging over dorsal median groove in right valve. Muscle scar field central. Scars consisting of a round frontal scar and a vertical row of four ovate adductor scars.

Dimension.—Measurements of specimens from sample 135-3, Yabu Member of Yabu Formation are given below.

Sp	N	Me	$\bar{X}$	Sd	V	OR
LV ♂	2	L	0.620			0.62
		H	0.425			0.42–0.43
LV ♀	4	L	0.573	0.015	2.62	0.56–0.59
		H	0.403	0.013	3.12	0.39–0.42
RV ♂	2	L	0.605			0.60–0.61
		H	0.400			0.39–0.41
RV ♀	2	L	0.560			0.56
		H	0.375			0.37–0.38

Remarks.—The species resembles *Loxoconcha tosaensis* Ishizaki, 1968, in carapace size and lateral outline, but differs distinctly in surface ornamentation.

Occurrence.—Abundant in Yabu Member and common in Kamiizumi Member of Yabu Formation; also common in Kiyokawa Formation.

Genus *Nipponocythere* Ishizaki, 1971

Nipponocythere bicarinata (Brady, 1880)

Pl. 13, figs. 1–4.

Cythere bicarinata: Puri and Hulings, 1976, p. 269, pl. 10, figs. 12, 13.

Nipponocythere delicata Ishizaki and Kato, 1976, p. 134, 136, pl. 3, figs. 2–6 [Type: Diluvium Furuya Mud, Shizuoka Pref.; IGPS-91742].

Remarks.—Puri and Hulings (1976, p. 269) designated a carapace (BMNH 80.38.50) as lectotype.

As shown in Pl. 13, figs. 1–4, surface ornamentation of this species varies from those with punctation on entire surface, through those with punctation on restricted areas of surface, to those with smooth surface. In original description of *Nipponocythere delicata* Ishizaki and Kato (1976), the authors wrote that *N. delicata* differs from *N. asamushiensis* (conspecific with *N. bicarinata* according to Hanai et al., 1977) "in having a thinner but more inflated carapace and a more restricted ditribution of lateral ornamentation with small pits." It is, however, highly probable that the surface ornamentation of type specimens of *N. delicata* falls within the range of variation of *N. bicarinata*.

Family Xestoleberididae Sars, 1928

Genus *Xestoleberis* Sars, 1866

Xestoleberis suetsumuhana n. sp.

Pl. 15, figs. 11, 12; text-figs. 16-5, 6.

Types.—Holotype, a left valve, UMUT-CA 9913 (Pl. 15, fig. 12; text-fig. 16-5. L, 0.64; H, 0.35), sample 138-5, Loc. 138, Kamiizumi Member of Yabu Formation.

Illustrated specimen.—A right valve, UMUT-CA 9914 (Pl. 15, fig. 11; text-fig. 16-6. L, 0.68; H, 0.37), sample 284-3, Yabu Member of Yabu Formation.

Diagnosis.—*Xestoleberis* moderate in size, characterized by flattened venter, and strong lateral inflation, yet with kidney-shaped lateral outline common to many species of *Xestoleberis*.

Description.—Viewed laterally, carapace moderate in size, kidney-shaped, and highest at middle. Anterior margin obliquely and narrowly rounded. Dorsal margin gently arched. Ventral margin sinuate in its anterior half and straight in its posterior half. Ventral part of posterior margin gently arched and dorsal part of posterior margin narrowly rounded. Posterior margin narrowly rounded. Posterior margin truncated vertically in its middle. Carapace inflated laterally and with strongly flattened venter.

Xestoleberis spot distinct, large and long. Normal pores of sieve type. Radial pore canals simple, short, evenly spaced along anterodorsal, ventral, and posterior margins, and dense along anteroventral margin. Marginal infold wide along anterior and mid-ventral margins.

Sp	Sa	Ho	N	Me	OR
RV	49–1	Toyonari M., Kioroshi F.	1	L	0.67
				H	0.36
RV	73–4	Kiyokawa F.	1	L	0.68
				H	0.34
RV	74–2	Kiyokawa F.	1	L	0.62
				H	0.31
LV	138–5	Kamiizumi M., Yabu F.	1	L	0.64
				H	0.35
LV	305–4	Yabu M., Yabu F.	1	L	0.66
				H	0.36
RV	284–3	Yabu M., Yabu F.	1	L	0.68
				H	0.37
RV A-1	6–1	Toyonari M., Kioroshi F.	1	L	0.52
				H	0.28
RV A-1	73–4	Kiyokawa F.	1	L	0.48
				H	0.26
RV A-1	74–2	Kiyokawa F.	1	L	0.53
				H	0.29
LV A-1	138–1	Kiyokawa F.	1	L	0.52
				H	0.29
RV A-1	305–2	Yabu M., Yabu F.	1	L	0.57
				H	0.31
RV A-1	305–5	Yabu M., Yabu F.	1	L	0.55
				H	0.28
RV A-1	284–3	Yabu M., Yabu F.	1	L	0.58
				H	0.30

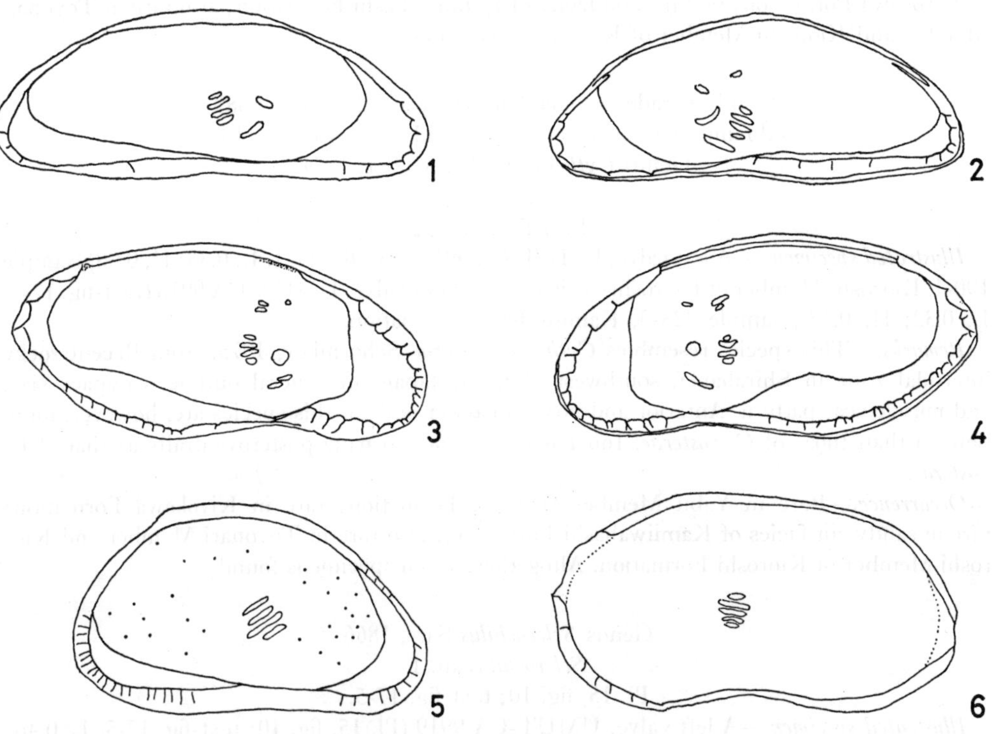

Text-fig. 17

Internal views. left row (odd number): left valves; right row (even number): right valves.

Figs. 1, 2. *Cytherois* sp. 1, CA 9917 (sample 190-1, Kioroshi Member of Kioroshi Formation). x 159. 2, CA 9918 (sample 228-3, Kamiiwahashi Formation). X 179.

Figs. 3, 4. *Paracytheroma* sp. 3, CA 9921 (sample 228-2, Kamiiwahashi Formation). x 107. 4, CA 9922 (sample 6-3, Toyonari Member of Kioroshi Formation). x 109.

Fig. 5. *Sclerochilus* sp. 1. CA 9919 (sample 308-1, Kiyokawa Formation). x 107.

Fig. 6. *Sclerochilus* sp. 2. CA 9920, a right immature valve (sample 272-1, Kioroshi Member of Kioroshi Formation). x 100.

Anterior vestibule crescent in lateral view and moderately deep. Hinge of *Xestoleberis* type merodont. Muscle scars located slightly anterior to middle, consisting of a vertical row of four large ovate adductor scars and a large round frontal scar.

Dimensions.—Measurements of pooled specimens are given.

Remarks.—This species resembles *Xestoleberis* sp. B reported by Ishizaki (1971, p. 42, pl. 6, fig. 16) from fine-grained sand bottom of Mutsu Bay, northern Honshu, in having similar outline and strongly flattened venter, but the species has less protrudent posteroventral margin. This species also resembles *Xestoleberis setouchiensis* Okubo, 1979, in having alaelike lateral inflation but is larger than *X. setouchiensis*.

Occurrence.—Rare in Yabu Member and Kamiizumi Member of Yabu Formation; rare in Kiyokawa Formation; rare in sand facies of Kamiiwahashi Formation; also rare in Toyonari Member and Kioroshi Member of Kioroshi Formation.

Family Paradoxostomatidae Brady and Norman, 1889
Subfamily Paradoxostomatinae Brady and Norman, 1889
Genus *Cytherois* G. W. Müller, 1884
Cytherois sp.
Text-figs. 17-1, 2.

Illustrated specimens.—A left valve, UMUT-CA 9917 (text-fig. 17-1. L, 0.38; H, 0.08), sample 190-1, Kioroshi Member of Kioroshi Formation; a right valve, UMUT-CA 9918 (text-fig. 17-2. L, 0.32; H, 0.12), sample 228-3, Kamiiwahashi Formation.

Remarks.—This species resembles *Cytherois zosterae* Schornikov, 1975, from Recent rocky intertidal zone in Shirahama, southwest Honshu, Japan, in general outline, carapace size, and muscle scar pattern. Anterior and posterior outer ends of this species are, however, more pointed than those of *C. zosterae*. Inner margin is not so wide posteroventrally as that of *C. zosterae*.

Occurrence.—Rare in Yabu Member of Yabu Formation; rare in Kiyokawa Formation; rare in sandy silt facies of Kamiiwahashi Formation; also rare in Toyonari Member and Kioroshi Member of Kioroshi Formation. Altogether eleven specimens found.

Genus *Sclerochilus* Sars, 1866
Sclerochilus sp. 1
Pl. 15, fig. 10; text-fig. 17-5.

Illustrated specimen.—A left valve, UMUT-CA 9919 (Pl. 15, fig. 10; text-fig. 17-5. L, 0.46; H, 0.27), sample 308-1, Kiyokawa Formation.

Remarks.—This species resembles *Sclerochilus meridionalis* G. W. Müller, 1908, from Recent Antarctic Sea in general outline, but differs in having smaller size of carapace. Muscle scars are located at a little anterior to the middle and consist of a vertical row of four horizontally elongate large adductor scars.

Occurrence.—Rare in Kamiizumi Member of Yabu Formation; also rare in Kiyokawa Formation. Altogether two left valves found.

Sclerochilus sp. 2
Text-fig. 17-6.

Illustrated specimen.—A right immature valve, UMUT-CA 9920 (text-fig. 17-6. L, 0.56; H, 0.30), sample 272-1, Kioroshi Member of Kioroshi Formation.

Remarks.—Although the specimens available are all immature forms, outline of carapace and muscle scar pattern suggest close affinity of this species to genus *Sclerochilus*.

Occurrence.—All immature valves. Rare in Toyonari Member and Kioroshi Member of Kioroshi Formation.

Family Cytheromatidae Elofson, 1939
Genus *Paracytheroma* Juday, 1907
Paracytheroma sp.
Pl. 15, figs. 3, 4, 7, 8, 13, 15; text-figs. 17-3, 4.

Illustrated specimens.—A left valve, UMUT-CA 9921 (Pl. 15, figs. 3, 13; text-fig. 17-3. L, 0.53; H, 0.26), sample 228-2, Kamiiwahashi Formation; a carapace, UMUT-CA 9922 (Pl. 15, fig. 4. L, 0.55; H, 0.24; W, 0.21), sample 6-3; a right valve, UMUT-CA 9923 (Pl. 15, figs.

7, 15; text-fig. 17-4. L, 0.54; H, 0.25), sample 6-3, Toyonari Member of Kioroshi Formation; a left valve, UMUT-CA 9924 (Pl. 15, fig. 8. L, 0.54; H, 0.24), sample 228-3, Kamiiwahashi Formation.

Remarks.—*Cytherois uranouchiensis* illustrated by Ishizaki (1968) resembles this species in having the following characteristics: marginal infold which is deep anteriorly and moderately deep along posterior and posteroventral margins; inner margin which is nearly straightly running from anterior cardinal angle to ventral margin and runs parallel to posterior cardinal angle to ventral margin and runs parallel to posterior contact margin and slightly projects near the middle of posetroventral margin; line of concrescenece which runs nearly parallel to outer margin throughout entire free margin; and deep anterior vestibule and shallow posteroventral vestibule. Normal pores of sieve type with a subcentral opening. Radial pore canals are short, simple, straight, and form pairs anteriorly. Muscle scars are located at a little anteriorly to the middle of carapace, and consist of an oblique row of four adductor scars, a round large frontal scar, two mandibular scars and a few dorsal scars. All of these characteristics suggest that this species as well as *Cytherois uranouchiensis* belong to genus *Paracytheroma.*

Occurrence.—Rare in Yabu Member of Yabu Formation; rare in Kiyokawa Formation; also rare in Toyonari Member and Kioroshi Member of Kioroshi Formation.

Acknowledgements

This study has been made under the direction of Professor Tetsuro Hanai of the University of Tokyo. I am much indebted to him for advice and encouragement which stimulated the present work. Gratitude is also expressed to Dr. Noriyuki Ikeya of Shizuoka University, and Drs. Masuoki Horikoshi, Kiyotaka Chinzei, Itaru Hayami, Tomowo Ozawa and Toshiyuki Yamaguchi of the University of Tokyo for their instructions during laboratory and field work. The manuscript was read by Professor T. Hanai and Dr. Paul M. Frydl. Part of photographs taken with scanning electron microscope were prepared by Mr. Takahiro Kamiya of the University of Tokyo.

ULTRASTRUCTURE AND PATTERN OF THE CARAPACE OF *BICORNUCYTHERE BISANENSIS* (OSTRACODA, CRUSTACEA)

Yutaka Okada

Abstract: The fine structures of the carapace of *Bicornucythere bisanensis* were studied; these include the cuticle, epidermal cells, subdermal cells, nervous tissue and spines. The correlation between epidermal cells and cuticular reticulation was made clear. Three-dimensional structure of setal sensilla (pores with setae) was examined by serial ultra-thin sections. Reticulation pattern and pore distribution pattern were described.

Introduction

An ostracod carapace has two special advantages in paleontological studies. One is its surface ornamentation which enables us to study cellular aspects of the ontogenetic and phylogenetic morphogeneses. This advantage has not been recognized yet. The other is the pores penetrating the carapace cuticle. Calcified parts of sense organs, which are usually preserved in the pores of fossil ostracods, have been utilized for analyses of paleoenvironments (Rosenfeld and Vesper, 1977), and enable us to investigate how the changes in sense organs correlate with phylogenetic development of ostracods.

Histological observations of ostracods are rare even at the light microscopic level; light microscopic studies—Cannon (1931, 1940), Rome (1947), Kesling (1951), Harding (1964), Kornicker (1969) and electron microscopic studies—Jørgensen (1970), Bate and East (1972, 1975), Puri (1974), and Andersson (1975)—have been done. Accurate function of various carapace characters (reticulatin, pores) cannot be inferred without a clear understanding of the three-dimensional structure of associated organs and tissues. The main purpose of this paper is to clarify the relation between an epidermal cell and a polygonal fossa of the ornamentation and to elucidate the three-dimensional structure of setal sensilla (pores with receptive setae). It is based on a histological study of an entire ostracod carapace utilizing the transmission electron microscope (TEM) and the scanning electron microscope (SEM).

The first intensive study of pores was made at the light microscopic level with a freshwater species by Rome (1947). Recently, most investigations have focused their attention on the pore surface owing to the development of SEM and classified pores into several types, which have been used in taxonomy, phylogenetical discussions, and environmental analysis (Hay and Sandberg, 1967; Sandberg and Plusquellec, 1969; Puri and Dickau, 1969; Puri, 1974). Puri (1974) used TEM to elucidate the internal structure of pores. However, more detailed observations are needed to estimate their functions, since sensory organs are usually composed of complex structures. Moreover, it is necessary to understand the distribution pattern of pores, since an individual responds to external stimuli after it has integrated information from its whole sensory system. Hanai (1970) found that a striking similarity of the pore canal patterns is observable when specimens of the same species are compared in spite of differences in sex, localities and horizons; even species of the same genus have a similar pattern of pore distribution. More than one hundred pores are usually distributed on a carapace of many ostracods, and some of them may have different functions. In this study, three-dimensional structures of identical pores of a species were examined at the EM level with seriall thin-sections.

Polygonal reticulation of the ostracod carapace is known in many species from Paleozoic to Recent. Reticulation patterns are used as a criterion for identification of many species, and their accurate illustration by SEM micrographs has been accumulated; Pokorný (1969), Liebau (1969, 1971, 1975), Benson (1972), and Al-Furaih (1977). Changes of reticulation patterns in phylogeny from Cretaceous to Recent were reported in Benson (1972). When the histological meanings and the ontogenetical changes of the reticulation are clarified, it will be useful to study the pattern formation and the cell lineage, both of which are important as the fundamental mechanisms controlling metazoan morphogeneses. Without understanding of the mechanism, their phylogenetical development is not fully comprehensible. In this context, Hanai (Hanai et al., 1977) used the term "organule", proposed by Lawrence (1966) in his study on the cell lineage of an insect, and emphasized the importance of "spatial patterns of organules for precisely placed pore canal opening" in systematic studies of cytheracean ostracods.

Spatial organization of cells is now intensively studied in two approaches. Some investigate how respective cells occupy their appropriate positions through their cell lineages, and the others study how the position affects the cell differentiation to form various patterns. A ridged cell lineage has been clarified with *Caenorhabditis elegans*, a species of Nematoda (Sulston and Horvitz, 1977; Deppe et al., 1978; Kimble and Hirsh, 1979); the existence of the genes affecting its cell divisions has also been recognized (Albertson, Sulston, and White, 1978). Various homeotic mutation genes are known to affect the morphogenesis of *Drosophila melanogaster*, and their operating processes are under investigation in many laboratories.

For the elucidation of multicellular metazoan evolution, it is important to understand the mechanisms controlling spatial organization of cells. This report shows the correspondence of each polygon to an underlying epidermal cell in a species with typical ornamentation. Post-embryonic changes in cell arrangement can be easily studied with the reticulation of larval and adult carapaces. The reticulated carapaces of this category are well preserved in fossil specimens of some species from Paleozoic to Recent. This raises the possibility to make a survey of the process of the changes in cell arrangement through both ontogeny and phylogeny.

Upon Hanai's suggestion (Hanai et al., 1977), a benthonic species *Bicornucythere bisanensis* (Okubo) was selected as an experimental animal. All specimens were dredged from the Aburatsubo Bay, Kanagawa Prefecture, Central Japan. For SEM observation, they were fixed in 5% glutaldehyde in 0.1 M sodium cacodylate buffer (pH 7.3) with 5% sucrose for 6 hours at 4° C, dehydrated in ethanol series, treated in amyl acetate, dried by critical-point method and coated with carbon and gold-palladium alloy. For TEM observation, they were fixed in the same glutaldehyde fixative for 24 hours at 4° C, decalcified in 5% EDTA (pH adjusted to 7.3 with NaOH) with 7% sucrose for

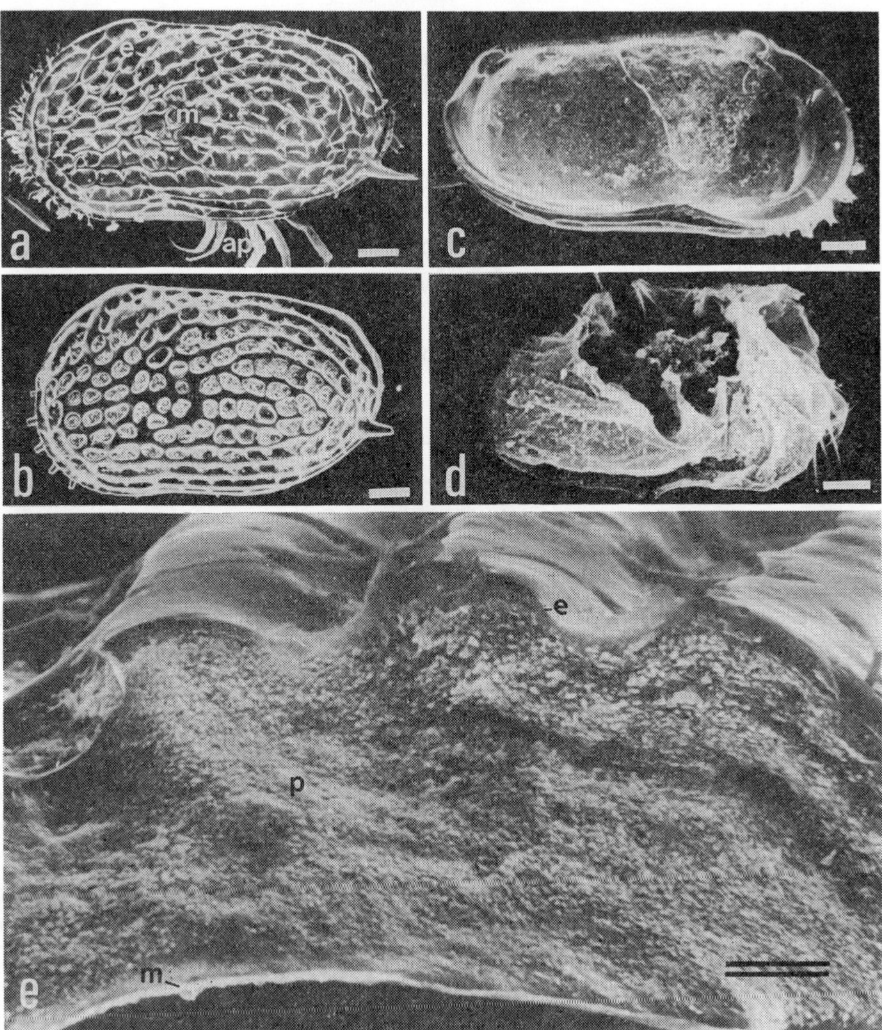

Text-fig. 1

a, b: Recticulation of the left valves. a: Recent *B. bisanensis* dredged from the Aburatsubo Bay. The soft parts of an ostracod are enclosed within the valves with some appendages (ap) extending out between them. Reticulation is somewhat modified by muscle attachments (m) and an underlying eye (e). Many sensory setae can be seen on the valves. b: fossil from the Anden Formation of the Oga Peninsula, northeast Japan. The reticulation pattern is nearly identical to that of fig. a. (scale bar = 100 μm). c, d: Correlation of carapace and soft part. c: Inner surface of a carapace, where inner lamella cuticle had been attached during drying except for an anterodorsal triangular area. d: specimen without outer lamella cuticle, which shows adepidermal surface of inner lamella cuticle and connecting area between carapace and soft part. (scale bar=100 μm) e: Fractured cuticle. (e) epicuticle of outer lamella. (m) membranous layer. (p) calcified procuticle of outer lamella.

Table 1
Structure of carapace integument.

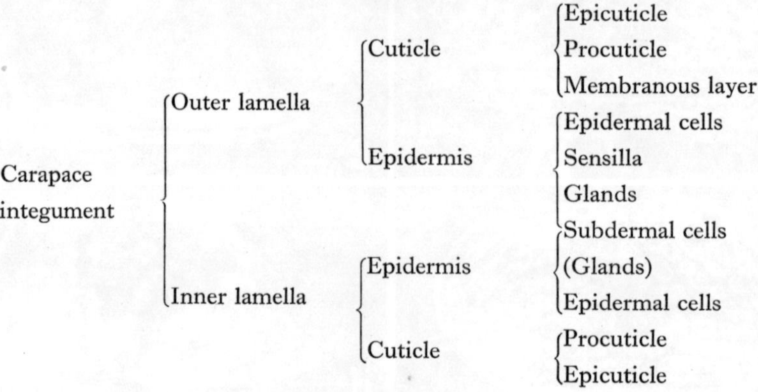

Carapace integument	Outer lamella	Cuticle	Epicuticle
			Procuticle
			Membranous layer
		Epidermis	Epidermal cells
			Sensilla
			Glands
	Inner lamella	Epidermis	Subdermal cells
			(Glands)
			Epidermal cells
		Cuticle	Procuticle
			Epicuticle

24 hours at 4° C, post-fixed in 1% osmium tetraoxide in 0.1 M cacodylate buffer (pH 7.3) with 5% sucrose for 3 hours at 4° C, dehydrated in ethanol series, and embedded in Spurr's resin. Thin sections were treated with uranyl acetate and lead citrate.

General structure

Histological outline: A carapace has two layers of cuticle: outer lamella cuticle and inner lamella cuticle (text-fig. 2a). Each cuticle is lined with a single layer of epidermal cells: outer epidermal cells and inner epidermal cells. The two kinds of cells have different compositions of organella and form cuticles of different structures. The outer lamella cuticle is thick and calcified. It consists of outer epicuticle, middle procuticle, and inner membranous layers. In the surface view, the outer cuticle is ornamented by reticulation (text-figs. 1a, b), which is composed of polygonal fossae reflecting the arrangement of the underlying epidermal cells. The inner lamella cuticle is thin and has a smooth surface, except for the cuticle near the free margin of the carapace where the lamella is calcified and increases its thickness. It consists of an epicuticle and a procuticle and lacks the membranous layer. There exist subdermal cells between inner and outer epidermal cell layers which lack the lining of obvious basement membranes. There is nervous tissue with sensilla. The nonbranching distal parts of four bipolar nerve cells pass through a pore in the outer lamella cuticle, and two of them penetrate its seta. Nearly three hundred pores with setae are scattered on an adult carapace. In addition to the setal pores, 32 secretory pores with ducts of exocrine glands, are distributed over an adult carapace.

The outer lamella cuticle is composed of two valves which are hinged together at the dorsal margin by a ligament, a thin strip of elastic cuticle. At the free margin of a valve, inner and outer layers of epidermal cells are in immediate connection. Outer and inner epidermal cells secrete cuticular substances before ecdysis and form outer and inner lamella cuticles, respectively. There are many spines near the anterior margin and a pair of spines at the posterodorsal area. Every spine is a cuticular pro-

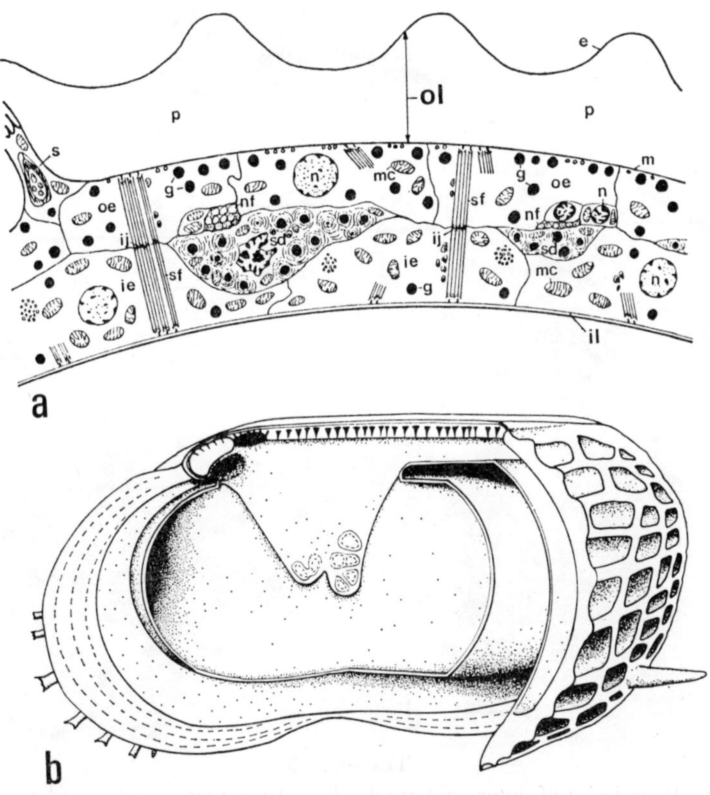

Text-fig. 2

a: General structure of carapace. (e) epicuticle. (g) granule. (ie) inner epidermal cell. (ij) intermediate junction. (il) inner lamella cuticle. (m) membranous layer. (mc) mitochondrion. (n) nucleus. (nf) nerve fiber. (oe) outer epidermal cell. (ol) outer lamella cuticle (p) procuticle. (s) sensillum. (sd) subdermal cell. (sf) supporting fibre. b: Schematic drawing of exoskeleton; an antero-left quarter is removed.

trusion containing an empty tube without any outlet.

The so-called soft part is enclosed within the bivalve carapace (text-fig. 2b). The soft part connects with the carapace at two triangular areas covering the median, middorsal and anterodorsal regions (text-figs. 1c, d). At the ventral corner, a cluster of adductor muscle fibers, which runs transversely in the soft part, attaches to the valves by a series of discrete structures. Eyes are situated at the anterodorsal corner.

Epidermis: Outer and inner epidermal cells, subdermal cells, and several kinds of cells of nervous tissue lie between the outer and inner lamella cuticle (pl. 16; text-fig. 2a). The features of these cells change through moltings. In this chapter, descriptions will be confined to adult calcified specimens.

Outer epidermal cells are large, about 30 to 50 μm in diameter and about 5 to 10 μm in thickness. Nuclei are round and light with sparse chromatin. Mitochondria are not abundant. The cells are connected to each other by apical desmosomes and septate

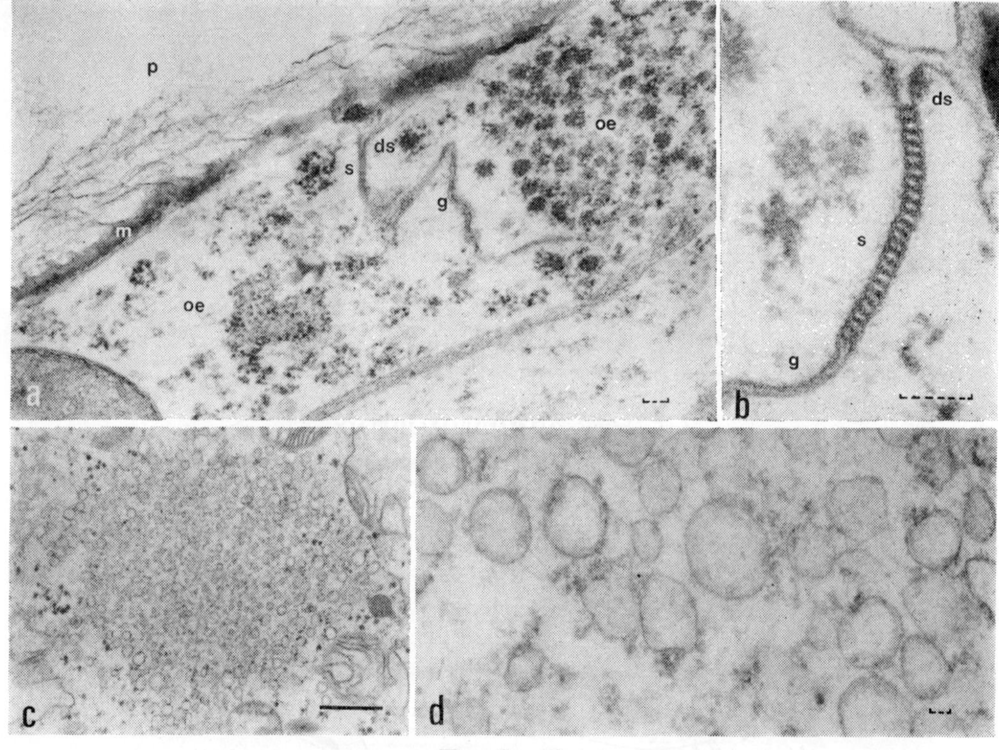

Text-fig. 3

a: Boundaries of outer epidermal cells. b: Septate junction. (ds) desmosomes. (g) gap-type junction. (m) membranous layer. (oe) outer epidermal cell. (p) procuticle. (s) septate junction. c, d: Multivesicle bodies.

junctions, and basal gap-type junctions (text-figs. 3a, b). They form a single layer which underlies the outer lamella cuticle.

Outer epidermal cells are characterized by many large granules, about 1 μm in diameter (pl. 18). They are often enclosed in single or double layered membranes. Some of them were found in r-ER. The granules can be classified into three types: (a) light granules, sometimes composed of subunits; (b) granules with cores of different electron density, and (c) dark homogeneous granules. These granules may be secreted by exocytosis.

Inner epidermal cells are also single-layered large cells with light round nuclei and underlie the inner lamella cuticle. Abundant large mitochondria characterize the cells. Numerous small vesicles, about 0.1 μm in diameter, aggregate to form multivesicle bodies, sometimes more than 10 μm in diameter (text-figs. 3c, d). The cells are connected to each other by apical desmosomes and widely distributed gap-type junctions (pl. 17, figs. g. h). No septate junction was observed between inner epidermal cells.

Subdermal cells are amoeboid. They are situated between the outer and inner epidermal cell layers. Nuclei also take amoeboid forms and contain much chromatin. Abundant r-ER characterizes the cells. Many layers of r-ER are arranged similar to a large Golgi apparatus in some specimens (pl. 18; pl. 19, fig. e). Large granules which

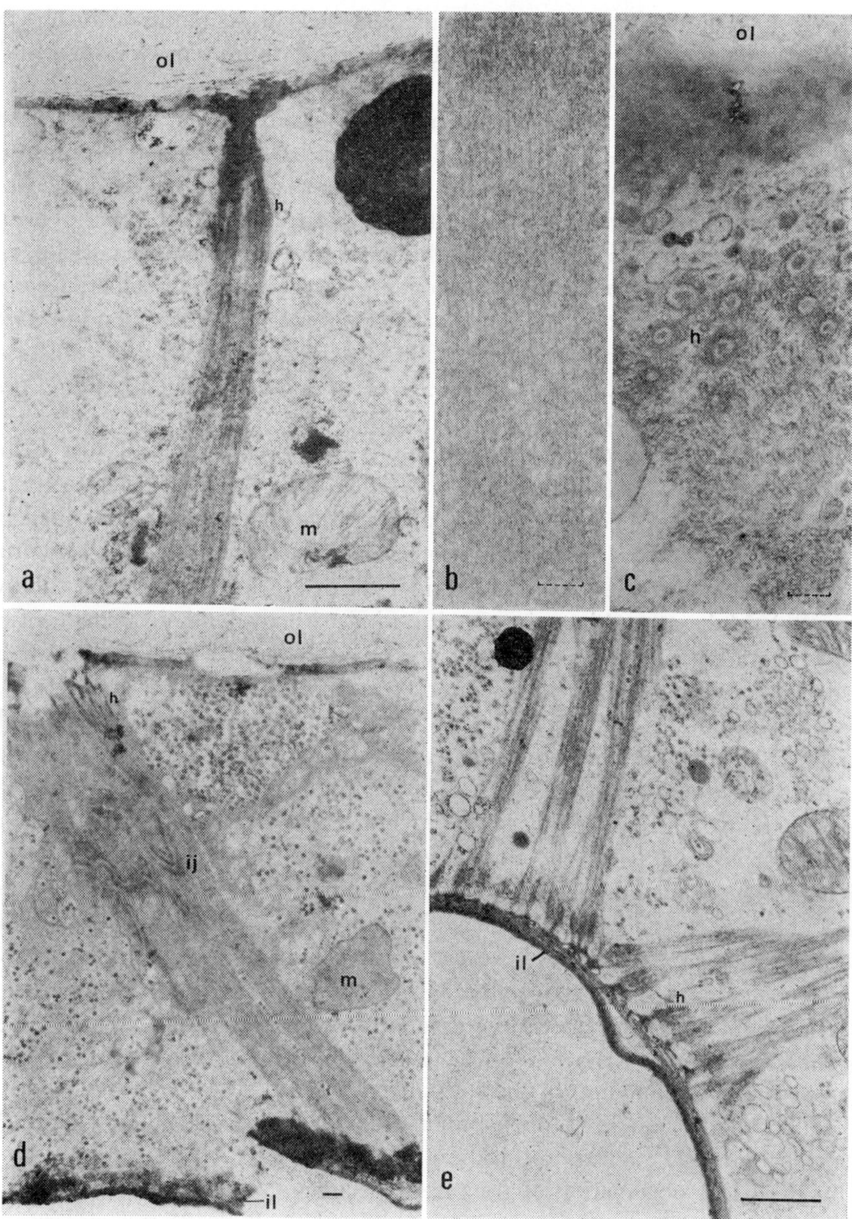

Text-fig. 4

Supporting fibres. a: Longitudinal section of the fiber in an outer epidermal cell. b: Longitudinal section of packed microtubules. c: Cross section of the fiber near outer lamella cuticle. d: Clusters of microtubules in outer and inner epidermal cells are longitudinally connected by the intermediate junction and bridge the epidermis from the outer lamella cuticle to the inner lamella cuticle. e: Fibers in an inner epidermal cell. (il) inner lamella cuticle. (ol) outer lamella cuticle. (h) conical hemidesmosomes. (ij) intermediate junction. (m) mitochondrion.

are similar to that observed in outer epidermal cells were numerously observed in some other specimens (pl. 19, figs. a–d). The cells are connected with outer and inner epidermal cells by gap-type junctions.

In the outer and inner epidermal cells, microtubules are packed into clusters at some portions of the cells (text-fig. 4). Each cluster of microtubules extends from conical hemidesmosomes in the apical cytoplasm of the inner and outer epidermal cells to the intermediate junction, where both cells are tightly connected to each other. The clusters appear to bridge the epidermis between the outer and inner lamella cuticles, running nearly perpendicular to the cuticles. Secondary lysosomes were sometimes observed around the clusters. This structure made of microtubules may correspond to *fibrilles de soutien* of Rome (1947) and supporting fibers of Kesling (1951). The clusters seem to restrict the displacement of epidermal cells, for they were frequently found throughout the moltings of ostracods.

Nervous tissue was observed between outer epidermal cells and just below the outer epidermal cell layer (text-fig. 2a). It extends from receptive setae to the central nervous system. Its detailed structures near receptive setae will be described later.

Cuticle at molt stage C: The feature of cuticle also changes through moltings. A molt cycle is usually divided into four periods (A to D) or more detailed subdivisions (A_1, A_2, etc.). Molt stage C is an intermolt period without dynamic changes of cuticle.

SEM observation of the fractured surface of outer lamella after etching treatment revealed that its cuticle consists of outer epicuticle, medium procuticle, and inner membranous layers when the lamella is fully formed (text-fig. 1e). Epicuticle and membranous layers could be discriminated from procuticle, because they are smooth layers and resist the etching of EDTA. The accurate thickness and detailed structure of these layers are not clear, because the layers curl somewhat at cut ends. Within a few days after ecdysis, when the cuticle is thin (about 5 μm at the bottom of fossae of reticulation), the membraneous layer is lacking and the inner surface of the cuticle is rough. The procuticle layer is filled with fine calcite grains and shows a laminated structure owing to partial delicacy to the etching reagent. The laminae are waving, affected by the shape of the reticulated surface made of ridges and fossae. Below the ridges, the height of the wave decreases gradually toward the inner surface of the cuticle. The granular structure of the procuticle is sometimes obscure near the membranous layer hidden under abundant organic substances.

The inner lamella cuticle is much thiner than the outer lamella cuticle and has no granular structure. The accurate thickness and detailed structure could not be recognized by the SEM observation of fractured specimens. The cuticle is sometimes attached to the outer lamella cuticle as one result of the collapse of the epidermis, when the fixation and the dehydration were not appropriately performed.

The three layered structure of outer lamella was ascertained in thin sections by the TEM observation, which revealed that the inner lamella cuticle consists of epicuticle and procuticle layers and always lacks a distinguisable membranous layer.

The epicuticle of both lamellae is composed of three distinct layers (pl. 17). The outermost layer is highly electron-dense, nearly equal to the plasma membrane in thickness. It is similar to the cuticulin layer (Locke, 1966) of the insect *Calpodes ethlius* in structure and forming process. This layer will be called the cuticulin layer in this paper. In places, the cuticulin layer is discontinuous and appears as a dotted line where it seems to form minute channels from the inside layer to the outside.

The middle layer of the epicuticle appears as a light zone and is three to five times thicker than the cuticulin layer. In this layer, a membraneous substructure could be observed, which is clear in some specimens and fragmental in other specimens.

The innermost layer (inner epicuticle) is much thicker and less electron dense than the cuticulin layer and probably corresponds to that observed as the epicuticle in the light microscope. In the outer lamella, its thickness is about 30 to 50 nm. An electron-denser membrane, thinner and lighter than the cuticulin layer, is usually at the outermost part of the inner epicuticle. Many granules (about 5 nm in diameter) attach to the bottom of the inner epicuticle. In the inner lamella, the inner epicuticle is about 40 to 100 nm in thickness and almost homogeneous without the outstanding substructure of the electron denser membrane.

The epicuticle of outer lamella forms many minute protrusions which are seemingly like the microvilli in the surface view. They protrude about 70 nm above the surface. They are also covered with a cuticulin layer and filled inside with inner epicuticle. They do not seem to affect the lower cuticle in structures.

Procuticle is the thickest layers in both lamella: 3 to 15 μm thick in the outer lamella and about 0.2 μm thick in the inner lamella. Its foundamental structure seems to be formed by piling many undulating membranes (sheets or belts), each of which fuses partially with the adjacent ones. In thin sections, they appear as a slightly compressed interlocking lattice, which does not consist of fibers as reported on carapace procuticle of *Cypridopsis vidua* and *Heterocypris incongruens* (Bate and East, 1972, 1975) but of membranes. Because they appear as belts in most specimens as obliquely sectioned membranes. The membranes are thinner than the plasma membranes. This structure made of the membranes is named the piled membrane structure. It is typically developed in the innermost part of procuticle of outer lamella just above the membraneous layer (text-fig. 3a), where calcite grains were sometimes obscure in SEM micrographs. It is also observed near the teeth of the hinge, around the setal pores (pls. 22, 23, 29, 30; text-figs. 14, 15), in the calcified inner lamella at the free margin (pl. 17, fig. i).

In the outer lamella, the procuticle often shows fragmental membranes, except for the innermost part where the original structure might be well preserved throughout the calcification of the cuticle. Such fragmental membranes are abundant in the narrow uppermost zone just beneath the epicuticle, and they are attached by many micro-granules about 5 nm in diameter (pl. 17, fig. a). They are, as a whole, arranged to form the lamination gently waving as observed in the fractured surface by SEM. The structure of fragmental membranes may result from the increase of the accumulatin of calcite granules in the space between membranes.

In the inner lamella, most procuticle shows a laminated structure. Each lamina is parallel to the surface and recognized by the difference of electron density (pl. 17, figs. g, h). Its detailed structure is obscure, owing to tightly packed organic material. The piled membrane structure was clearly observed in calcified parts of the inner lamella near the carapace free margin, where the procuticle is thicker than other parts of the lamella (pl. 17, fig. i).

The inner membraneous layer, which was observed only in fully formed outer lamella cuticle of larvae and adults, is an amorphous dense layer (pl. 18; text-figs. 1e, 3a). It attaches to the procuticle with an uneven surface and to the epidermis with a rather smooth surface. The cell membranes sometimes become indistinguishable when they closely contact to the layer. This layer was not observed in uncalcified specimens just after ecdysis (at molt stage A) nor in calcified specimens with thin carapaces fixed within a few days after ecdysis (at molt stage B).

Newly formed cuticle: A new cuticle is formed under the old cuticle by epidermal cells before ecdysis (at molt stage D). The new cuticle has the reticulation whose polygon has initially a similar size to that of the old cuticle. The polygon here consists of a rough surfaced center with many microvilli-like protrusions and surrounding smooth-srufaced belts with a few protrusions. They correspond to the fossa and the ridges of the old cuticle respectively. Though both surfaces have many furrows, they are rather prominant on the belts. The furrows across the belts may guarantee the dilatation of each polygon after ecdysis.

In cross-sections, the belt appears as a T-shaped projection of the cuticle with a smooth surfaced top. Rough surfaced cuticle with microprotrusions is situated between two T-shapes. The cuticle of both types already consists of epicuticle and procuticle layers. No significant difference could be detected between the newly formed epicuticle and that at molt stage C. The procuticle contains many small grains (about 5 nm in diameter), membranes which probably correspond to those of the piled membrane structure of later stages, and larger grains (about 50 nm in diameter) which seem to originate the grains observed in typical Golgi apparatus. T-shaped projections were always found above the boundary of adjacent outer epidermal cells.

The root of a T-shape is buried in a depression of the cuticle before ecdysis. Only its top part could be seen as a belt in the surface view. It becomes a true T-shaped projection projecting the root and the depressed part after ecdysis (at molt stage A). It gradually swells and increases its width. The T-shaped ridge changes to the ridge whose top and sides are convex and concave respectively. The ridge further increases the width at its base and finally takes the form of a smooth wavy ridge.

Reticulation

Reticulation pattern and cell-polygon correspondence: The carapace of many ostracod species is ornamented by polygonal fossae and ridges which form a species-dependent reticulation pattern, which is important in ostracod taxonomy. The number and arrangement of fossae are compared by numbering every fossa. These have been thought to be extremely consistent in each species.

Topotypes and published illustrations of adult *Bicornucythere bisanensis* (Okubo), an ostracod species found commonly in shallow marine environments near Japan, were examined and no difference in its reticulation pattern could be detected in spite of varieties in environments such as water temperature and salinity. More than fifty adult specimens from one population in a dredged sample from the Aburatsubo Bay did not show differences in the pattern (text-figs. 1a, 8). Variants of the pattern were rarely found after rough cultivation for four weeks under an abnormal condition —namely, high salinity, excessive aeration, and fluctuating room temperatures. They

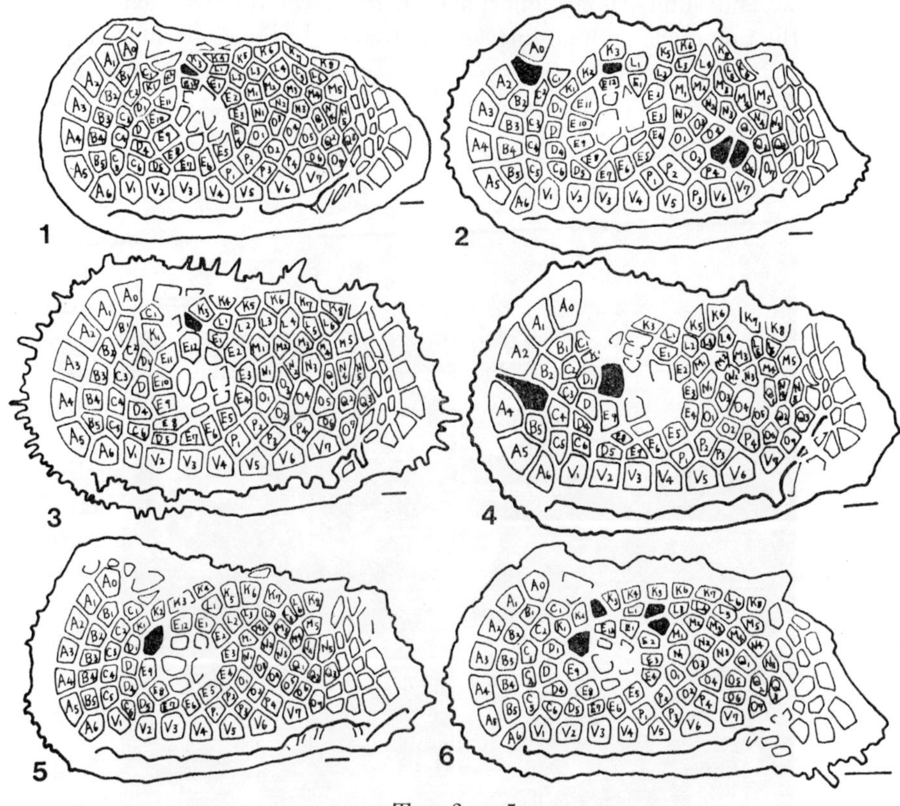

Text-fig. 5

Reticulation patterns of five species of *Agrenocythere* and one species of *Oertliella*, traced from the SEM micrographs of Benson (1972). Each polygonal fossa is compared by numbering method. Noncorrespondent fossae are colored black. As Benson has already pointed out the consistency of reticulation pattern in the phylogeny of these species, these illustrations show a consistent arrangement of fossae. Remarkable changes are the following. The fossa below K3 is increasing in every Recent species. Either E10 or E11 is lacking in Eocene species. Scale bars are 100 μm. 1. *Agrenocythere americana*, Recent, Gulf of Mexico. 2. *A. hazelae*, right valve, Recent, East Pacific. 3. *A. spinosa*, Recent, Mozambique Channel. 4. *A. hazelae*, Miocene, Trinidad. 5. *A. gosnoldia*, right valve, Eocene, northeast North America. 6. *Oertliella ducassae*, right valve, Eocene, Biarritz (France).

lack one polygon in one valve of their carapaces and have bilaterally nonsymmetrical reticulation patterns.

Among four fossil specimens which were found from the Pleistocene Anden Formation (Okada, 1979), one variant (text-fig. 1b) was found to lack one fossa in anterior area of the carapace, meaning that the evolution of the reticulation pattern was quite conservative. Similar conservativeness is known in the phylogeny of *Agrenocythere* group (text-fig. 5). Correspondence of a cuticular polygon to an epidermal cell is postulated in many arthropods and is clearly shown in the insect *Calpodes ethlius* (Locke,

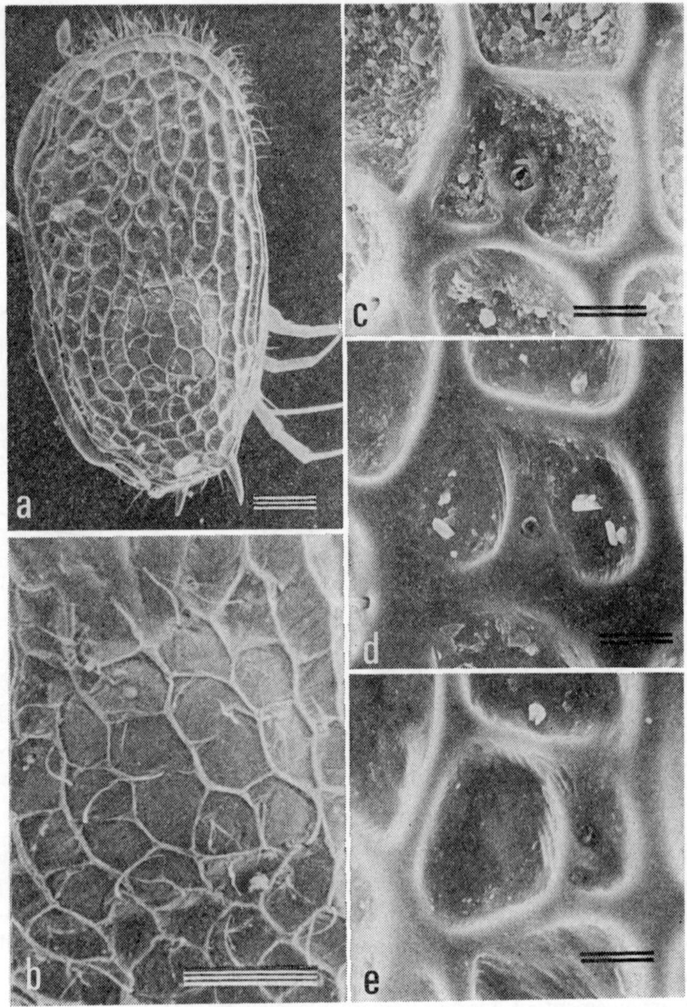

Text-fig. 6

a, b: Single nodes were occasionally formed on carapaces during the laboratory cultivation. Their size varies from sample to sample. Ridges on the nodes are sometimes obscure and two or more adjacent fossae in longitudinal rows are almost fused. c–e: Modification of reticulation by pores at the corresponding positions of different specimens. Fossae appear to be divided into two parts (d, e).

1966). If each polygonal fossa of ostracods represents its underlying epidermal cell, a stable cell arrangement could be deduced.

In order to examine the relation between the polygons and the underlying epidermal cells, TEM observations were made on carapaces at molt stage C (pl. 16, fig. c) and at molt stage D. In the latter case, the cuticle was thin and its relationship to epidermal cells could be studied easily. Boundaries of two epidermal cells of the outer lamella were distinguished by apical desmosomes and adjacent septate junctions. In the specimens

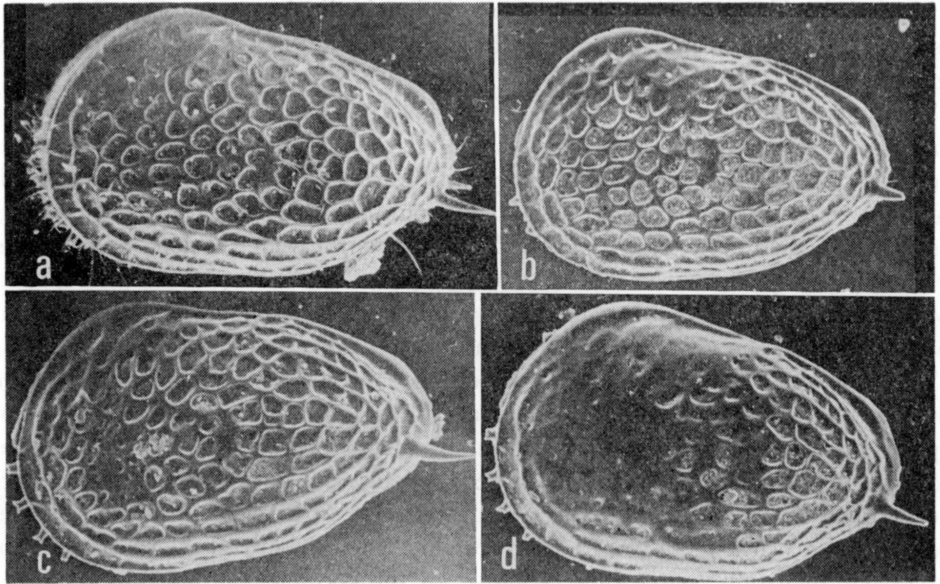

Text-fig. 7
Obscured reticulation at (−2) stage. Reticulation is obscure at antero-dorsal margin (a), just behind the eye (b), and just before the frontal muscle attachment (c). In the specimen whose obscured area extends widely to the posterior region behind the adductor muscle attachments, fossae around the attachments are not flattened (d).

just before ecdysis, a cuticular projection which usually takes a T-shape was always found above each boundary. SEM observation of the newly formed cutcile revealed a difference between a fossa with rough surface and the smooth surrounding belts formed by the T-shaped cuticle. Time-sequential observations of the T-shaped cuticles showed that they develop into the ridges of the reticulation after ecdysis. These observations point out clearly that each polygon corresponds to one epidermal cell. The result of this species suggests the possibility of the correspondence in certain ostracods with polygonal reticulation, from Paleozoic to Recent. Fossil ostracods may indicate a conservative nature of cell arrangment in ostracod evolution.

Modification of reticulation: The outer epidermal cells, which secrete the outer lamella cuticle, seem to be affected by various factors. The specialization of cells can be recognized by modification of reticulation. The boundaries of outer epidermal cells are not always clearly represented by ridges of the reticulation. Some types of modfication are related to the existence of organs, the functional specialization of epidermal cells, or some accidental cause. The fossae which are situated above the eyes are not formed and show a smooth surface in most specimens. Many ridges widen around the pores which are situated in ridges. Single small mounds are formed around the pores in fossae. Such mounds are usually connected with adjacent ridges by single short ridges, occasionally double short ridges at opposite sides of the pores (text-figs. 6c–e.) Muscle attachment changes ridges into mounds. Ridges are not clear near the sockets at hingements. Two or more adjacent fossae in longitudinal rows are sometimes almost fused on the nodes which are occasionally formed in the laboratory cultivation (text-figs. 6 a, b).

The reticulation is further modified on a larger scale by the flattening of the carapace surface, which may be useful to consider the relation between species with reticulated carapaces and those with smooth carapaces. This type of modification was called "obsolete reticulation" or "obscured reticulation" by Ishizaki (1972, 1975), who discussed the relation between this modification and environmental factors. Some features of this complex phenomenon will be described in the following without assumption of its meaning. The measurement suitable to represent the degree of the extension of the obscured reticulation may be devised on the basis of these observations.

The obscureness often occurs at anterior parts, especially in younger molt stages. The reticulation pattern in the obscured area emerges at later molt stages. The area expands at various degrees in different samples of the same stages. When it is narrow, the obsured reticulation occurs at three regions: anterodorsal margin, just behind the eye, and just before the frontal muscle attachment (tex-figs. 7a–c). In many specimens the obscureness coexists in two or three regions. When the area is wider, a continuous obscured reticulation covers these regions. In the specimens whose obscured areas extend widely to the posterior regions behind the adductor muscle attachments, fossae around the attachments are often clearly observed (text-fig. 7d).

There are intermediate stages between well-developed reticulation and completely obscured reticulation. The fossae at the intermediate stages can be classified into two types. One type shows shallow fossae with a rather smooth surface, and the other shows fossae with several pits like the second-order reticulation. The obscured fossae at intermediate stages are often observed around completely obscured regions.

Spine

There are a pair of spines at posterodorsal area of the carapace (text-figs. 1a, b) and about eight pairs along anterior margins of valves. Spines are cone-shaped protrusions of cuticle. The posterior spines are usually unbranching, about 80 μm in length and about 30 μm in diameter at their base. The anterior spines are variable in number (often eight), shape, and size. Two dorsal and two venteral spines are usually small, sometimes vestigial, while four middle spines are large and often have two branches with minor projections near their crotches. The spines are 10 to 25 μm in length and 5 to 15 μm in diameter at their base. Every spine has similar inner structure (text-figs. 8 a, b).

Spines already exist as almost flattened protrusions with many furrows before ecdysis under the old carapace (pl. 20 figs. c, d). They are completely covered with epicuticle. In the calcified specimens, calcite grains are arranged concentrically (pl. 20, fig. a). The center of the cone is empty and contains no protoplasm. In sections, the wall of the empty space appears as a circle formed by a concentration of amorphous substance without membrane structure (pl. 20, fig. b). The circle decreases its diameter toward the tip, and the empty cavity does not open outside. Toward the epidermis, it ends abruptly near the base of the spine and does not have an outlet (text-figs. 8a, b).

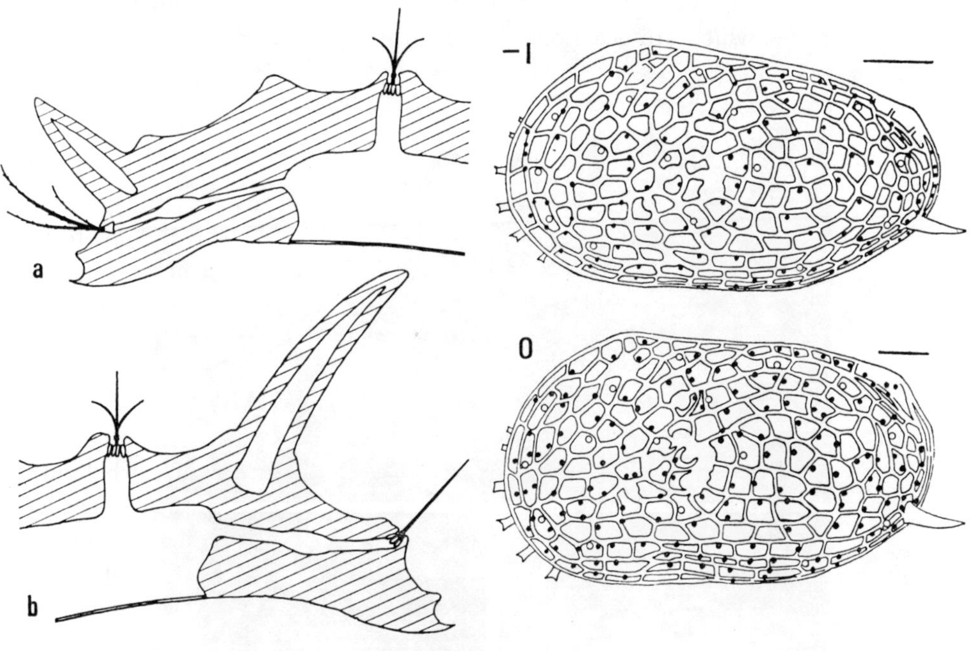

Text-fig. 8

a, b: Longitudinal sections of anterior (a) and posterior (b) margin with spines. Positions of sensillum pores are also indicated.

−1, 0: Distribution of pores except those at the margin. Number of setal pores (●) increases at the last molt, (−1) to (0). Number of secretory pores (○) does not change at the molt.

The spines have been worn at their tips, and calcite grains are exposed in some specimens (pl. 20, fig. a). Occasionally, distal parts of spines have been cut off, probably by some accident in their life, and the spines show their cavities. Such damages might become fatal if the empty cavity was continuous with its body cavity.

Before entering into this histological study, various functions were expected for the spines. The EM observations, however, have not revealed the meaning of this conspicuous morph. Ethological study may be needed for the understanding of its existence.

Pores

Histological survey of setal sensilla (pores with setae) of ostracod carapaces was difficult, and no literature showed their three-dimensional structures.

About two hundred and fifty or more setae could be found over an adult carapace of this species (text-figs. 8–1, 0). When every seta has a different inner structure, a curious schema may possibly be created by the usual method in which the observation of many different portions of different specimens is composed to make one model of a structure. Therefore, examination of the identical seta and its pore was the first con-

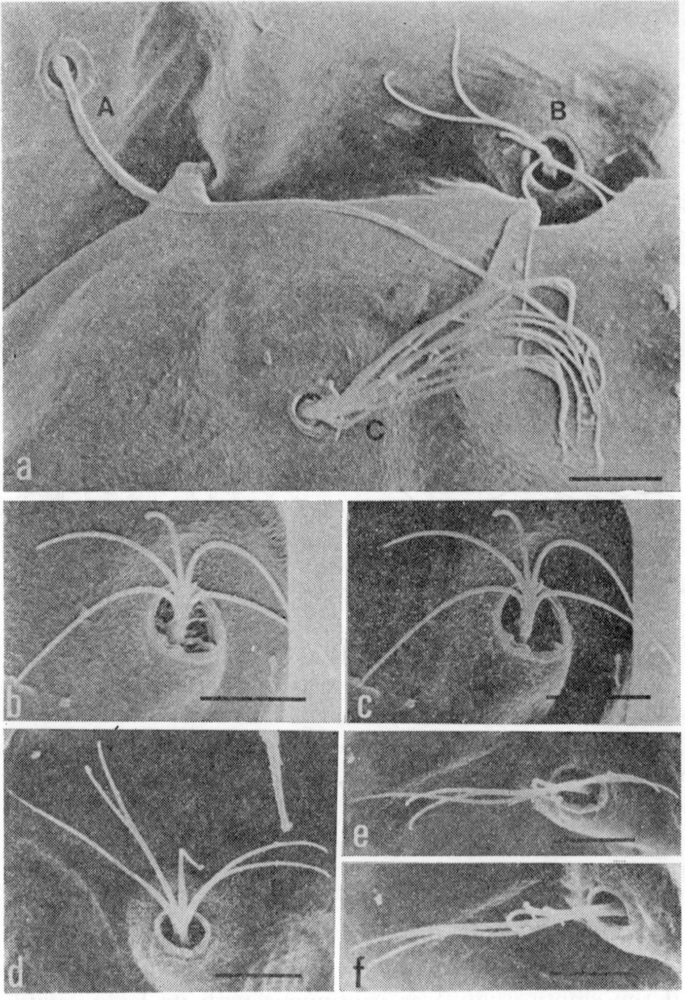

Text-fig. 9

Three types of setae may be sensible of different stumuli. SEM micrographs (a-c) show inefficiency of the dichotomic classification of pores into the sieve and simple types. b-c: Low and high contrast micrographs of the same pore with a B-type seta. d-f: A seta observed from three angles. B-type setae have branches extending in a plane.

sideration in this study.

Some serial thin sections (about 0.3 μm thick) and a few thick sections (about 1 μm thick) were cut alternately. The identical pore was distinguished in thick sections with a light micrscope and then the pore, and its seta was examined at EM level consecutively. In the most successful case, 46 thin sections interrupted 14 times by 29 thick sections were observed about a single sensilum. Some of these sections are shown in plates 21 to 27 and text-figures 14, 15a, and 16a. The three-dimensional structure of each pore and seta was compared and their schemata were drawn to show their basic struc-

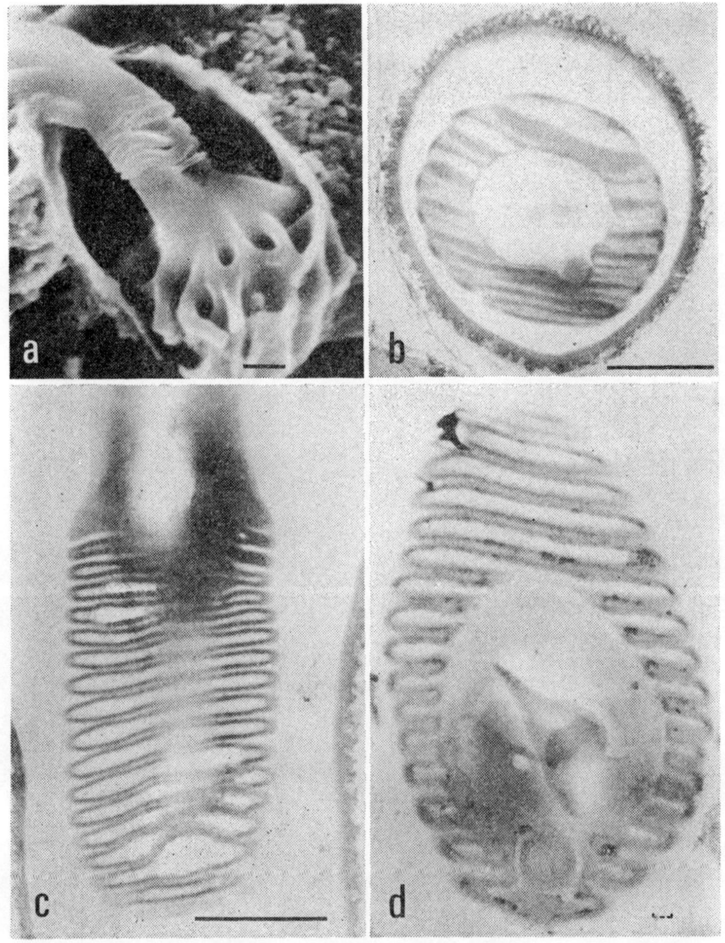

Text-fig. 10

a: Circular fold and multiple pouch structure. b–d: Circular fold sectioned at various angles. Two cilia eccentrically penetrate the fold.

tures.

Pores in the carapaces of ostracods have traditionally been classified into sieve-type pores and simple-type pores by surface view, and into normal pores and radial pores (or marginal pores) by the position where they penetrate. The former classification seems to need reexamination. Regarding the latter classification, radial pores have been defined as the pores running along the plane called the marginal zone. However, this is not reasonable in this species for the following reason. No pore runs along the marginal zone, though many pores appear to run radially in the marginal area of the carapace when they are observed from the lateral side with a light microscope. Furthermore, the pores in this species include secretory pores without setae. This type of pore has been called the Ben-type pore in our laboratory.

In this article, for convenience sake, the term normal pore will be used for the setal

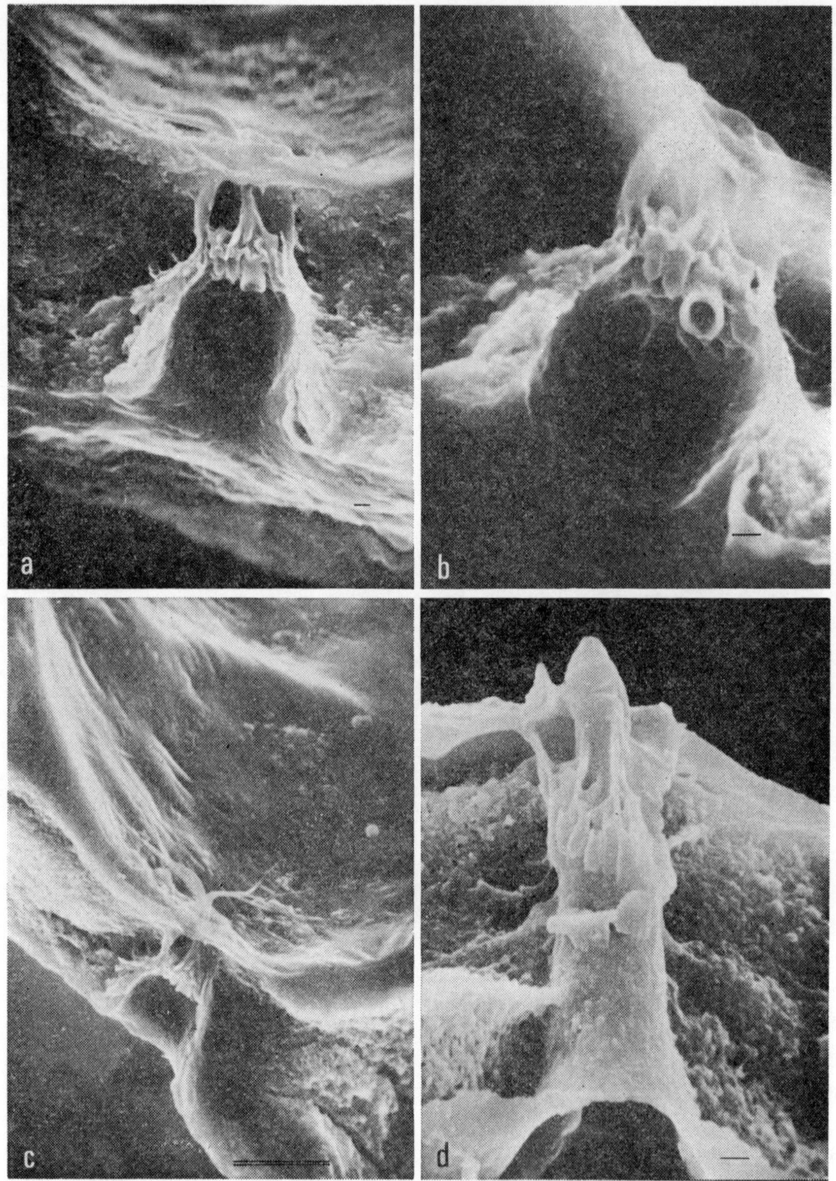

Text-fig. 11

Cuticular structure of normal pores. Epicuticle at the surface of the distal part of pores, multiple pouch structure near the middle part, setal cuticle with opened bottom and membranous layer at the surface of the proximal part can be seen as smooth continuous membranes. A pore was observed from various directions (a to c).

pore whose openings can be seen in the lateral view of a carapace. Two rows of setal pores along the anterior margin of the carapace (text-fig. 16c) have a peculiar structure which is distinct from the structure of normal pores. These pores will be referred to as anterior marginal pores. Pores near the posterior margin of the carapace have essentially the same structure as that of normal pores.

About one hundred normal pores with setae near the adductor muscle attachment, fifty anterior marginal pores, and twenty pores near the posterior margin were examined.

Cuticular structure of normal pores: SEM observation of setal length, the presence of ramifications, and opening of their pores revealed three types of seta: A, B, and C-types (text-fig. 9). A-type is a long seta (about 90 μm in length when measured from the surface of carapace) without apparent branches, and its pore has a narrow opening (about 3 μm in diameter). B-type is a short ramifying seta (about 20 μm in length) with 5 to 6 branches near the base; its pore has a wide opening (about 5 μm in diameter). C-type is a middle-sized seta (about 40 μm in length) with many branches which diverge at various portions of the seta; its pore has a narrow opening (about 3 μm in diameter). The diameter of setae near the base is about 1 μm, 0.7 μm and 1 μm in A-type, B-type, and C-type. The distribution pattern of normal pores with setae of each type was almost constant at least in 4 adult specimens and every corresponding pore with a seta was distinguishable in 10 examined samples. The seta comes out nearly vertically from carapace like a branching tree when observed with light microscope and TEM, though they look winding when dried by critical point method.

Setae of the three types have essentially the same cuticular structure in the pores. Every seta becomes thicker just beneath the opening of its pore and forms a unique part which may correspond to the circular reinforcement of Puri (1974). Observations with SEM and analysis of many TEM micrographs of preparations sectioned at various angles revealed that this is a cuticular fold which repeatedly surrounds the trunk of a seta, giving the appearance of a screw (pl. 20, fig. e; text-fig. 10). This structure is called a circular fold in this paper. It is about 3 μm in length, and 1.5 μm in diameter, and the pitch of the fold is about 0.2 μm.

Below the circular fold, the setal cuticle is surrounded by many deep tubular depressions with round bottoms, which represent the structure herein named the multiple pouch structure (pl. 20, fig. e; pl. 22; text-figs. 10a, 11, 14). Each depression is about 10 μm in depth and 0.5 μm in diameter. About 10 to 30 depressions surround a seta.

The cuticles of setae, circular folds, and multiple pouch structures have the same structure as the epicuticle of the carapace surface and have a distinct cuticulin layers. Moreover, these cuticles are continuous with each other and the boundaries between these structures are indistinguishable from each other at the boundary of the structures.

The inner surface of the proximal part of a pore is lined with the membranous layer (pl. 20, fig. e; text-figs. 11, 15). The layer attaches to the multiple pouch structure near the base of each pouch.

In fossil specimens, the pores for B-type setae are distinguishable from others. The problems of preservation of pores are mentioned in Okada (1979) with other species.

Internal morphology of normal pores: In spite of the difference in forms of the setae, no significant difference was detected in the internal morphology of about one hundred

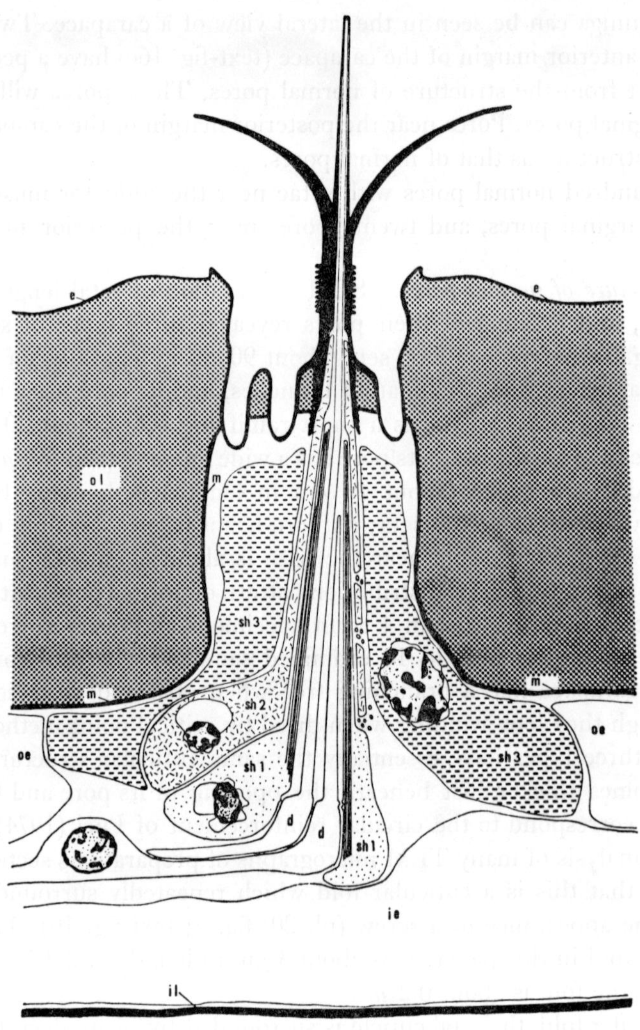

Text-fig. 12

Structure of setal sensillum. (d) dendrite. (e) epicuticle. (ie) inner epidermal cell. (il) inner lamella cuticule. (m) membranous layer. (oe) outer epidermal cell. (ol) outer lamella cuticule. (sh 1 to 3) sheath cells.

normal pores (text-fig. 12). Four ciliated cells (bipolar sensory neurons) and three sheath cells (cells enveloping dendrites) protrude into every pore. The nuclei of the four ciliated cells are situated between outer and inner epidermal cell layers near the inner opening of each pore. Each neuron has a nonbranching cilium. Two cilia are long and penetrate one branch of a seta where they are packed in a narrow tube (pl. 21; text-fig. 13). Two other cilia are short and never reach the setal cuticle. Near the ciliary base, the microtubules are arranged in the usual 9+0 pattern in two cilia, whereas they are arranged in an uncommon 9+0 pattern in the other two modified cilia (pl. 25). In the latter case, nine pairs of microtubules without apparent arms are closely

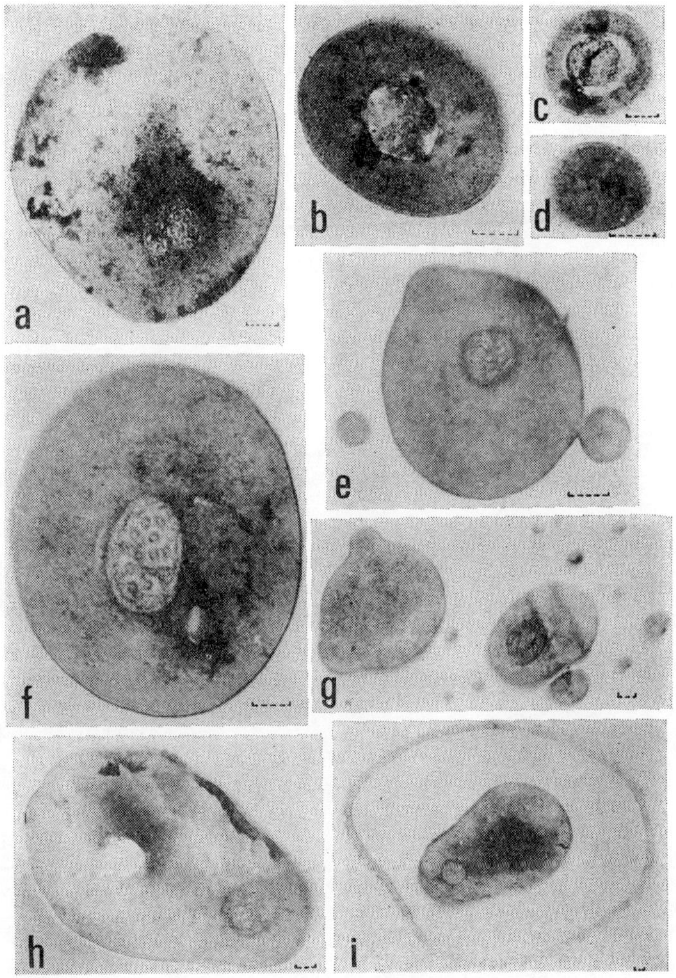

Text-fig. 13

Cross sections of setae at distal part. a–d: Setae of normal pore. d: A branch without cilia. e, f: Setae of pores near posterior margin. g: Setae of anterior marginal pores. Two long cilia penetrate one branch of each seta where they are packed in a narrow tube. h-i: Setae above the multiple pouch structure.

arranged in a smaller circle, and the ciliary matrix is denser. At the base of cilia of both types, two minor center tubules were sometimes observed which may correspond to the center microtubules of 9+2 pattern (pl. 26; pl. 28, fig. c). Near the same region, microtubules appear to be connected by an electron-dense substance which tubularly encloses the center. A striated rootlet begins at the basal endings of microtubules in the usual type cilia and lines the cell membrane of its dendrite at the plane where it adheres to an innermost sheath cell (sh-1) with a desmosome-like connection. At the distal part of cilia, the microtubules vary in their number and do not make the 9+0 arrangement.

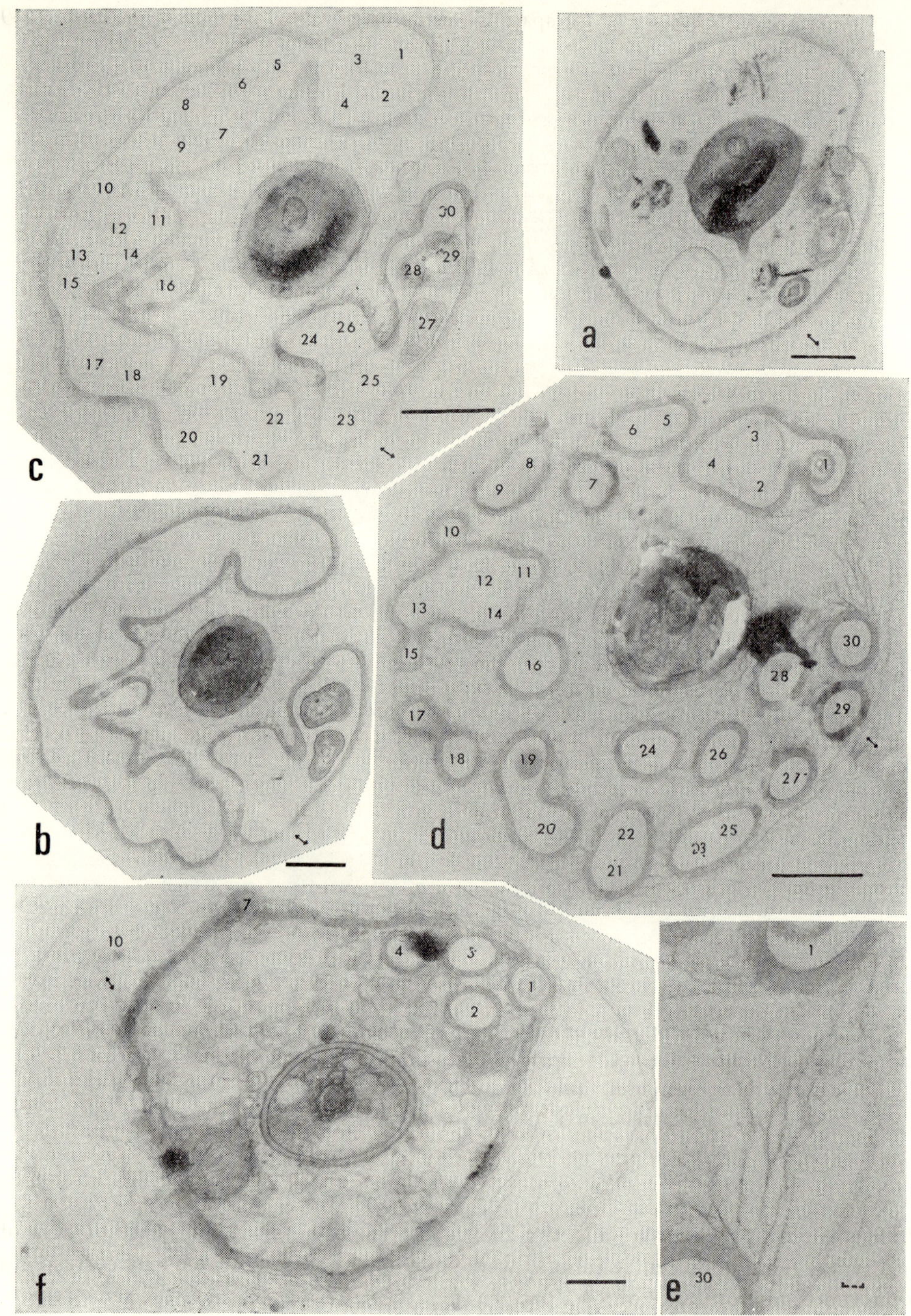

Text-fig. 14

Distal (a) to proximal (f) parts of multiple pouch structure. Tubular depressions of multiple pouch structure are numbered. Sections between fig. d and f are illustrated in pl. 22. e: Piled membrane structure around tubular depressions of fig. d.

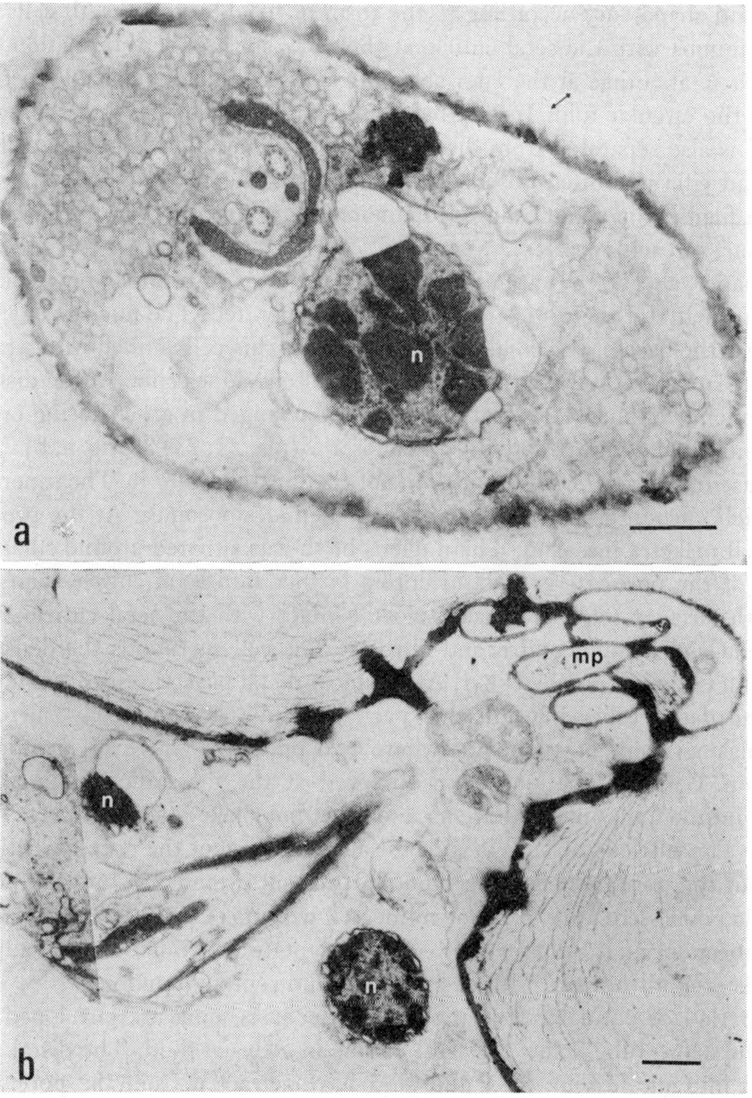

Text-fig. 15
a: Nuclei of the sh-3 cell near the proximal end of a pore. The sh-1 cell does not form a complete tube there. b: Oblique section of a sensillum showing ciliary bases, nuclei (n) of sh-2 and sh-3, and the multiple pouch structure (mp).

Serial thin sections of a sensilum suggest that the short cilia probably have the usual 9+0 pattern (pls. 22 to 25).

An electron-dense tube, which will correspond to the dense tube of Guthrie (1966) or the cuticular sheath of Anderson (1975), encloses cilia nearly from the base of cilia to the top of the seta (pls. 21 to 23; text-figs. 13, 14). The tube shows a quite different appearance from the cuticle of carapces and setae. It has no lamellate structure. Its thickness and shape vary according to the form of neighboring sheath cells. This tube is not continuous with the setal cuticle at the early stage of seta formation, and seems to fuse with setal cuticle at the later stage. It is often indistinguishable in the finished seta above the circular fold. In and below the circular fold, it stands apart from setal cuticle and is also separated from the adjacent cell membranes of sheath cells. Between the tube and cilia was often observed a filamentous sheath which runs parallel to the tube. Each filament is about 5 nm in diameter. They are bundled into bunches in the proximal part of each pore.

In this paper, sheath cells are called sh-1, sh-2 and sh-3 from center to margin (text-fig. 12). Chromatin of their nuclei are highly agglomerated. The nucleus of sh-1 is often situated near the base of the cilia. The cytoplasm of this cell is filled with straight intracellular fibrillae (about 5 nm in diameter) where it envelopes cilia. At the distal portion, the cell branches into about 10 parts, which are arranged in an enclosing layer around cilia and connected with each other by desmosome (pls. 22; 23; 28, fig. a, b). They reach near the base of the setal cuticle and some of them extend into it. The inner cell membranes which encounter the dense tube make hemidesmosomes. At the proximal portion, the cell makes a mesaxon. The nucleus of sh-2 is situated around ciliary base, but not inside of the pore. Its cytoplasm enters farther into setal cuticle than sh-1, ends just beneath circular fold, and makes close contact with the setal cuticle. It contains few organella such as mitocondria and fibrillae. The cell surrounds the dense tube and the enclosing layer of branched sh-1. At the most distal part, the sh-2 cell incompletely encloses the dense tube. Beneath this part, the tube is completely surrounded by sh-2 cytoplasm except at one point where two opposing sh-2 cell membranes form the mesaxon. Septate junctions were observed at the mesaxon. Proximally, sh-2 is often split into cytoplasmic digitations and does not make a sheath where sh-1 is non branching. The nucleus of sh-3 is situated near the base of the pore, sometimes in the basal part of the pore (text-fig. 15). Though this cell extends up into the pore toward the multiple pouch structure, it is not connected with its cuticle. The cytoplasm is often rich in mitocondoria. It adheres to sh-2 with septate junctions and to adjacent epidermal cells around the pore with desmosomes and septate junctions.

At the period of cuticular formation before ecdysis, cilia are enveloped with only sh-2 and the dense tube below the setal cuticle in ecdysial fluid. The distal part of the cilium runs into an old seta. Sh-1 and sh-3 have retracted from the pore. New setal cuticle is formed around the sh-2 cell. The tip of a new seta is situated just below the old setal cuticle. Supposing the cilia is sensible of the external stimulations even at the time of cuticular formation, the conduction of stimuli could be possible only with the sheath of sh-2 and the dense tube.

New setal cuticle was also observed around a filamentous cytoplasmic projection without cilia. Many micrographs clearly show that the cytoplasmic projection is a dendroid part of sh-2. The three types of setae therefore certainly originate from three types of sh-2 cells. Three types of setae should be sensitive to different stimuli owing to the different setal forms, while excitation in every pore may be the same because of the internal uniform structure. The ostracod may be sensitive to various external stimuli by the differentiation of sh-2 cells.

Pores near posterior margin: Setae in the pores near the posterior margin also have

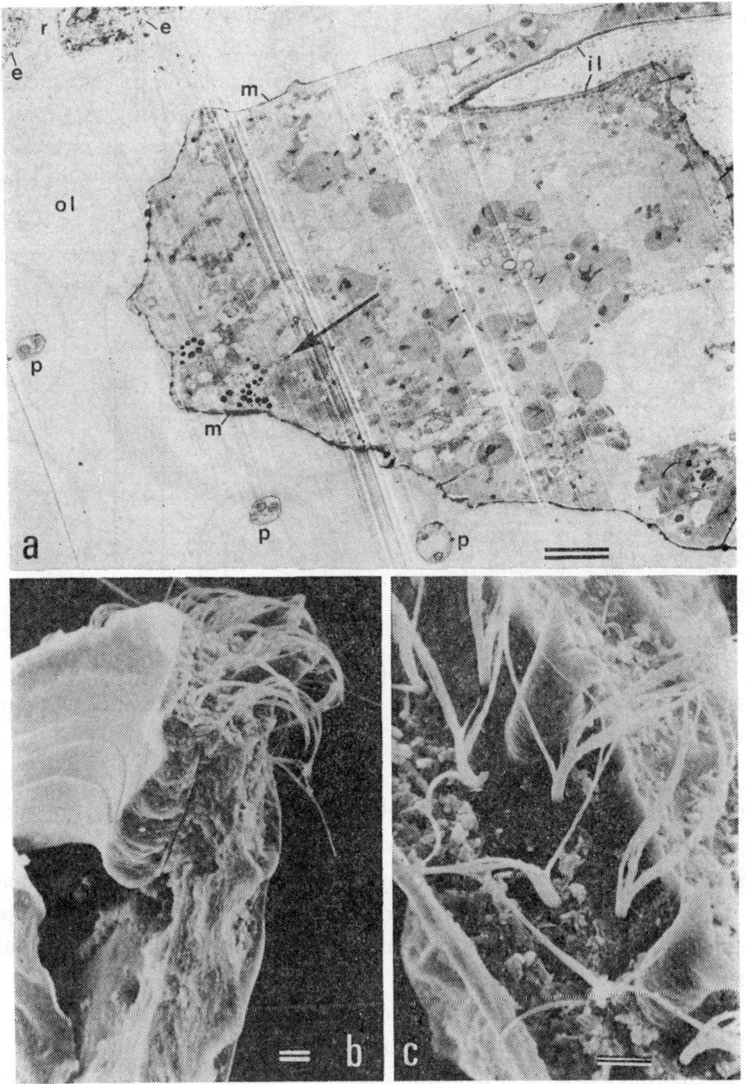

Text-fig. 16

a: Lower magnification micrograph of a section in the series. An arrow indicates the position of the sensillum in pl. 27, fig. c. Other pores (p) are also shown. (e) epicuticle. (il) inner lamella cuticle. (ol) outer lamella cuticle. (r) reticulated surface. b-c: Anterior marginal setae and anterior spines. b: Fractured specimen showing a long pore. c: Surface view showing two rows of anterior marginal setae.

circular folds and are surrounded by multiple pouch structures (pl. 29; text-fig. 17d). However, the number of depressions of the multiple pouch structure is smaller, about six to ten. Four ciliated cells and three sheath cells protrude into every pore, and two long cilia penetrate one branch of a seta, as in the case of normal pores. The nature of

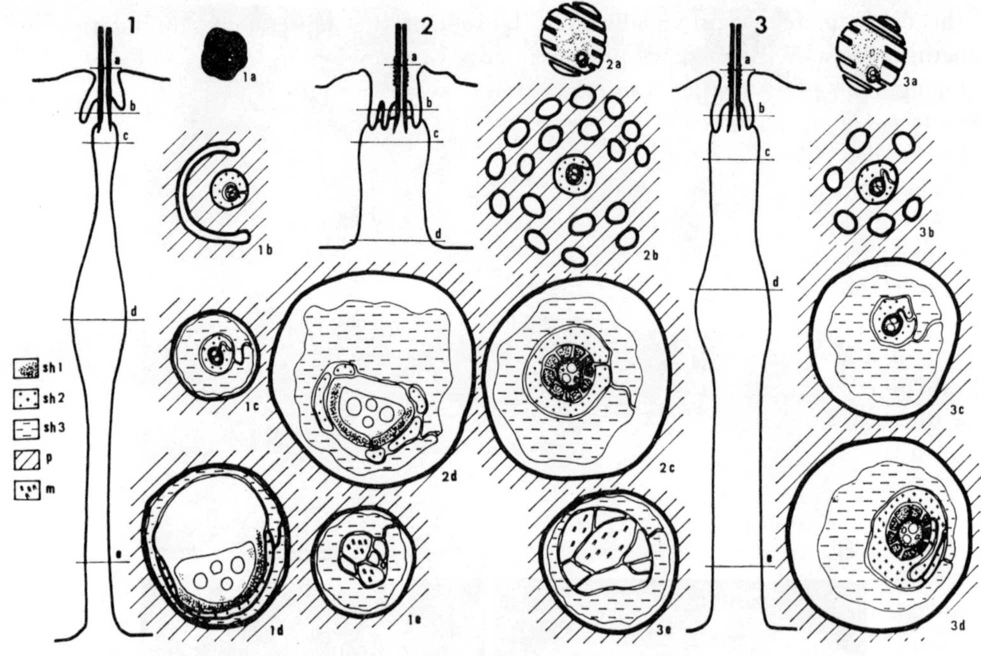

Text-fig. 17

Schemata of pores at three regions. 1. anterior marginal pore. 2. normal pore. 3. pore near posterior margin.

these cells, the connection between these cells, an electron-dense tube, and a filamentous sheath are essentially similar to that observed in the normal pores.

The pores are long (about 60 μm) as they penetrate the thick procuticle near the marginal zone. The structures observed in the normal pore are present in the distal one-fourth of each pore, and the structures in the epidermis under the normal pores are situated in the proximal three-fourths.

The bases of the cilia are present at the portion of one-third of the pore from the opening, where sh-1 becomes a wider tube and where the diameter of the pore also increases. The diameter is largest near the middle portion of the pore where two ciliated cells and sh-1 are connected by desmosomes. The pore of the portion is cylindroid. The major axis and minor axis are about 5 μm and 3 μm.

The proximal narrower portion of the pore contains a sh-3 cell which loosely sheathes four dendrite and a few filamentous cytoplasms. The cytoplasms are as thin as cilia and seem to be the continuations of sh-1 and sh-2 cells. Two dendrites are usually thick (about 0.5 μm in diameter) and contains microtubules and mitochondria.

Anterior marginal pores: The cuticular structure of anterior marginal pores is different from that of normal pores and pores near the posterior margin. The basal part of the seta is not round in cross section but ellipsoid with an uneven circumference. They are not surrounded by hellical folds. Cuticular depression around a seta does not form the multiple pouch structure but forms a skirt-form depression which surrounds a seta as a circular structure in cross section (pl. 30; text-fig. 17a). The diameter of the circle increases proximally. The bottom of the depression is undulating and sometimes shows a fragmentary portion in cross section.

Though the pores are longer and narrower than the normal pores (text-fig. 16b), the internal morphology of the pores is essentially the same as that described above. The base of the cilia is situated at the portion of one-fourth of a pore from the opening. The diameter of the pores is larger at about one-third of the pores from the opening, where two ciliated cells and sh-1 are connected by desmosomes. The anterior marginal pores are close to each other, as the density of pores is highest at anterior margin in the carapace. Two adjacent pores sometimes fuse with each other near the widest portion (pl. 30, fig. b). Just beneath the portion, nucleus of a sh-1 cell was observed in a pore. The proximal narrower portion of the pore contains the sheath cell (sh-3), four dendrites, and a few filamentous cytoplasm, as in the case of the pores near the posterior margin.

Acknowledgments

I wish to express my deep gratitude to Professor Tetsuro Hanai of the Geological Institute, University of Tokyo, under whose direction this study was carried out. Professor Tetsuro Hanai gave me helpful advice and read this manuscript. I would like to also express my appreciation to Professor Kenjiro Wake, Drs. Itaru Hayami, Kiyotaka Chinzei, Noriyuki Ikeya, Tomowo Ozawa, Toshiyuki Yamaguchi, and Masumi Ichikawa, and Osamu Tachikawa, Norimasa Nishida, and Paul Frydl for discussions and technical advice.

This article is a part of the author's doctoral thesis written in 1979. I thank to Professor Tetsuro Hanai, Professor Toshio Kimura, Professor Takeo Mizuno, Dr. Kiyotaka Chinzei and Dr. Itaru Hayami for reading the thesis at the doctoral examination in 1980.

REFERENCES CITED

Albertson, D. G., J. E. Sulston and J. G. White. 1978. Cell cycling and DNA replication in a mutant blocked in cell division in the nematode *Caenorhabditis elegans*. *Dev. Biol.*, v. 63 no. 1, p. 165–178, 8 fig., 3 tables.

Al-Furaih, A. F. 1977. Cretaceous and Palaeocene species of the ostracod *Hornibrookella* from Saudi Arabia. *Palaeontology*, v. 20, pt. 3, p. 483–502, pl. 53–58, 9 text-fig.

Andersson, A. 1975. The ultrastructure of the presumed chemoreceptor aethetasc "Y" of a cypridid ostracode. *Zooloica Scripta*, v. 4, no. 4, p. 151–158, 16 fig.

Aoki, N., T. Awano, A. Fukuzawa, O. Horiguchi, N. Ikeda, S. Kinosaki, K. Koike, T. Koike, Y. Mori, Y. Sugiyama, M. Suzuki, and F. Yoshimura. 1962. On the molluscan fossils of the Narita Formation, Boso Peninsula. *J. geol. Soc. Japan*, v. 68, no. 6, p. 341–346, 2 text-fig. [in Japanese]

——— and K. Baba. 1971. Molluscan fossil assemblages and their stratigraphic positions in the Semata, Kamiizumi and Narita Formations of the Kisarazu-Ichihara area, Boso Peninsula. *J. geol. Soc. Japan*, v. 77, no. 3, p. 137–151, 2 text-fig. [in Japanese]

Asano, K. 1936. Foraminifera of the Numa coral bed. *J. geol. Soc. Japan*, v. 43, no. 519, p. 921. [in Japanese]

Athersuch, J. 1979. On *Acanthocythereis hystrix* (Reuss). *Stereo-atlas of ostracod shells*, v. 6, pt. 2, p. 133–140.

——— and G. Ruggieri. 1975. On *Urocythereis phantastica* Athersuch and Ruggieri sp. nov. *Stereo-atlas of ostracod shell*, v. 2, pt. 4, p. 223–230.

Bate, R. H. and B. A. East. 1972. The structure of the ostracode carapace. *Lethaia*, v. 5, no. 2, p. 177–194, 11 fig.

——— and ———. 1975. The ultrastructure of the ostracode (Crustacea) integument. *In* F. M. Swain [ed.] Biology and paleobiology of Ostracoda. *Bull. Am. Paleontol.*, v. 65, no. 282, p. 529–547, 4 pl.

Benson, R. H. 1972. The *Bradleya* problem, with description of two new psychrospheric ostracode genera, *Agrenocythere*, and *Poseidonamicus* (Ostracoda: Crustacea). *Smithson. Contr. Paleobiology*, no. 12, 138 p., 14 pl., 67 text-fig.

———. 1974. The role of ornamentation in the design and function of the ostracode carapace. *Geoscience and Man*, v. 6, p. 47–57, 1 pl., 4 text-fig.

———. 1975*a*. Morphologic stability in Ostracoda. *In* F. M. Swain [ed.] Biology and paleobiology of Ostracoda. *Bull. Am. Paleontol.*, v. 65, no. 282, p. 13–46, 23 text-fig.

———. 1975*b*. Discussion. [In A. Bertels, "Ostracode ecology during the Upper Cretaceous and Cenozoic in Argentina"]. *In* F. M. Swain [ed.] Biology and paleobiology of Ostracoda. *Bull. Am. Paleontol.*, v. 65, no. 282, p. 341.

Bertels, A. 1969. "Rocaleberidinae", nueva subfamilia (Ostracoda, Crustacea) del limite Cretacico-Terciario de Patagonia Septentrional (Argentina). *Ameghiniana*, v. 6, no. 2, p. 116–171, 5 pl.

———. 1975. Ostracode ecology during the Upper Cretaceous and Cenozoic in Argentina. *In* F. M. Swin [ed.] Biology and paleobiology of Ostracoda. *Bull. Am. Paleontol.*, v. 65, no. 282, p. 317–351, 5 pl., 2 text-fig.

———. 1976. Evolutionary lineages of some Upper Cretaceous and Tertiary ostracodes of Argentina. *In* G. Hartmann [ed.] Evolution of post-Paleozoic Ostracoda. *Abh. Verh.*

naturwiss. Ver. Hamb., n. ser., 18/19 (supl.), p. 175–190, 3 pl., 1 text-fig.

Brady, G. S. 1868. A monograph of the Recent British Ostracoda. *Linn. Soc. London Trans.*, v. 26, no. 2, p. 353–495.

———. 1880. Report on the Ostracoda dredged by H. M. S. *Challenger*, during the years 1873–1876. Report of Scientific Results of Voyage H. M. S. *Challenger*, Zoology, v. 1, pt. 3, p. 1–184, pl. 1–44.

———. 1898. On new or imperfectly known species of Ostracoda, chiefly from New Zealand. *Trans. zool. Soc. London*, v. 14, pt. 8, p. 429–452, pl. 43–47.

Cannon, H. G. 1931. On the anatomy of a marine ostracod, *Cypridina* (*Doloria*) *levis* Skogsberg. *Discovery Rept.*, v. 2, p. 435–482, pl. 6, 7, 12 text-fig.

———. 1940. On the anatomy of *Gigantocypris mülleri*. *Discovery Rept.*, v. 19, p. 185–244, pl. 39–52, 16 text-fig.

Carbonnel, G. 1969. Les Ostracodes du Miocène Rhodanien. Part I, II. *Doc. Lab. Géol. Fac. Sci. Lyon*, no. 32 (1, 2), p. 1–469, 16 pl., 48 tex-fig.

Cronin, T. M. and H. Khalifa. 1979. Middle and Late Eocene Ostracoda from Gebel El Mereir, Nile Valley, Egypt. *Micropaleontology*, v. 25, no. 4, p. 397–411, 2 pl., 2 text-fig.

Deppe, U., E. Schierenberb, T. Cole, C. Krieg, D. Schmitt, B. Yoder, and G. von Ehrenstein. 1978. Cell lineages of the embryo of the nematode *Caenorhabditis elegans*. *Proc. natl. Acad. Sci. U.S.A.*, v. 75, no. 1, p. 376–380, 4 fig.

Doruk, N. 1973. On *Mutilus keiji* Ruggieri. *Stereo-atlas of ostracod shells*, v. 1, pt. 2, p. 117–120.

———. 1974. On *Urocythereis favosa* (Roemer). *Stereo-atlas of ostracod shells*, v. 2, pt. 1, p. 33–44.

Eguchi, M. and R. Mori. 1973. A study of fossil corals from Tateyama City and its environs and Recent coral fauna of Chiba Prefecture, central Japan. *Bull. Tokyo Coll. Domest. Sci.*, no. 13, p. 41–57, 5 pl., 2 text-fig., 5 tables. [in Japanese]

Gründel, J. 1969. Neue taxonomische Einheiten der Unterklasse Ostracoda. *Neues Jahrb. Geol. Paläontol. Mon.*, no. 6, p. 353–361.

———. 1973. Zur Entwickung der Trachyleberididae (Ostracoda) in der Unterkreide und in der tiefen Oberkreide. Teil 1: Taxonomie. *Z. geol. Wiss. Berlin* 1, v. 1, no. 11, p. 1463–1474, 6 fig.

———. 1976. Neue taxonomische Einheiten der Cytherocopina Gründel 1967 (Ostracoda). *Z. geol. Wiss. Berlin*, v. 4, no. 9, p. 1295–1304.

Guillaume, M. C. 1976. A taxonomic revision of two species of the family Leptocytheridae, *Leptocythere pellucida* (Baird) and *Leptocythere castanea* (Sars), with a description of a new species, *Leptocythere psammophila*. *In* G. Hartmann [ed.] Evolution of post-Paleozoic Ostracoda. *Abh. Verh. naturwiss. Ver. Hamb.*, n. ser., 18/19 (supl.), p. 325–330, 6 pl., 1 text-fig.

Guthrie, D. M. 1966. The function and fine structure of the cephalic airflow receptor in *Schistocerca gregaria*. *J. Cell. Sci.*, v. 1, no. 4, p. 463–470, 15 fig.

Hamada, T. 1963. Some problems on the Numa coral bed in Chiba Prefecture. *Chigaku Kenkyu*, spec. issues, p. 94–119, 3 pl., 10 text-fig., 2 tables. [in Japanese]

Hanai, T. 1957*a*. Studies on the Ostracoda from Japan, I. Subfamily Leptocytherinae, n. subfam. *J. Fac. Sci. Univ. Tokyo*, sec. 2, v. 10, pt. 3, p. 431–468, pl. 7–10, 2 text-fig.

———. 1957*b*. Studies on the Ostracoda from Japan, II. Subfamily Pectocytherinae n. subfam. *J. Fac. Sci. Univ. Tokyo*, sec. 2, v. 10, pt. 3, p. 469–482, pl. 11, 6 text-fig.

———. 1957*c*. Studies on the Ostracoda from Japan, III. Subfamilies Cytherurinae G. W. Müller (emend. G. O. Sars, 1925) and Cytheropterinae n. subfam. *J. Fac. Sci. Univ. Tokyo*, sec. 2, v. 11, pt. 1, p. 11-36, pl. 2–4, 9 text-fig.

———. 1959. Studies on the Ostracoda from Japan: Historical review with bibliographic index of Japanese Ostracoda. *J. Fac. Sci. Univ. Tokyo*, sec. 2, v. 11, pt. 4, p. 419–439.

———. 1961. Studies on the Ostracoda from Japan: Hingement. *J. Fac. Sci. Univ. Tokyo*, sec. 2, v. 13, pt. 2, p. 345–377, 14 text-fig.

———. 1970. Studies on the ostracod subfamily Schizocytherinae, Mandelstam. *J. Paleontol*, v. 44, no. 4, p. 693–729, pl. 107, 108, 19 text-fig., 5 tables.

———. 1974. Notes on the taxonomy of Japanese Cypridnids. *Geoscience and Man*, v. 6, p. 117–126, 4 text-fig.

———, N. Ikeya, K. Ishizaki, Y. Sekiguchi, and M. Yajima. 1977. Checklist of Ostracoda from Japan and its adjacent seas. *Univ. Mus., Univ. Tokyo, Bull.* no. 12, vi+242 p., 4 pl., 2 text-fig.

———, ——— and M. Yajima. 1980. Checklist of Ostracoda from Southeast Asia. *Univ. Mus., Univ. Tokyo, Bull.* no. 17, 242 p., 4 text-fig.

Harding, J. P. 1965 [1964]. Crustacean cuticle with reference to the ostracod carapace. *Publ. staz. zool. Napoli*, v. 33, supl., p. 9–31, 28 fig.

——— and P. C. Sylvester-Bradley. 1953. The ostracod genus *Trachyleberis*. *Brit. Mus. nat. Hist., Bull. Zool.*, v. 2, no. 1, 15 p., 2 pl., 25 text-fig.

Hartmann, G. 1959. Zur Kenntnis der lotischen Lebensbereiche der pazifischen Küste von El Salvador unter besonderer Berücksichtigung seiner Ostracoden fauna. (III. Beitrag zur Fauna El Salvador). *Kieler Meeresforsch.*, v. 15, no. 2, p. 187–241, 22 text-fig.

———. 1964. Zur Kenntnis der Ostracoden des Roten Meeres. *Kieler Meeresforsch.*, v. 20, spec. pub., p. 35–127, 62 pl.

———. 1966. Ostracoda. *In* H. G. Bronns's Klassen und Ordnungen des Tierreichs. 5. Bd. Arthropoda, 1. Abt. Crustacea, 2. Buch, 4. Teil, 1. Lieferung, 216 p., 121 text-fig.

———. 1978. [in Hartmann-Schröder, G. and G. Hartmann, 1978] Zur Kenntnis des Eulitorals der australischen Küsten unter besonder Berücksichtigung der Polychaeten und Ostracoden. *Mitt. hamb. zool. Mus. Inst.*, v. 75, p. 63–219, pl. 1–14, 673 text-fig.

——— and H. S. Puri. 1974. Summary of neontological and paleontological classification of Ostracoda. *Mitt. hamb. zool. Mus. Inst.*, v. 70, p. 7–73.

Hay, W. W. and P. A. Sandberg. 1967. The scanning electron microscope, a major breakthrough for micropaleontology. *Micropaleontology*, v. 13, no. 4, p. 407–418, 2 pl., 1 table.

Hazel, J. E. 1967. Classification and distribution of the recent Hemicytheridae and Trachyleberididae (Ostracoda) off northeastern North America. *U.S. geol. Surv., Prof. Paper* 564, 49 p., 11 pl., 2 text-fig.

———. 1970. Atlantic continental shelf and slope of the United States, ostracode zoogeography in the southern Nova Scotian and northern Virginian faunal Provinces. *U.S. geol. Surv., Prof. Paper*, 529-E, p. i-v, E1–E21, 69 pl., 11 fig., 3 tables.

Herrig, E. 1977. Ostracoden aus dem Plio/Pleistozän der Sozialistischen Republik Vietnam. Teil II. *Z. geol. Wiss.*, v. 5, no. 10, p. 1253–1267, 2 pl., 6 text-fig.

Holden, J. C. 1967. Late Cenozoic ostracodes from the drowned terraces in the Hawaiian Islands. *Pacific Sci.*, v. 21, no. 1, p. 1–50, 37 text-fig.

Horiguchi, O. and K. Ohara. 1972. Diatom fossil assemblages from the Anegasaki and Narita Formations of the Kisarazu-Chiba area, Boso Peninsula. *J. geol. Soc. Japan*, v. 78, no. 6, p. 281–287, 2 text-fig.

Hoshino, M. 1967. Radiocarbon age of Numa coral bed. *Earth Science* (*Chikyu Kagaku*), v. 21, no. 6, p. 38–39, 1 text-fig. [in Japanese]

Howe, H. V. 1961. Family Hemicytheridae Puri, 1953. *In* R. C. Moore [ed.] Treatise on invertebrate paleontology, pt. Q Arthropoda 3, Crustacea, Ostracoda. *Geol. Soc. Am.* Lawrence: University of Kansas Press. p. Q300–Q306.

Howe, R. C. and H. J. Howe. 1975. Species determination of molts from the Shubuta Clay of Mississippi. *In* F. M. Swain [ed.] Biology and paleobiology of Ostracoda. *Bull. Am. Paleontol.*, v. 65, no. 282, p. 61–75, 1 pl., 7 text-fig.

Hu, C. H. 1976. Studies on the Pliocene ostracodes from the Cholan Formation, Miaoli district, Taiwan. *Proc. geol. Soc. China*, no. 19, p. 25–51, 3 pl., 20 text-fig.

———. 1977*a*. Studies on ostracodes from the Pleistocene Toukoshan Formation in the Miaoli area, Taiwan. *Proc. geol. Soc. China*, no. 20, p. 80–107, 4 pl., 19 text-fig.

———. 1977*b*. Studies on ostracodes from the Toukoshan Formation (Pleistocene), Miaoli district, Taiwan. *Petroleum Geology of Taiwan*, no. 1, p. 181–217, 27 text-fig.

——— and Y. N. Cheng. 1977. Ostracodes from the Late Pleistocene Lungkang Formation near Miaoli, Taiwan. *Mem. geol. Soc. China*, no. 2, p. 191–205, 3 pl., 13 text-fig.

——— and L. C. Yang. 1975. Studies on Pliocene ostracodes from the Chinshui Shale, Miaoli district, Taiwan. *Proc. geol. Soc. China*, no. 18, p. 103–114, 2 pl.

——— and K. Y. Yeh. 1978. Ostracod faunas from Pleistocene Liushuang Formation in the Taiwan area, Taiwan. *Proc. geol. Soc. China*, no. 21, p. 151–162, 3 pl. 6 text-fig.

Ikebe, N. 1936. Pleistocene shell bed of Toyonari, Tiba Prefecture, with description of two new species of *Odostomia* and a note on *Arcopagia* (*Merisca*) *serricostata* (Tokunaga). *The Venus*, v. 4, no. 4, p. 189–205, 8 text-fig. [in Japanese]

Ikeya, N. 1970. Population ecology of benthonic Foraminfera in Ishikari Bay, Hokkaido, Japan. *Rec. oceanogr. Works Japan*, v. 10, no. 2, p. 173–191, 10 fig., 1 table.

———. 1971. Species diversity of recent benthonic Foraminifera off the Pacific coast of north Japan. *Rep. Fac. Sci. Shizuoka Univ.*, v. 6, p. 179–201, 7 text-fig., 1 table.

———. 1977. Ecology of Foraminifera in the Hamana Lake region on the Pacific coast of Japan. *Rep. Fac. Sci. Shizuoka Univ.*, v. 11, p. 131-159, 12 text-fig., 1 table.

———, T. Hanai and Y. Okada. 1976. Culture methods of benthonic Foraminifera. *Marine Sciences*, v. 8, no. 2, p. 128–134, 3 text-fig. [in Japanese]

——— and T. Handa. 1972. Surface sediments in Hamana Lake, the Pacific coast of Japan. *Rep. Fac. Sci. Shizuoka Univ.*, v. 7, p. 129–148, 14 text-fig.

Imamura, A. 1925. Change of the coast line in Boso Peninsula. *Rep. Imp. Earthquake Invest. Comm.*, v. 100, no. 2, p. 91–93, 4 fig. [in Japanese]

Ishizaki, K. 1966. Miocene and Pliocene ostracodes from the Sendai area, Japan. *Tohoku Univ., Sci. Rep.*, 2nd ser. (Geol.), v. 37, no. 2, p. 131–163, pl. 16–19, 1 text-fig.

———. 1968. Ostracodes from Uranouchi Bay, Kochi Prefecture, Japan. *Tohoku Univ., Sci. Rep.*, 2nd ser. (Geol.), v. 40, no. 1, p. 1–45, pl. 1–9, 17 text-fig.

———. 1969. Ostracodes from Shinjiko and Nakanoumi, Shimane Prefecture, western Honshu, Japan. *Tohoku Univ., Sci. Rep.*, 2nd ser. (Geol.), v. 41, no. 2, p. 197–224, pl. 24–26, 16 text-fig.

———. 1971. Ostracodes from Aomori Bay, Aomori Prefecture, northeast Honshu, Japan. *Tohoku Univ., Sci. Rep.*, 2nd ser. (Geol.), v. 43, no. 1, p. 59–97, pl. 1–7, 8 fig.

———. 1972. Morphological variation of *Leguminocythereis hodgii* (Brady), Ostracoda (Crustacea), from Japan. Prof. Jun-ichi Iwai memorial volume, p. 697–707, pl. 7, 8, 5 text-fig., 4 tables.

———. 1975. Morphological variation in *Leguminocythereis*? *hodgii* (Brady), Ostracoda (Crustacea) from Japan. *In* F. M. Swain [ed.] Biology and paleobiology of Ostracoda. *Bull. Am. Paleontol.*, v. 65, no. 282, p. 245–262, 2 pl., 5 text-fig., 3 tables.

——— and M. Kato. 1976. The basin development of the diluvium Furuya Mud basin, Shizuoka Prefecture, Japan, based on faunal analysis of fossil ostracodes. *In* Y. Takayasu and T. Saito [eds.] Progress in micropaleontology, p. 118–143, 4 pl., 9 text-fig.

Jørgensen, N. O. 1970. Ultrastructure of some ostracods. *Bull. geol. Soc. Denmark*, v. 20, pt. 1, p. 79–92, 7 pl., 1 fig., 1 table.

Kanamori, H. 1973. Mode of strain release associated with major earthquakes in Japan. *Ann. Rev. Earth Planet. Sci.*, v. 1, p. 213–239, 19 fig., 1 table.

Key [Keij], A. J. 1953. Preliminary note on the Recent Ostracoda of the Snellius Expedition. *Proc. K. Ned. Akad. Wet.*, ser. B, v. 56, no. 2, p. 155–168, 2 pl., 1 text-fig.

———. 1954. Some Recent Ostracoda of Manila (Philippines). *Proc. K. Ned. Akad. Wet.*, ser. B, v. 57, no. 3, p. 351–363, 3 pl., 2 text-fig.

Keij, A. J. 1979a. On two new *Baltrella* species (Ostracoda, Bythocytheridae) from the western Indopacific. *Proc. K. Ned. Akad. Wet.*, ser. B, v. 82, no. 1, p. 32–44, 2 pl., 3 text-fig.

———. 1979b. Review of the Indo-west Pacific Neogene to Holocene ostracode genus *Atjehella. Proc., K. Ned. Akad. Wet.*, ser. B. v. 82, no. 4, p. 449–464, 2 pl., 8 text-fig.

———. 1979c. Brief review of type species of genera from the Kingma collection.*In* N. Krstić [ed.] Proceedings of VII international symposium on ostracodes. Taxonomy, biostratigraphy and distribution of ostracodes. The Servian Geological Society, Beograd. p. 59–62, 2 pl.

Kesling, R. V. 1951. The morphology of ostracod molt stages. *Illinois Biological Monographs*, v. 21, no. 1–3, p. 1–126, 46 pl., 5 charts, 36 fig., 8 tables.

Kigoshi, K., H. Aizawa and N. Suzuki. 1969. Gakushuin natural radiocarbon measurements VII. *Radiocarbon*, v. 11, no. 2, p. 295–326.

Kikuchi, T. 1972. A characteristic trace fossil in the Narita Formation and its paleogeographical significance. *J. geol. Soc. Japan*, v. 78, no. 3, p. 137–144, 1 pl., 6 text-fig. [in Japanese]

Kilenyi, T. I. 1971. Some basic questions in the paleoecology of ostracods. In H. J. Oertli [ed.] Paléoécologie d'Ostracodes, Pau, 1970. *Bull. Centre Rech. PAU-SNPA*, v. 5, supl., p. 31–44, 3 fig.

Kimble, J. and D. Hirsh. 1979. The postembryonic cell lineages of the hermaphrodite and male gonads in *Caenorhabditis elegans. Dev. Biol.*, v. 70, no. 2, p. 396–417, 18 fig.

Kobayashi, H., Y. Matsui and H. Suzuki. 1971. University of Tokyo radiocarbon measurements IV. *Radiocarbon*, v. 13, no. 1, p. 97–102.

Konishi, K. 1967. Rate of vertical displacement and dating of reefy limestones in the marginal facies of the Pacific Ocean. *Quaternary Res.* (Japan), v. 6, no. 4, p. 207–223, 8 text-fig., 6 tables. [in Japanese]

Kontrovitz, M. 1967. An investigation of ostracode preservation. *Quat. J. Florida Acad. Sci.*, v. 29, no. 3, p. 171–177, 3 fig.

———. 1975. A study of the differential transportation of ostracodes. *J. Paleontol.*, v. 49, no. 5, p. 937–941, 3 text-fig., 2 tables.

——— and M. J. Nicolich. 1979. On the response of ostracode valves and carapaces to water currents. *In* N. Krstić [ed.] Proceedings of the VII International Symposium on ostracodes, Belgrade, p. 269–272, 2 fig., 1 table.

Kornicker, L. S. 1969. Relationship between the free and attached margins of the myodocopid ostracod shell. *In* J. W. Neale [ed.] The taxonomy, morphology and ecology of Recent Ostracoda. p. 109–135, 7 pl., 12 fig.

——— and F. E. Caraion. 1974. West African myodocopid Ostracoda (Cylindroleberididae). *Smithson. Contr. Zool.*, no. 179, 78 p., 43 fig.

——— and I. G. Sohn. 1971. Viability of ostracode eggs egested by fish and effect of digestive fluids on ostracode shells—ecologic and paleoecologic implications. *In* H. J. Oertli [ed.] Paléoécologie des Ostracodes, Pau, 1970. *Bull. Centre Rech. Pau-SNPA*, v. 5, supl., p. 125–135, 1 pl., 3 tables.

Kozima, N. 1958*a*. Geological study of the Kioroshi district, Chiba Prefecture, Japan. The studies on the Narita Group (1). *J. geol. Soc. Japan*, v. 64, no. 4, p. 165–171, 4 text-fig.

[in Japanese]

———. 1958*b*. On the mode of occurrence of the fossil shells in the Narita Formation of the Kioroshi district, Chiba Prefecture, Japan. The studies on the Narita Group (2). *J. geol. Soc. Japan*, v. 64, no. 5, p. 213–221, pl. 3, 4; 3 text-fig. [in Japanese]

———. 1959. Geological studies of the environs of Imba-numa, Chiba Prefecture. The studies on the Narita Group (3). *J. geol. Soc. Japan*, v. 65, no. 10, p. 595–605, 4 text-fig. [in Japanese]

———. 1962. On the Narita Group in the district from south of Imba-numa to Ōami Shirasato-machi, Chiba Prefecture, Japan. The studies on the Narita Group (4). *J. geol. Soc. Japan*, v. 68, no. 12, p. 676–686, 6 text-fig. [in Japanese]

———. 1963. Geological history and sedimentation of the Narita Group from the vicinity of Tega-numa to Ōami Shirasato-machi, Chiba Prefecture, Japan. The studies on the Narita Group (5). *J. geol. Soc. Japan*, v. 69, no. 4, p. 172-183, 2 pl., 4 text-fig. [in Japanese]

———. 1966*a*. Geological structure of the southern coast of Tokyo Bay, Japan. Studies on the Narita Group (6). *J. geol. Soc. Japan*, v. 72, no. 4, p. 205–212, 5 text-fig. [in Japanese]

———. 1966*b*. On the fossil shells in the Narita Group along the southeast coast of Tokyo Bay, Japan. Studies on the Narita Group (7). *J. geol. Soc. Japan*, v. 72, no. 12, p. 573–584, 4 text-fig. [in Japanese]

Kuroda, T. and K. Hatanaka. 1979. Palynological study of the late Quaternary in the coastal plain along Hakata Bay, in Fukuoka City, northern Kyushu, Japan. *Quaternary Res.* (Japan), v. 18, no. 2, p. 53–68, 2 pl., 3 fig., 2 tables.

Langer, W. 1971. Rasterelektronenmikroscopische Beobachtungen über den Feinbau von Ostracoden-Schalen. *Paläontol. Z.*, v. 45, no. 3/4, p. 181–186, pl. 22, 23; 2 text-fig.

Lawrence, P. A. 1966. Development and determination of hairs and bristles on the milk-weed bug, *Oncopeltus fasciatus* (Lygaeidae, Hemiptera). *J. Cell Sci.*, v. 1, no. 4, p. 475–498, 15 fig., 1 table.

Liebau, A. 1969. Homologisierende Korrelation von Trachyleberididen-Ornamenten (Ostracoda, Cytheracea). *Neues Jahrb. Geol. Paläontol.*, Mh. 7, p. 390–402, 4 fig.

———. 1971. Homologe Skulpturmuster bei Trachyleberididen und verwandten Ostrakoden. Dissertation an der Technischen Universität Berlin, D 83, 118 p., 32 fig. (English translation, 1977).

———. 1975*a*. Comment on suprageneric taxa of the Trachylebeirdidae s. l. (Ostracoda, Cytheracea). *Neues Jahrb. Geol. Paläontol.*, *Abh.*, v. 148, no. 3, p. 353–379, 3 text-fig.

———. 1975*b*. The left-right variation of the ostracode ornament. *In* F. M. Swain [ed.] Biology and paleobiology of Ostracoda. *Bull. Am. Paleontol.*, v. 65, no. 282, p. 77–86. 1 pl., 5 text-fig.

———. 1977. Carapace ornamentation of the Ostracoda Cytheracea: Principles of evolution and functional significance. *In* H. Löffler and D. Danielopol [eds.] Aspects of ecology and zoogeography of Recent and fossil Ostracoda. Saalfelden, 1976. The Hague: Dr. W. Junk b. v. Publishers. p. 107–120, 2 pl.

———. 1978. Die Evolution der Trachyleberididen-Poren—Differentiation eines Sinnesorgansystems. *Neues Jahrb. Geol. Paläontol.*, Abh., v. 157, no. 1/2, p. 128–133, 2 text-fig.

Locke, M. 1966. The structure and formation of the cuticulin layer in the epicuticle of an insect, *Calpodes ethlius* (Lepidoptera, Hesperiidae). *J. Morphol.*, v. 118, no. 4, p. 461–494, 37 fig.

Maddocks, R. F. 1966. Distribution pattern of living and subfossil podocopid ostracodes in the Nosy Be area, Northern Madagascar. *Paleontol. Contr. Univ. Kans.*, Arthropoda, Paper 12, 77 p., 63 fig., 11 tables.

Mandelstam, M. I., G. F. Shneider, Z. V. Kuznetsova, and F. I. Katz. 1957. New genera of Ostracoda in the families Cypridae and Cytheridae. *All-Union paleontol. Soc., Ann. Mosc.*, v. 16, p. 165–193, 4 pl. [in Russian]

Matsuda, T., Y. Ota, M. Ando, and N. Yonekura. 1978. Fault mechanism and recurrence time of major earthquakes in the southern Kanto district, Japan, as deduced from coastal terrace data. *Geol. Soc. Am. Bull.*, v. 89, no. 11, p. 1610–1618, 12 fig., 1 table.

Matsushima, Y. 1979. Littoral molluscan assemblages during the postglacial Jomon transgression in the southern Kanto, Japan. *Quaternary Res.* (Japan), v. 17, no. 4, p. 243–265, 9 text-fig., 6 tables. [in Japanese]

McKenzie, K. G. 1967. Recent Ostracoda from Port Phillip Bay, Victoria. *Proc. r. Soc. Victoria*, v. 80, pt. 1, p. 61–106, pl. 11–13, 10 text-fig.

Morkhoven, F. P. C. M. van. 1962. 1963. Post-Paleozoic Ostracoda: Their morphology, taxonomy and economic use. vol. I, general, vii+204 p., 79 fig. Vol. II, generic descriptions, 478 p., 763 fig., Elsevier, Amsterdam.

Motomura, I. 1932. A statistical method in animal synecology. *Zool. Mag. Japan*, v. 44, no. 528, p. 379–383, 3 fig. [in Japanese]

Müller, G. W. 1894. Die Ostracoden des Golfes von Neapel und der angrenzenden Meersabschnitts. *Naples Stn. Zool. Fauna Flora, Golfes Neapel, Monographie* 31, p. 1–404.

Nakata, T., M. Koba, T. Imaizumi, W. R. Jo., H. Matsumoto, and T. Suganuma. 1980. Holocene marine terraces and seismic crustal movements in the southern part of Boso Peninsula, Kanto, Japan. *Geogr. Rev. Japan*, v. 53, no. 1, p. 29–44, 7 text-fig., 1 table. [in Japanese]

Naruse, Y. and A. Sugimura. 1953. The Tomoegawa shell bed. *J. geol. Soc. Japan*, v. 59, no. 690, p. 92. [in Japanese]

Naumann, E. 1877. Ohshima Kazan-ki [Ohshima volcanoes, translated by I. Wada], Gakugeishirin, Tokyo, p. 1–48.

Neale, J. W. 1965 [1964]. Some factors influencing the distribution of Recent British Ostracoda. *Publ. Stn. Zool. Napoli*, suppl., p. 247–307, 11 text-fig.

———. 1974. The genus *Finmarchinella* Swain, 1963 (Crustacea: Ostracoda) and its species. *Bull. Br. Mus. nat. Hist.*, (*Zool.*), v. 27, no. 2, p. 83–94, 2 pl., 2 text-fig.

——— and H. V. Howe. 1975. The marine Ostracoda of Russian Harbour, Novaya Zemlya and other high latitude faunas. *In* F. M. Swain [ed.] Biology and paleobiology of Ostracoda. *Bull. Am. Paleontol.*, v. 65, no. 282, p. 381–431, 7 pl., 10 text-fig.

Nomura, S. 1932. Molluscs from the raised beach deposits of the Kanto region. *Sci. Rep. Tohoku Imp. Univ.*, 2nd ser., v. 15, p. 65–141, 1 map.

Nonaka, T., T. Muramatsu, K. Suzuki, and T. Hirai. 1973. Effects of the development works on the shallow water area of the Hamana-ko Bay: Circulation of sea water. *Suisan Doboku*, v. 10, no. 1, p. 47–51. [in Japanese]

Ohara, S. 1968. The type Semata Formation. *J. Coll. Arts Sci. Chiba Univ.*, v. 5, no. 2, p. 303–318, 2 text-fig. [in Japanese]

Ohmert, W. 1968. Die Coquimbinae, eine neue Unterfamilie der Hemicytheridae (Ostracoda) aus dem Pliozän von Chile. *Mitt. Bayer. Staatssaml. Paläontol. hist. Geol.*, v. 8, p. 127–165, 38 text-fig.

———. 1971. Die Ostracodengattungen *Palaciosa* und *Caudites* aus dem Pliozän von Chile. *Mitt. Bayer. Staatssamml. Paläontol. hist. Geol.*, v. 11, p. 87–116, pl. 9, 10; 21 text-fig.

Oinomikado, T. 1937. Shell bed from Iwatomi-machi and Sakura-machi, Chiba Prefecture. *Chikyu*, v. 27, no. 5, p. 323–335, 1 text-fig. [in Japanese]

Okada, Y. 1979. Stratigraphy and Ostracoda from Late Cenozoic strata of the Oga Peninsula, Akita Prefecture. *Trans. Proc. palaeontol. Soc. Japan*, n. s., no. 115, p. 143–173, pl. 21–23, 18 text-fig.

Okubo, I. 1977. Fourteen species of the genus *Paradoxostoma* Fischer, 1855 in the Inland Sea, Japan (Ostracoda). *Publ. Seto Mar. biol. Lab.*, v. 24, nos. 1/3, p. 99–131, 20 fig., 1 table.

———. 1979a. Three species of *Xestoleberis* (Ostracoda) from the Inland Sea of Japan. *Proc. Japan. Soc. syst. Zool.*, no. 16, p. 9–16, pl. 1.

———. 1979b. Six species of marine Ostracoda from the Inland Sea of Seto. *Res. Bull. Okayama Shujitsu jun. Coll.*, no. 9, p. 143–157, 7 fig.

———. 1980. Taxonomic studies on Recent marine podocopid Ostracoda from the Inland Sea of Japan. *Publ. Seto Mar. biol. Lab.*, v. 25, no. 5/6, p. 389–443, 23 fig., 3 tables.

Omatsola, M. E. 1970. On structure and morphologic variation of normal pore system in Recent cytheroid Ostracoda (Crustacea). *Acta. Zool.*, v. 51, p. 115–124, 3 pl., 1 text-fig.

———. 1971. *Campylocythereis*, a new genus of the Campylocytherinae (Ostr. Crust.) and its muscle scar variation. *In* H. J. Oertli [ed.] Paléoécologie d'Ostracodes Pau 1970. *Bull. Centre Rech., Pau-SNPA*, v. 5, supl., p. 101–123, 4 pl., 3 text-fig.

Omoto, K. 1976. Tohoku University radiocarbon measurements III. *Sci. Rep. Tohoku Univ.*, ser. 7, v. 26, no. 1, p. 135–157.

———. 1978. Tohoku University radiocarbon measurements IV. *Sci. Rep. Tohoku Univ.*, ser. 7, v. 28, no. 1, p. 101–116.

Omura, H. 1926. The change of elevation of land caused by the great earthquake of September 1st, 1923. *Bull. Earthquake Res. Inst. Univ., Tokyo Imp.*, v. 1, p. 65–68, 1 map. [in Japanese]

Oyama, K. and S. Ishiyama. 1968. A problem for estimating sedimentary condition as an example of horizons in Kami-izumi, Chiba Prefecture. *Bull. geol. Surv. Japan*, v. 19, no. 9, p. 569–574, 1 text-fig. [in Japanese]

Phleger, F. B. 1951. Ecology of Foraminifera, northwest Gulf of Mexico. Pt. 1, Foraminiferal distribution. *Geol. Soc. Am., Mem.* 46, p. 1–88, 2 pl., 33 fig., 30 tables.

———. 1960. Ecology and distribution of Recent Foraminifera. Baltimore: Johns Hopkins University Press, 297 p.

Plusquellec, P. L. and P. A. Sandberg. 1969. Some genera of the ostracode subfamily Campylocytherinae. *Micropaleontology*, v. 15, no. 4, p. 427–480, 10 pl., 14 text-fig.

Pokorný, V. 1955. Contribution to the morphology and taxionomy of the subfamily Hemicytherinae Puri, 1953. (Crust., Ostracoda). *Acta Univ. Carolinae*, 3 *Geol.*, p. 1–35, 19 text-fig.

———. 1968. *Radimella* gen. n., a new genus of the Hemicytherinae (Ostracoda, Crust.). *Acta Univ. Carolinae, Geol.*, 1968, no. 4, p. 359–373, 12 fig.

———. 1969. The genus *Radimella* Pokorný, 1969 (Ostracoda, Crustacea) in the Galapagos Islands. *Acta Univ. Carolinae, Geol.* 1969, no. 4, p. 293–334, 10 pl., 39 fig.

———. 1971. The genus *Caudites* Coryell and Fields, 1937 (Ostracoda, Crust.) in the Galapagos Islands. *Acta Univ. Carolinae, Geol.* 1971, no. 4, p. 267–302, 6 pl., 20 text-fig.

Puri, H. S. 1952. Ostracode genus *Cytherideis* and its allies. *J. Paleontol.*, v. 26, no. 6, p. 902–914, pl. 130, 131, 14 text-fig.

———. 1974. Normal pores and the phylogeny of Ostracoda. *Geoscience and Man*, v. 6, p. 137–151, 13 pl., 1 text-fig.

———, G. Bonaduce and J. Malloy. 1965 [1964]. Ecology of the Gulf of Naples. *Publ. Stn. Zool. Napoli*, v. 33, supl., p. 87–199, 67 text-fig.

——— and B. E. Dickau. 1969. Use of normal pores in taxonomy of Ostracoda. *Trans. Gulf Coast Assoc. Geol. Soc.*, v. 19, p. 353–367, 6 pl.

——— and N. C. Hulings. 1976. Designation of lectotypes of some ostracods from the *Challenger* Expedition. *Bull. Br. Mus. nat. Hist.*, (*Zool.*), v. 29, no. 5, p. 249–315, 27 pl., 14 text-fig.

Reyment, R. A. 1966. Preliminary observations on gastropod predation in the western Niger delta. *Paleogeogr. Paleoclim. Paleoecol.*, v. 2, no. 2, p. 81–102, 4 fig., 3 tables.

———. 1967. Paleoethology and fossil drilling gastropods. *Kansas Acad. Sci., Trans.*, v. 70, p. 33–50, 2 fig., 3 tables.

Rome, D. R. 1944. Note sur la présence de Scolopoides chez les Ostracodes. *Bull. Mus. r. Hist. nat. Belg.*, v. 20, no. 2, p. 1–8, 2 text-fig.

———. 1947. *Herpetocypris reptans* Baird (Ostracoda). Etude Morphologique et Histologique. I. Morphologie Externe et Système Nerveux. *La Cellule*, v. 51, pt. 1, p. 51–152, 15 pl., 20 fig.

Rosenfeld, A. and B. Vesper 1977. The variability of the sieve pores in Recent and fossil species of *Cyprideis torosa* (Jones, 1890), as an indicator for salinity and paleosalinity. *In* H. Löffler and D. Danielopol [eds.] Aspects of ecology and zoogeorgraphy of Recent and fossil Ostracoda, p. 55–67, 1 pl., 4 fig. The Hague: Dr. W. Junk by Publishers.

Ruggieri, G. 1975. Contributo all consenze del genere *Aurila* (Ostracoda, Podocopa) con particolare riguardo ai suoi rappresentanti Pleistocene italiano. [A contribution to knowledge of the genus *Aurila* (Ostracoda, Podocopa), with special reference to its representative in the Italian Pleistocene]. *Boll. Soc. Paleontol. Ital.*, v. 14, no. 1, p. 27–46, 1 pl., 15 text-fig.

Sandberg, P. A. 1970. Scanning electron microscopy of freeze-dried Ostracoda (Crustacea). *Trans. Am. Microsc. Soc.*, v. 89, no. 1, p. 113–124, 14 fig.

——— and W. W. Hay. 1967. Study in microfossils by means of the scanning electron microscope. *J. Paleontol.*, v. 41, no. 4, p. 999–1001, pl. 131, 132.

——— and P. L. Plusquellec. 1969. Structure and polymorphism of normal pores in cytheracean Ostracoda (Crustacea). *J. Paleontol.*, v. 43, no. 2, p. 517–521, 1 text-fig.

Schäfer, W. 1962. Aktuo-paläontologie nach Studien in der Nordsee. Senkenberg-Buch 41. W. Kramer. 568 p., 39 pl. 277 fig.

Schornikov, E. I. 1975. Ostracod fauna of the intertidal zone in the vicinity of the Seto Marine Biological Laboratory. *Publ. Seto Mar. biol. Lab.*, v. 22, no. 1, p. 1–30, 15 text-fig.

——— and S. V. Schaitarov. 1979. A new genus of ostracods from Far Eastern seas. *Biologija Morija*, v. 2, p. 41–47, pl. 3, 2 text-fig. [in Russian]

Scott, T. 1902. Observations on the food of fishes. *Ann. Rep. Fish Board Scotland* 1901, pt. 3, p. 486–538.

Siddiqui, Q. A. 1971. Early Tertiary Ostracoda of the family Trachyleberididae from West Pakistan. *Bull. Br. Mus. nat. Hist.*, (*Geol.*), supl., 9, 98 p., 42 pl., 7 text-fig.

——— and U. M. Grigg. 1975. A preliminary survey of the ostracodes of Halifax Inlet. *In* F. M. Swain [ed.] Biology and paleobiology of Ostracoda. *Bull. Am. Paleontol.*, v. 65, no. 282, p. 369–379, 2 pl., 2 text-fig.

Simpson, E. H. 1949. Measurement of diversity. *Nature*, no. 163, p. 688.

Sissingh, W. 1972. Late Cenozoic Ostracoda of the South Aegean Island Arc. *Utr. Micropaleontol. Bull.*, no. 6, 187, p., 12 pl.

———. 1973. *Carinovalva* n. g. (Ostracoda), and comments on the ostracode genus *Lixouria* Uliczny (1969). *Proc. K. Ned. Akad. Wet.*, ser. B. v. 76, no. 2, p. 143–147, 2 pl.

Sohn, I. G. 1975. Discussion. [*In* A. Leibau "The left-right variation of the ostracode ornament."] *In* F. M. Swain [ed.] Biology and paleobiology of Ostracoda. *Bull Am. Paleontol.*, v. 65, no. 282, p. 86.

——— and L. S. Kornicker. 1969. Significance of Cretaceous nodules in myodocopid ostracod carapaces. *In* J. W. Neale [ed.] The taxonomy, morphology and ecology of Recent Ostracoda. p. 99–108, 3 pl.

Sugihara, S. 1970. Geomorphological developments of the western Shimōsa upland in Chiba Prefecture, Japan. *Geogr. Rev. Japan*, v. 43, no. 12, p. 703–718, 12 text-fig.

———, F. Arai and H. Machida. 1978. Tephrochronology of the Middle to Late Pleistocene sediments in the northern part of the Boso Peninsula, central Japan. *J. geol. Soc. Japan*, v. 84, no. 10, p. 583–600, 6 text-fig. [in Japanese]

Sugimura, A. and Y. Naruse. 1954, 1955. Changes in sea level, seismic upheavals, and coastal terraces in the southern Kanto region, Japan. I, II. *Japan J. Geol. Geogr.*, v. 24, p. 101–113, v. 26, nos. 3/4, p. 165–176, 8 fig., 4 tables.

Sulston, J. E. and H. R. Horvitz. 1977. Post-embryonic cell lineage of the nematode, *Caenorhabdis elegans*. *Dev. Biol.*, v. 56, no. 1, p. 110–156, 27 fig., 3 tables.

Swain, F. M. 1963. Pleistocene Ostacoda from the Gubik Formation, Arctic coastal plain, Alaska. *J. Paleontol.*, v. 37, no. 4, p. 798–834, 5 pl., 13 text-fig.

—— and J. C. Kraft. 1975. Biofacies and microstructure of Holocene Ostracoda from tidal bays of Delaware. *In* F. M. Swain [ed.] Biology and paleobiology of Ostracoda. *Bull. Am. Paleontol.*, v. 65, no. 282, p. 601–622, 5 pl., 2 text-fig.

Sylvester-Bradley, P. C. 1948. The ostracode genus *Cythereis*. *J. Paleontol.*, v. 22, no. 6, pl. 222, 1 text-fig.

——— and R. H. Benson. 1971. Terminology for surface features in ornate ostracodes. *Lethaia*, v. 4, no. 3, p. 249–286, 48 text-fig.

Triebel, E. 1941. Zur Morphologie und Ökologie der fossilen Ostracoden. *Senckenbergiana*, v. 23, p. 294–400, 15 pl.

Uliczny, F. 1969. Hemicytheridae und Trachyleberididae (Ostracoda) aus dem Pliozän der Insel Kephalinia (Westgriechenland). Zur Erlangung der Doktorwürde der Hohen Naturwissenschaftlichen Fakulat der Ludwig Maximilian Universität zu München, 152 p., 18 pl.

———. 1971. Zur Revision des Genotypes von *Incongruellina* (*Lixouria*) Uliczny, 1969 (Crustacea, Ostracoda). *Neues Jahrb. Geol. Paläontol. Mon.*, p. 734–740, 4 text-fig.

Utinomi, H. 1971. Scleractinian corals from Kamae Bay, Oita Prefecture, northeast of Kyushu, Japan. *Publ. Seto Mar. biol. Lab.*, v. 19, no. 4, p. 203–229.

Valentine, P. C. 1976. Zoogeography of Holocene Ostracoda off western North America and paleoclimatic implications. *U. S. geol. Surv.*, *Prof. Paper* 916, 47 p., 14 pl., 19 text-fig.

Valicenti, V. H. 1977. Some Hemicytherinae from the Tertiary of Patagonia (Argentina) and their morphological relationship and stratigraphical distribution. *In* H. Löffler and D. Danielopol [eds.] Aspects of ecology and zoogeography of Recent and fossil Ostracoda, p. 93–106, 2 pl., 2 text-fig. Saalfelden 1976. The Hague: Dr. W. Junk b. v. Publishers.

Wagner, C. W. 1957. Sur les Ostracodes du Quaternaire Recent des Pays-Bas et leur Utillisation dans l'Etude geologique des Depots Holocenes. Diss Univ. de Paris, 259 p., 50 pl., 26 text-fig.

Yabe, H. and T. Sugiyama. 1932. A study of Recent and semifossil corals of Japan. *Sci. Rep. Tohoku imp. Univ.*, 2nd ser., v. 14, no. 2A, p. 119–133, pl. 37–39, 1 text-fig.

——— and ———. 1935. Geological and geographical distribution of coral reefs in Japan. *J. Paleontol.*, v. 9, no. 3, p. 183–217, pl. 21, 5 text-fig., 4 tables.

Yajima, M. 1978. Quaternary Ostracoda from Kisarazu near Tokyo. *Trans. Proc. palaeontol. Soc. Japan*, n. s., no. 112, p. 371–409, pl. 49, 50, 10 text-fig.

Yasuda, Y. 1978. Vegetational history and paleogeography of the Kawachi Plain for the last 13,000 years. *Quaternary Res.* (Japan), v. 16, no. 4, p. 211–229, 8 text-fig. [in Japanese]

Yokota, K. 1978. Holocene coastal terraces on the southeast coast of the Boso Peninsula. *Geogr. Rev. Japan*, v. 51, no. 5, p. 349–364, 7 text-fig., 4 tables. [in Japanese]

Yokoyama, M. 1911. Climatic changes in Japan since the Pliocene epoch. *J. Coll. Sci., Imp. Univ. Tokyo*, v. 32, artic. 5, p. 1–16, 1 pl.

———. 1924. Molluscs from the coral-bed of Awa. *J. Coll. Sci.*, *Imp. Univ. Tokyo*, v. 45, artic. 1, p. 1–62, pl. 1–5.

Yonekura, N. 1975. Quaternary tectonic movements in the outer arc of southwest Japan with special reference to seismic crustal deformations. *Bull. Dept. Geogr., Univ. Tokyo,* no. 7, p. 19–71, 18 fig., 2 tables.

Yoshikawa, T. and M. Saito. 1954. Topography of the shallow sea floor in the vicinity of Chikura Harbour on the southwestern coast of Boso Peninsula, on the Pacific coast of Japan. *Bull. geogr. Inst. Tokyo Univ.,* no. 3, p. 40–50, 5 fig. [in Japanese]

Yule, G. U. 1944. The statistical study of literary vocabulary. Cambridge: At the University Press. 306 p.

INDEX

HANAI, T. [ed.] 1982. STUDIES ON JAPANESE OSTRACODA

PLATES 1–30

Explanation of Plate 1

Figs. 1–6, Cyprididae n. sp.

1a, b. Stereo pair of dorsal view; 2a, b. stereo pair of left lateral veiw; 3. right lateral view. ×60 (IGSU-O-8). 4. Interior lateral view of left valve. ×60 (IGSU-O-9). 5. Interior lateral view of right valve. ×60 (IGSU-O-10). 6. A normal pore canal opening and sensory hair with a thickned rim on dorsomedian surface of right valve. ×3600 (IGSU-O-8).

Figs. 7–10, *Physocypria* sp.

7a, b. Stereo pair of dorsal view; 8a, b. stereo pair of right lateral view; 9. interior lateral view of right valve. ×65; 10. adductor muscle scars on right valve. ×468 (IGSU-O-11).

Figs. 11–13, *Cypridopsis vidua* (O. F. Müller)

11a, b. Stereo pair of dorsal view; 12a, b. stero pair of left lateral view; 13. interior lateral view of left valve. ×55 (IGSU-O-12).

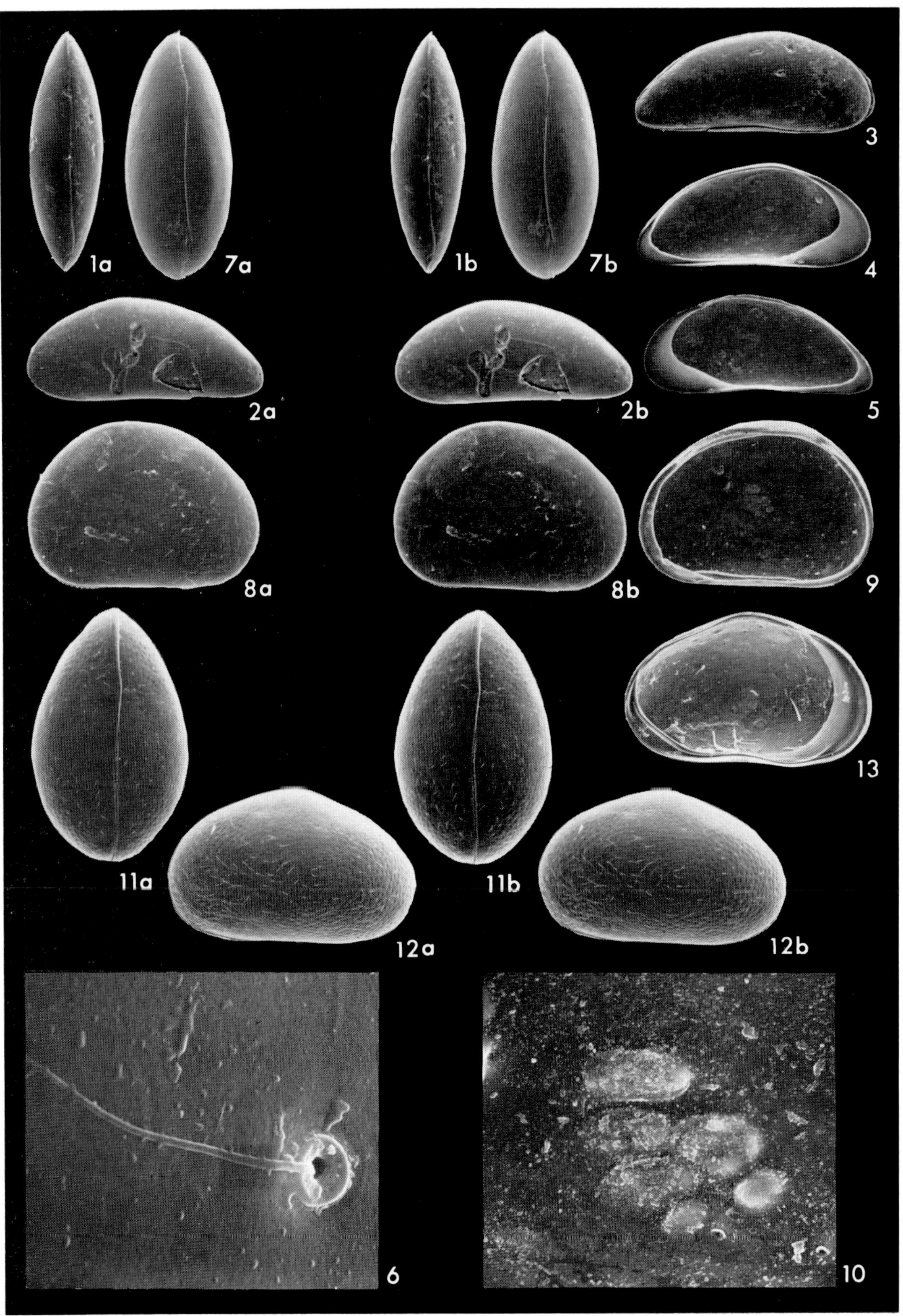
1a
7a
1b
7b
3
4
2a
2b
5
8a
8b
9
11a
11b
13
12a
12b
6
10

Explanation of Plate 2

Figs. 1–5, *Pontocythere sekiguchii* Ikeya and Hanai, n. sp.

1a, b. Stereo pair of dorsal view; 2a, b. stereo pair of left lateral view; 3. right lateral view. ×65. 4. Interior lateral view of left valve. ×98 (Holotype, IGSU-O-13). 5. Interior view of right valve. ×100 (IGSU-O-14).

Figs. 6–12, *Pontocythere minuta* Ikeya and Hanai, n. sp.

6a, b. Stereo pair of dorsal view; 7a, b. stereo pair of left lateral view; 9. right lateral view. ×99 (Holotype, IGSU-O-15). 9. Interior lateral view of left valve. ×67 (IGSU-O-16). 10. Irregularly shaped sieve plate without normal pore opening on central surface of right valve; 11. a normal pore canal opening with sieve-like rim on central surface of right valve. ×6600. 12. Three punctae (normal pore opening, irregular shaped sieves, and simple pits), dorsomedian surface of right valve. ×66 (Holotype, IGSU-O-15).

Figs. 13–17, *Neocytherideis punctata* Ikeya and Hanai, n. sp.

13a, b. Stereo pair of dorsal view; 14a, b. stereo pair of left valve; 15. right lateral view. ×90 (IGSU-O-17). 16. Interior lateral view of left valve; 17. interior lateral view of right valve. ×90 (Holotype, IGSU-O-18).

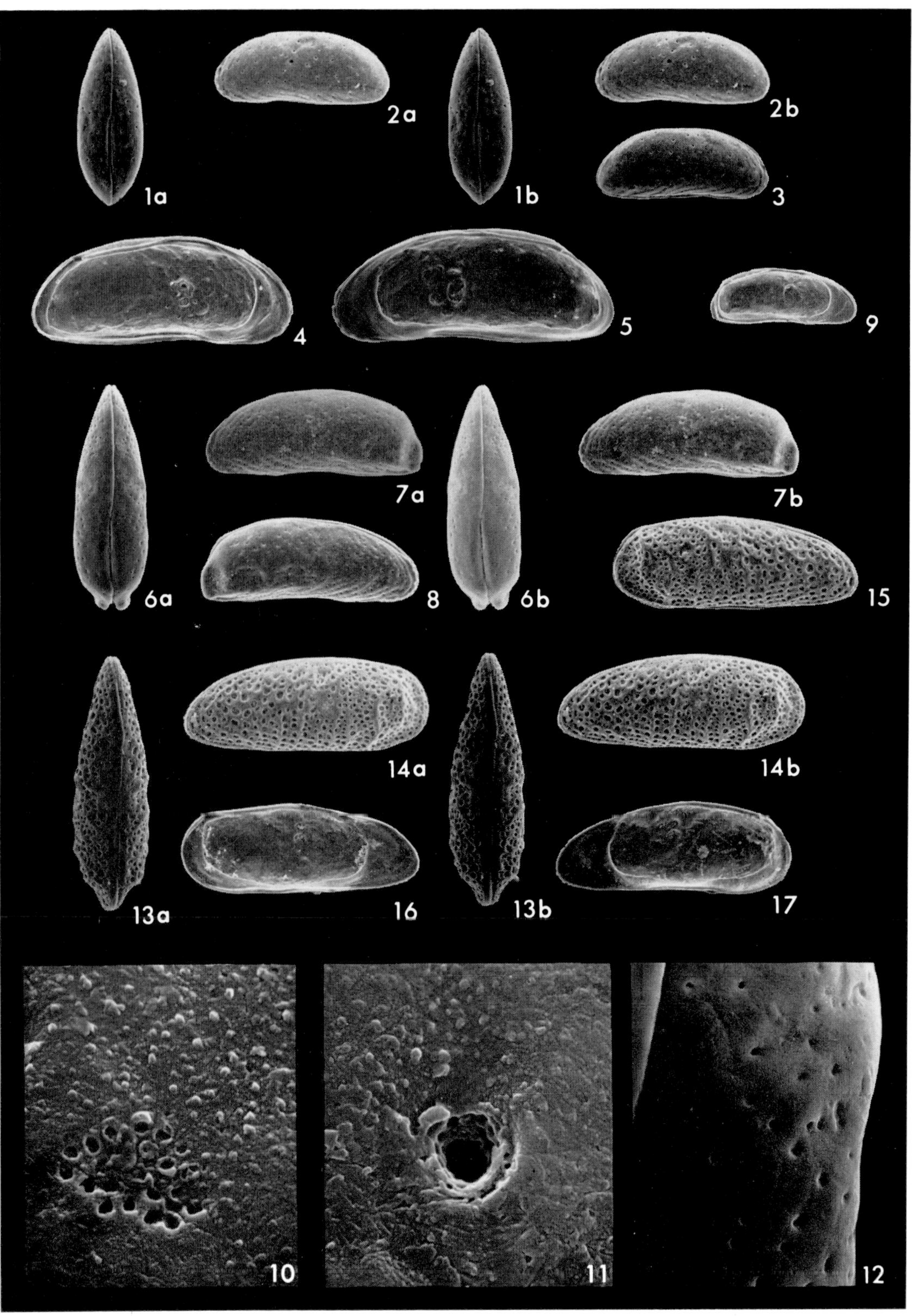
1a
2a
1b
2b
3
4
5
9
6a
7a
8
6b
7b
15
13a
14a
16
13b
14b
17
10
11
12

Explanation of Plate 3

Figs. 1–5, *Cornucoquimba rugosa* Ikeya and Hanai, n. sp.
1a, b. Stereo pair of left lateral view; 2a, b. stereo pair of dorsal view; 3. right lateral view; 4. interior lateral view of left valve; 5. interior lateral view of right valve. ×65 (Holotype, IGSU-O-19).

Figs. 6–10, *Bythocythere maisakensis* Ikeya and Hanai, n. sp.
6a, b. Stereo pair of dorsal view; 7. right lateral view; 8a, b. stereo pair of left lateral view; 9. interior lateral view of right valve; 10 interior lateral view of left valve. ×82 (Holotype, IGSU-O-20).

Fig. 11, *Pontocythere minuta* Ikeya and Hanai, n. sp.
Hinge and adductor muscle scars on left valve. ×468 (IGSU-O-16).

Fig. 12, *Cypridopsis vidua* (O. F. Müller)
Adductor muscle scars on right valve. ×264 (IGSU-O-12).

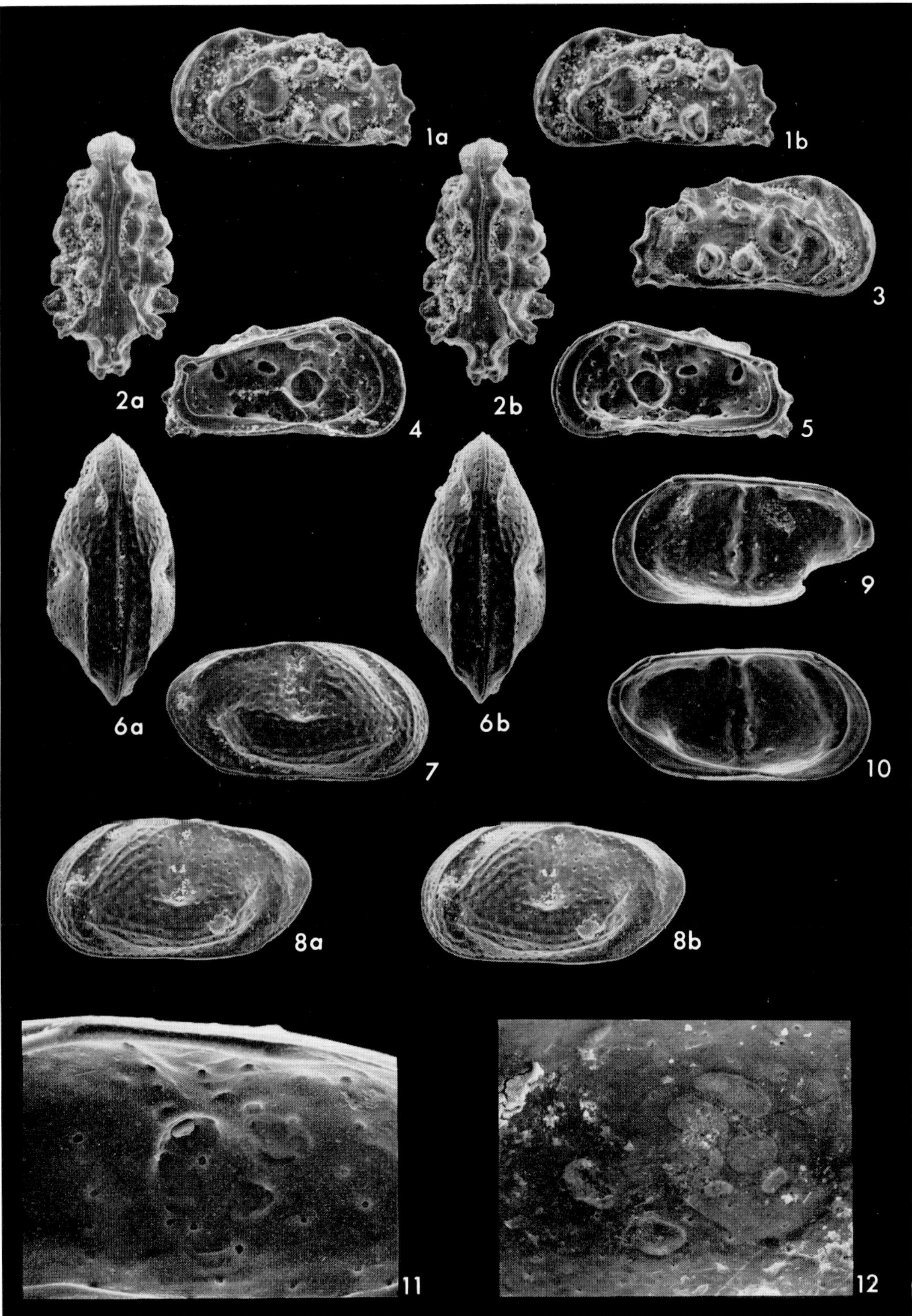
1a
1b
2a
2b
3
4
5
6a
6b
7
8a
8b
9
10
11
12

Explanation of Plate 4

Figs. 1–4, *Phlyctocythere hanamensis* Ikeya and Hanai, n. sp.
1a, b. Stereo pair of dorsal view; 2a, b. stereo pair of right lateral view; 3. left lateral view; 4. interior lateral view of right valve. ×87 (Holotype, IGSU-O-21).

Figs. 5, 6. *Bythoceratina elongata* Ikeya and Hanai, n. sp.
5a, b. Stereo pair of left lateral view; 6. interior lateral view of left valve. ×170 (Holotype, IGSU-O-22).

Fig. 7, *Actinocythereis* sp.
7a, b. Stereo pair of left lateral view. ×89 (IGSU-O-23).

Fig. 8, *Trachyleberis* sp.
8a, b. Stereo pair of right lateral view. ×70 (IGSU-O-24).

Fig. 9, *Neocytheretta* sp.
9a, b. Stereo pair of left lateral view. ×81 (IGSU-O-25).

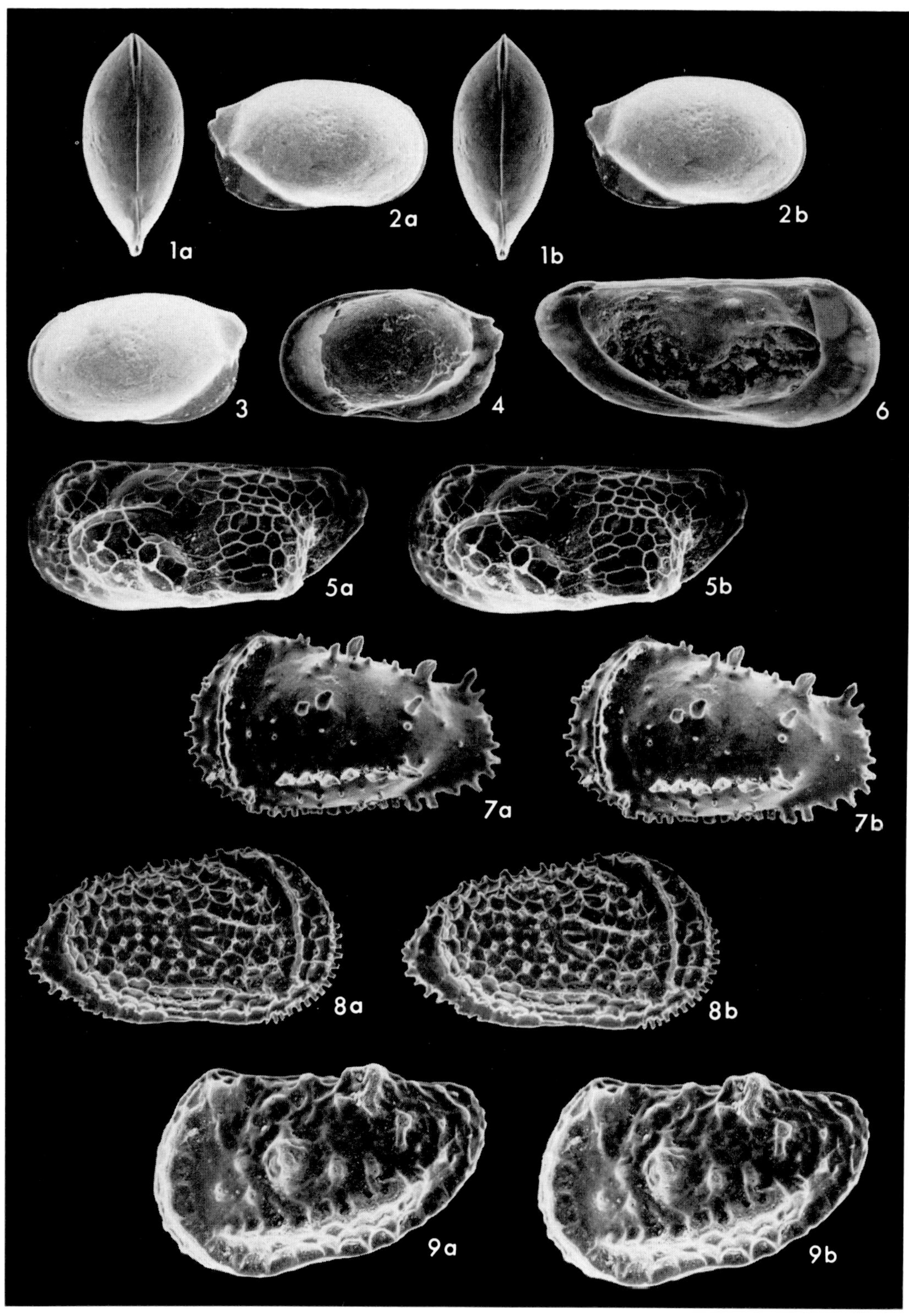
1a
2a
1b
2b
3
4
6
5a
5b
7a
7b
8a
8b
9a
9b

Explanation of Plate 5

Figs. 1–3, *Platymicrocythere* sp.

1a, b. Stereo pair of left lateral view. ×132 (IGSU-O-49, male). 2a, b. Stereo pair of dorsal view. ×117 (IGSU-O-28, female). 3. Right lateral view. ×132 (IGSU-O-49, male).

Figs. 4, 5, *Semicytherura elongata* Ikeya and Hanai, n. sp.

4a, b. Stereo pair of right lateral view; 5. interior lateral view of right valve. ×125 (Holotype, IGSU-O-29).

Fig. 6, *Semicytherura* sp.

6a, b. Stereo pair of left lateral view. ×195 (IGSU-O-30)

Figs. 7–9, *Callistocythere* sp.

7a, b. Stereo pair of right lateral view. ×115; 8. muscle scars. ×860; 9. a normal pore opening on the ridge of dorsocental area of carapace, exterior view of right valve. ×15000 (IGSU-O-32).

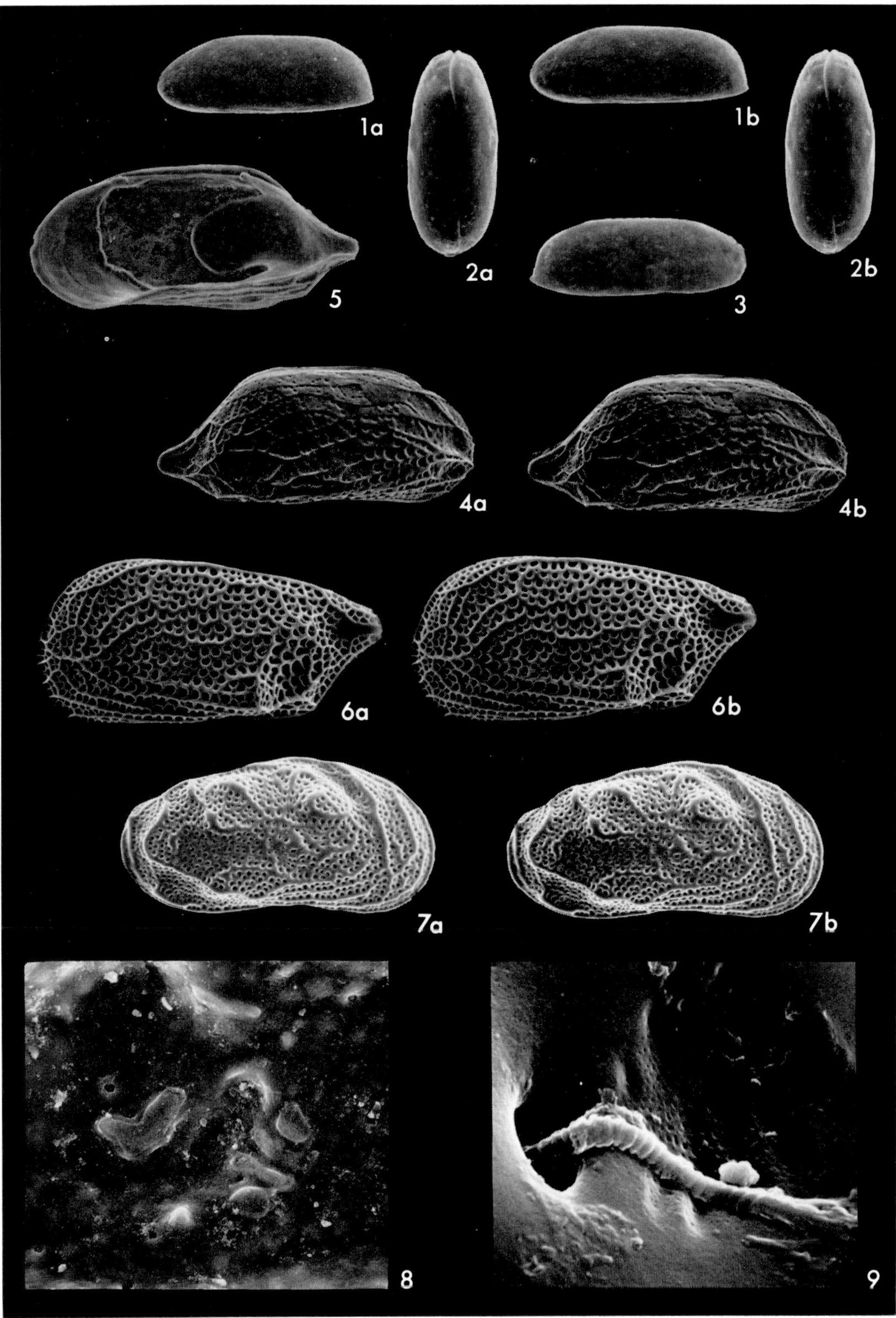
1a
1b
2a
2b
3
5
4a
4b
6a
6b
7a
7b
8
9

Explanation of Plate 6

Figs. 1–4, *Cytheroma*? sp.

1a, b. Stereo pair of dorsal view. ×200 (IGSU-O-33). 2a, b. Stereo pair of right lateral view; 3. interior lateral view of right valve. ×200 (IGSU-O-51). 4. A sieve type normal pore canal and sensory hair on dorsomedian surface of right valve. ×3600 (IGSU-O-33).

Figs. 5–10, *Neopellcuistoma inflatum* Ikeya and Hanai, n. gen., n. sp.

5a, b. Stereo pair of dorsal view; 6a, b. stereo pair of left lateral view; 7 right lateral view; 8. interior lateral view of left valve. ×66 (Holotype, IGSU-O-35). 9. Interior lateral view of right valve. ×66 (IGSU-O-36). 10. Hinge and adductor muscle scars on left valve. ×160 (Holotype, IGSU-O-35).

Fig. 11, *Neocytherideis punctata* Ikeya and Hanai, n. sp.

Dorsal view of anterodorsal area of complete carapace. ×544 (IGSU-O-17)

Figs. 12, 13, *Phlyctocythere hamanensis* Ikeya and Hanai, n. sp.

12. Dorsal view of anterodorsal area of complete carapace; 13. dorsal view of posterodorsal area of complete carapace. ×396 (Holotype, IGSU-O-21).

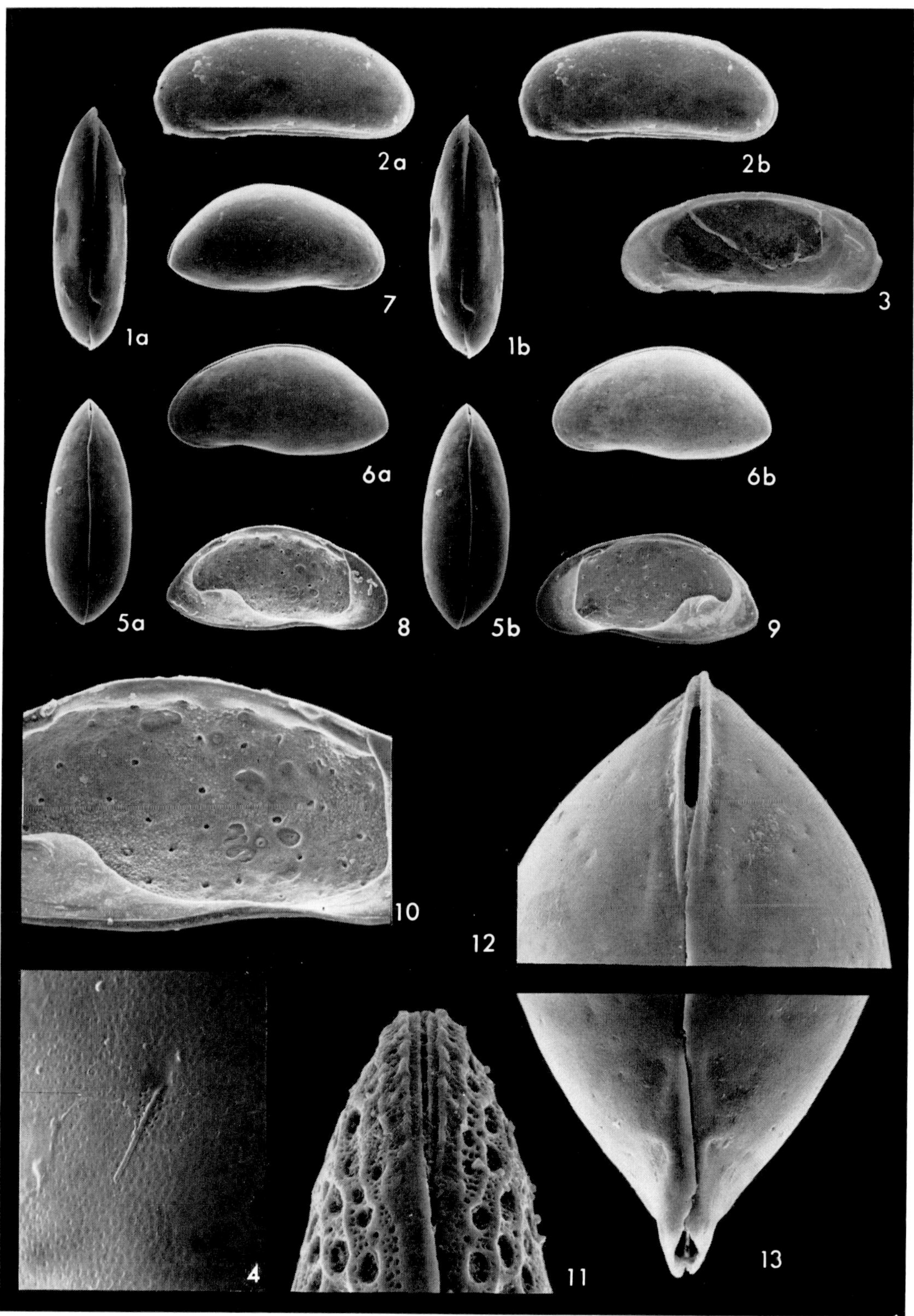
1a
1b
2a
2b
3
4
5a
5b
6a
6b
7
8
9
10
11
12
13

Explanation of Plate 7

Fig. 1, *Pontocythere sekiguchii* Ikeya and Hanai, n. sp.
A sieve-type normal pore canal on central area of carapace, exterior lateral view of left valve. ×5640 (Holotype, IGSU-O-13).

Fig. 2, *Neocytherideis punctata* Ikeya and Hanai, n. sp.
Two type normal pore canals on dorsocentral area of carapace, exterior lateral view of left valve, ×2680 (IGSU-O-17).

Fig. 3, *Cornucoquimba rugosa*, Ikeya and Hanai, n. sp.
A sieve-type normal pore canal on central area of carapace, interior lateral view of right valve. ×2680 (Holotype, IGSU-O-19).

Fig. 4, *Bythoceratina elongata*, Ikeya and Hanai, n. sp.
A sieve-type normal pore canal and muri on posteroventral area of carapace, exterior lateral view of left valve. ×2400 (Holotype, IGSU-O-22).

Fig. 5, *Semicytherura elongata*, Ikeya and Hanai, n. sp.
A normal pore canal opening with a thickened rim, on anteroventral area of carapace, exterior lateral view of right valve. ×5640 (Holotype, IGSU-O-29).

Fig. 6, *Neopellucistoma inflatum* Ikeya and Hanai, n. gen., n. sp.
A sieve-type normal pore canal on dorsocentral area of carapace, exterior lateral view of right valve. ×2680 (Holotype. IGSU-O-35).

Fig. 7, *Physocypria* sp.
A normal pore canal opening with a thickend rim, on dorsocentral area of carapace, exterior view of right valve. ×6600 (IGSU-O-11).

Fig. 8, *Platymicrocythere* sp.
A normal pore canal opening with a thickend rim, on dorsocentral area of carapace, exterior view of left valve. ×8600 (IGSU-O-49).

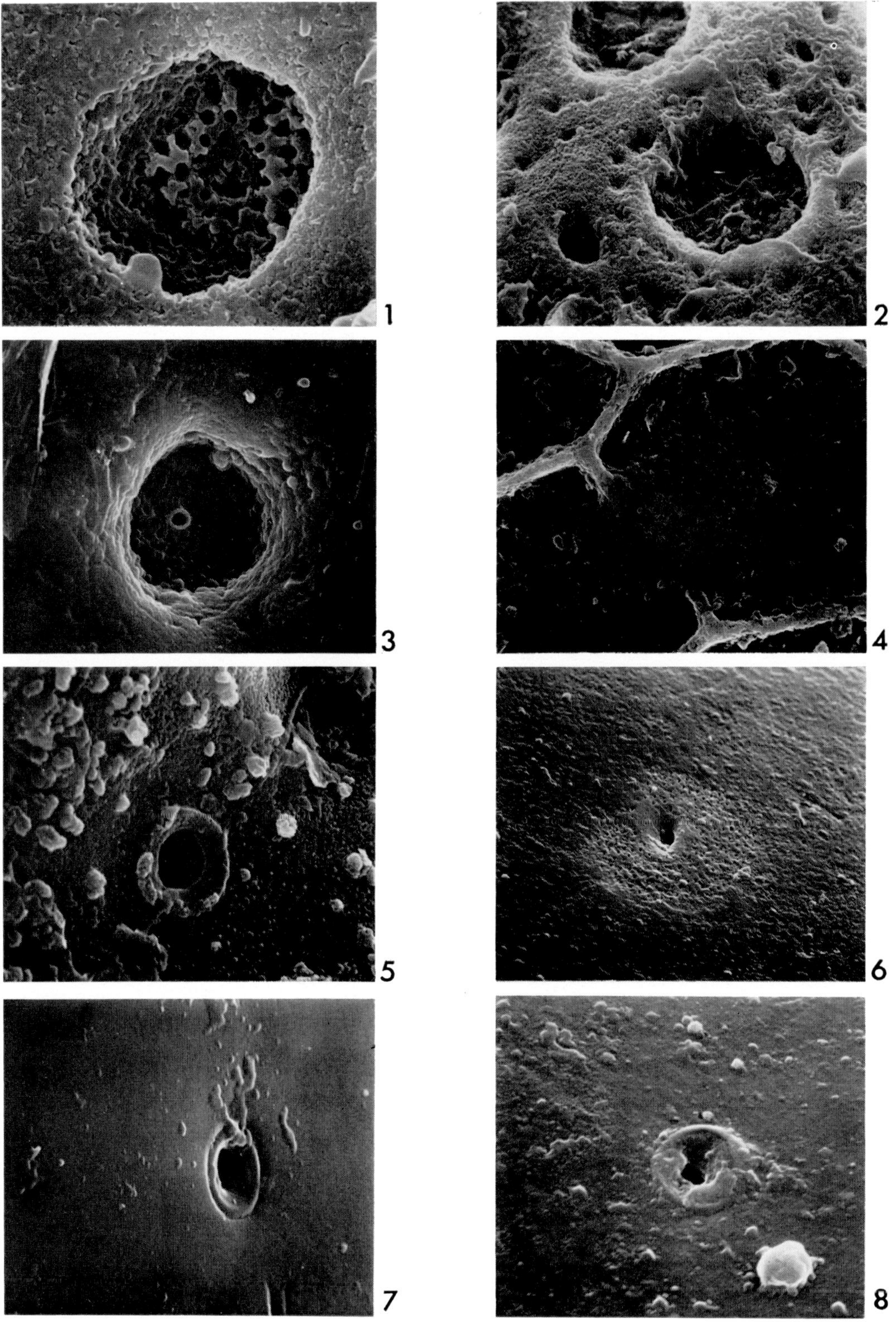

IKEYA and HANAI, Ecology of Recent Ostracods

Explanation of Plate 8

Figs. 1, 2, *Trachyleberis straba* n. sp.

1. Lateral view, left valve, female (UMUT-CA 15020), detail of the sinuous anterior ridge. ×220.
2. Internal view, right valve, female (UMUT-CA 15021), detail of the muscle scar. ×340.

Figs. 3–7, *Argilloecia lunata* n. sp.

3. Lateral view, left valve, female (UMUT-CA 15002). ×88. 4. Lateral view, left valve, male (UMUT-CA 15003). ×85. 5. Lateral view, right valve, female (UMUT-CA 15004). ×80. 6. Internal view, left valve, male (UMUT-CA 15001). ×83. 7. Internal view, right valve, female (UMUT-CA 15005). ×83.

Figs. 8–14, 16, 17, *Semicytherura sabula* n. sp.

8. Internal view, left valve, female (UMUT-CA 15024). ×136. 9. Dorsal view, carapace, male (UMUT-CA 15025). ×136. 10. Dorsal view, carapace, female (UMUT-CA 15023). ×136. 11. Internal view, right valve, female (UMUT-CA 15026). ×133. 12. Internal view, right valve, male (UMUT-CA 15027). ×133. 13. Lateral view, right valve, male (UMUT-CA 15028) ×133. 14. Lateral view, left valve, male (UMUT-CA 15029). ×133. 16. Lateral view, right valve, female (UMUT-CA 15040). ×135. 17. Lateral view, left valve, female (UMUT-CA 15041). ×129.

Figs. 15, 18–20, *Nipponocythere hastata* n. sp.

15. Internal view, left valve, female (UMUT-CA 15033). ×105. 18. Internal view, right valve, female (UMUT-CA 15034). ×105. 19. Lateral view, right valve, female (UMUT-CA 15032). ×105. 20. Lateral view, left valve, female (UMUT-CA 15035). ×107.

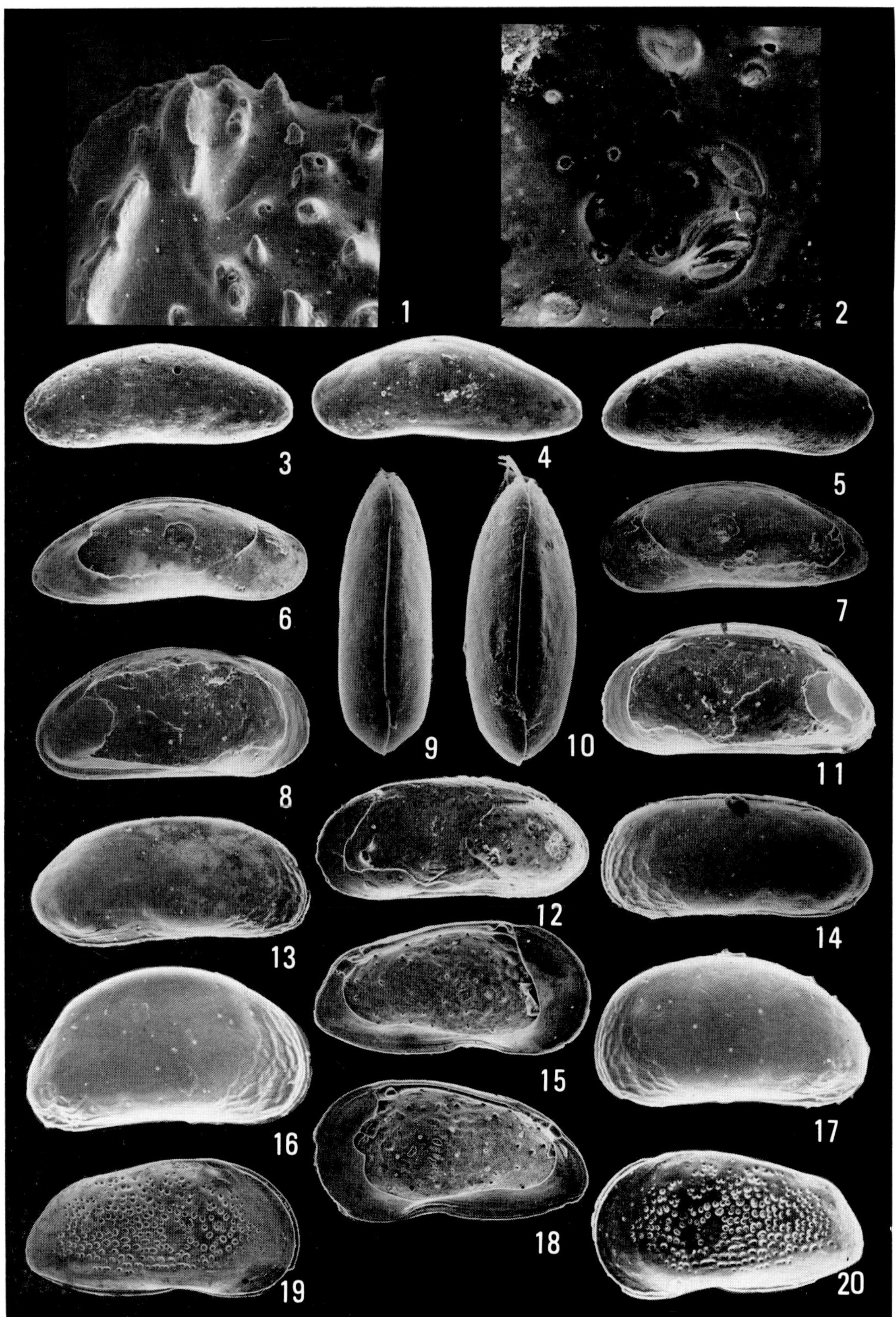
1
2
3
4
5
6
7
8
9
10
11
12
13
14
15
16
17
18
19
20

Explanation of Plate 9

Figs. 1–3, *Callistorcythere numaensis* n. sp.

1. Lateral view, right valve (UMUT-CA 15007) ×100. 2. Lateral view, left valve (UMUT-CA 15008). ×100. 3. Internal view, left valve (UMUT-CA 15009). ×104.

Figs, 4–9, *Callistocythere tateyamaensis* n. sp.

4. Lateral view, right valve, male (UMUT-CA 15011). ×83. 5. Lateral view, left valve, female (UMUT-CA 15010). ×82. 6. Internal view, left valve, female (UMUT-CA 15012). ×83. 7. Lateral view, right valve, female (UMUT-CA 15013). ×78. 8. Lateral view, left valve, male (UMUT-CA 15043). ×84. 9. Internal view, right valve, female (UMUT-CA 15044). ×80.

Figs. 10, 11, *Aurila* sp. A

10. Lateral view, right valve (UMUT-CA 15016). ×69. 11. Lateral view, left valve (UMUT-CA 15017). ×63.

Fig. 12, *Neomonoceratina* sp.

Lateral view, left valve (UMUT-CA 15014). ×107.

Figs. 13, 14. *Aurila* sp. B.

13. Lateral view, right valve (UMUT-CA 15018). ×57. 14. Lateral view, left valve (UMUT-CA 15019). ×57.

Fig. 15, *Hemicythere* sp.

Lateral view, right vale (UMUT-CA 15015). ×83.

Figs. 16–18, *Trachyleberis straba* n. sp.

16. Internal view, left valve, male (UMUT-CA 15022). ×57. 17. Internal view, right valve, female (UMUT-CA 15021). ×62. 18. Lateral view, left valve, female (UMUT-CA 15020) ×62.

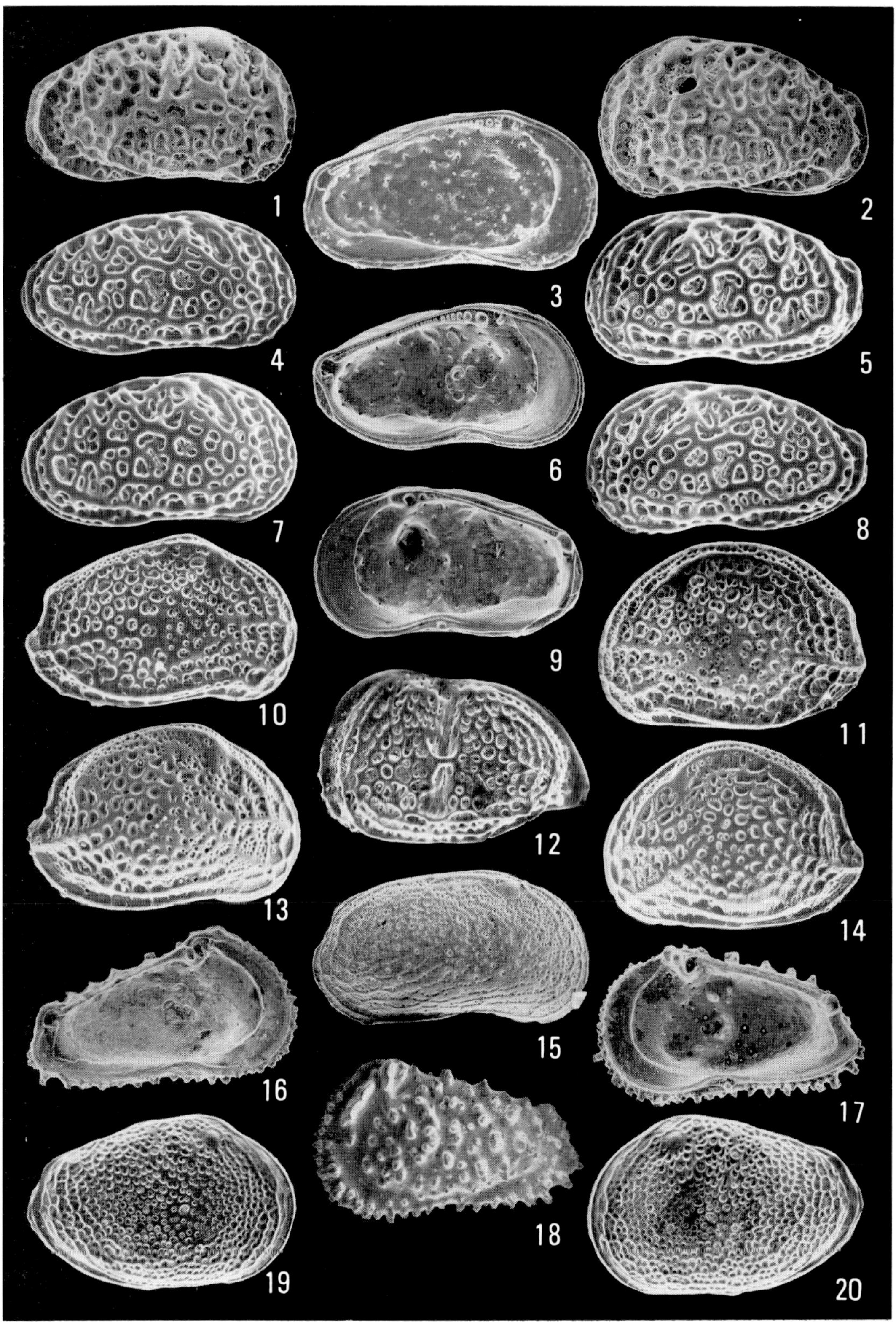
1
2
3
4
5
6
7
8
9
10
11
12
13
14
15
16
17
18
19
20

Explanation of Plate 10

Figs. 1–6. *Neocytherideis aoi* n. sp.

1. Lateral view of right male valve (CA 9807, sample 189–3, Kioroshi Member of Kioroshi Formation). ×85. 2. Inner view of right female valve (CA 9806, sample 189–3, Kioroshi Member of Kioroshi Formation). ×85. 3. Lateral view of right female valve (CA 9810, sample 120–1, Yabu Member of Yabu Formation). ×85. 4. Lateral view of left female valve (holotype, CA 9805, sample 189–3, Kioroshi Member of Kioroshi Formation). ×85. 5. Dorsal view of male carapace (CA 9808, sample 273–1, Kioroshi Member of Kioroshi Formation). ×80. 6. Dorsal view of female carapace (CA 9808, sample 120–1, Yabu Member of Yabu Formation). ×80.

Figs. 7, 8. *Pontocythere* sp. 1

7. Lateral view of right valve (CA 9811, sample 190–2, Kioroshi Member of Kioroshi Formation). ×100. 8. Lateral view of left valve (CA 9812, sample 190–2, Kioroshi Member of Kioroshi Formation). ×100.

Figs. 9, 12. *Munseyella oborozukiyo* n. sp.

9. Lateral view of right valve (holotype, CA 9819, sample 66–2, Toyonari Member of Kioroshi Formation). ×80. 12. Lateral view of left valve (CA 9820, sample 66–2, Toyonari Member of Kioroshi Formation). ×80.

Figs. 10, 11. *Pontocythere* sp. 2.

10. Lateral view of right valve (CA 9813, sample 228–3, Kamiiwahashi Formation). ×80. 11. Lateral view of left valve (CA 9814, sample 228–3, Kamiiwahashi Formation). ×80.

Figs. 13, 14, 16–19. *Eucythere yugao* n. sp.

13. Lateral view of right valve (CA 9816, sample 189–3, Kioroshi Member of Kioroshi Formation). ×80. 14. Lateral view of left valve (holotype, CA 9815, sample 189–3). ×80. 16. Inner view of left valve (holotype, CA 9815). ×80. 17. Inner view of right valve (CA 9816). ×80. 18. Normal pore canal opening on exterior central surface of right valve (CA 9816). ×900. 19. Normal pore canal opening on inner central surface of left valve (holotype, CA 9815). ×1260.

Fig. 15. *Munseyella* sp.

Right valve view of carapace (CA 9821, sample 305–4, Yabu Member of Yabu Formation). ×110.

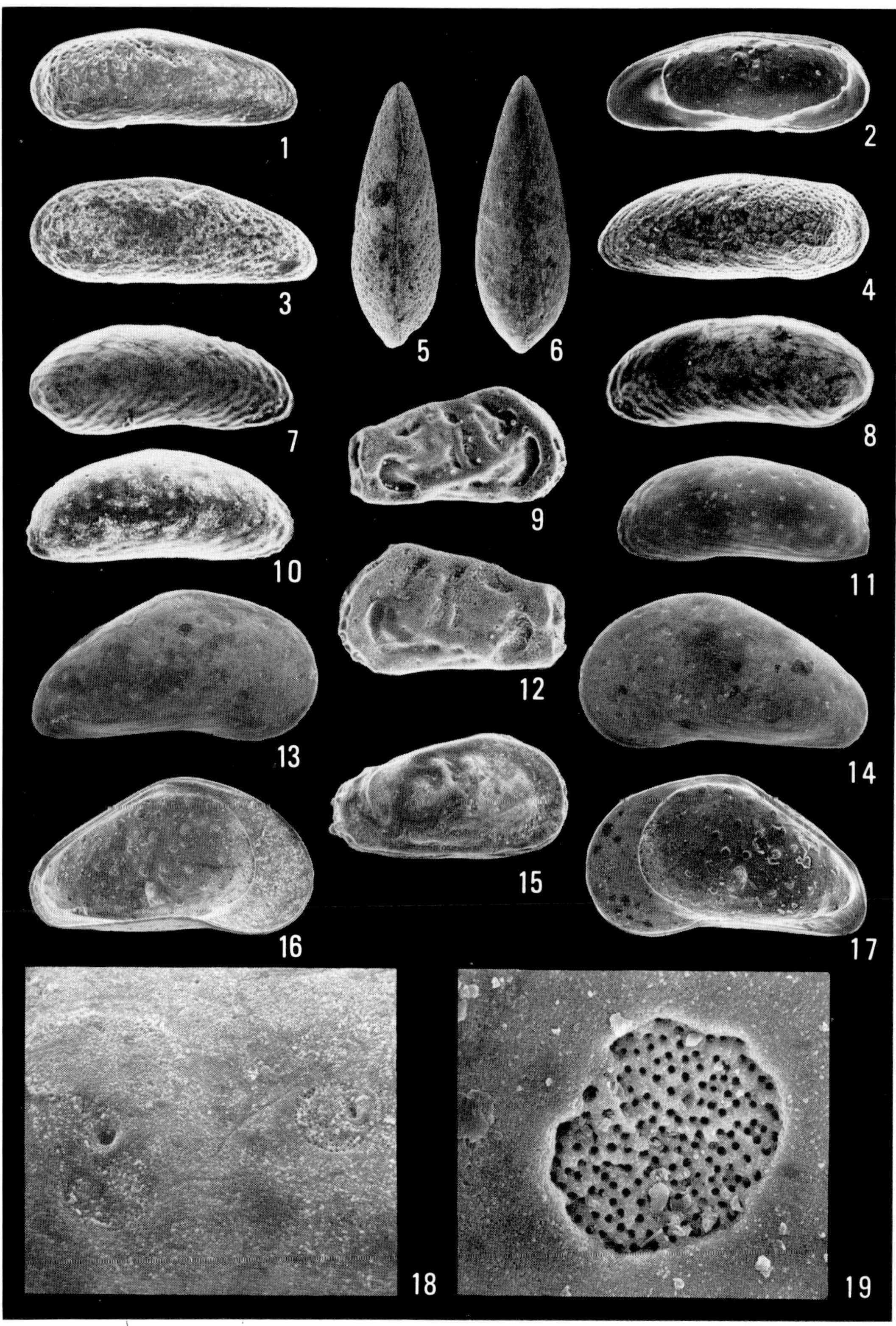
1
2
3
4
5
6
7
8
9
10
11
12
13
14
15
16
17
18
19

Explanation of Plate 11

Figs. 1–4. *Loxoconcha hanachirusato* n. sp.

1. Lateral view of left male valve (CA 9897, sample 189–3, Kioroshi Member of Kioroshi Formation). ×85. 2. Lateral view of right male valve (holotype, CA 9895, sample 189–3). ×90. 3.Lateral view of left female valve (CA 9898, sample 189–3). ×86. 4. Lateral view of right female valve (CA 9896, sample 189–3). ×86.

Figs. 5, 6. *Callistocythere hotaru* n. sp.

5. Lateral view of left female valve (holotype, CA 9826, sample 312–1, Kiyokawa Formation). ×85. 6. Lateral view of right female valve (CA 9827, sample 312–1). ×85.

Fig. 7. *Callistocythere rugosa* Hanai, 1957

Lateral view of left male valve (CA 9828, sample 73–4, Kiyokawa Formation). ×88.

Figs. 8, 9. *Callistocythere* sp.

8. Lateral view of right immature valve (CA 9831, sample 120–1, Yubu Member of Yabu Formation). ×85. 9. Lateral view of left immature valve (CA 9832, sample 120–1). ×85.

Figs. 10–13. *Neomonoceratina microreticulate* Kingma, 1948

10. Lateral view of left male valve (CA 9835, sample 228–3, Kamiiwahashi Formation). ×90. 11. Dorsal view of male carapace (CA 9836, sample 228–3). ×90. 12. Lateral view of left female valve (CA 9837, sample 228–3). ×90. 13. Dorsal view of female carapace (CA 9838, sample 228–3). ×90.

Figs. 14, 15. *Schizocythere asagao* n. sp.

14. Lateral view of right female valve (CA 9834, sample 135–3, Yabu Member of Yabu Formation). ×75. 15. Lateral view of left female valve (holotype, CA 9833, sample 305–3, Yabu Member of Yabu Formation). ×75.

Figs. 16, 17. *Loxoconcha tamakazura* n. sp.

16. Lateral view of right valve (CA 9904, sample 120–1, Yabu Member of Yabu Formation). ×88. 17. Lateral view of left valve (holotype, CA 9903, sample 120–1). ×88.

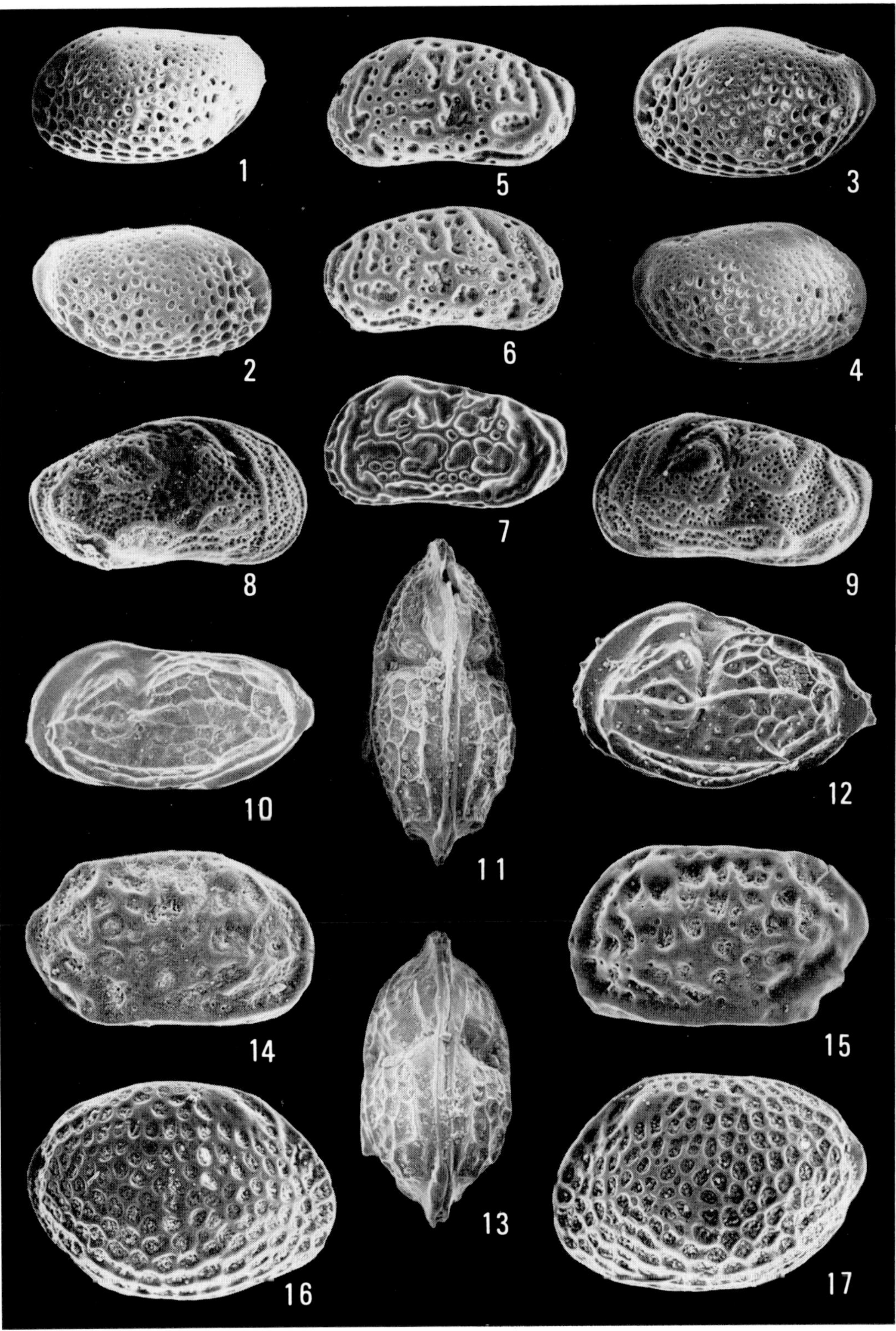

YAJIMA, Late Pleistocene Ostracoda

Explanation of Plate 12

Figs. 1, 2. *Rocaleberis*? sp.

1. Lateral view of right valve (CA 9846, sample 74–2, Kiyokawa Formation). ×40. 2. Inner view of right valve (CA 9846). ×40.

Figs. 3, 4. *Cletocythereis rastromarginata* (Brady, 1880)

3. Right valve view of carapace (CA 9844, sample 74–2, Kiyokawa Formation). ×55. 4. Dorsal view of carapace (CA 9844). ×55.

Fig. 5. *Acanthocythereis*? sp.

Lateral view of left partly broken valve (CA 9843, sample 135–3, Yabu Member of Yabu Formation). ×47.

Figs. 6, 7, 10. *Campylocythereis? ukifune* n. sp.

6. Lateral view of right male valve (CA 9858, sample 120–2, Yabu Member of Yabu Formation). ×49. 7. Lateral view of left male valve, (CA 9857, sample 284–3, Yabu Member of Yabu Formation). ×54. 10. Lateral view of left female valve (holotype, CA 9856, sample 284–3). ×57.

Figs. 8, 9, 11. *Trachyleberis*? *tosaensis* Ishizaki, 1968

8. Lateral view of right immature valve (CA 9841, sample 203–3, Kamiiwahashi Formation). ×56. 9. Lateral view of right female valve (CA 9839, sample 279–2, Kioroshi Member of Kioroshi Formation). ×58. 11. Lateral view of right male valve, (CA 9840, sample 189–3, Kioroshi Member of Kioroshi Formation). ×52.

Fig. 12. *Stigmatocythere*? sp.

Lateral view of right valve (CA 9845, sample 120–1, Yabu Member of Yabu Formation). ×68.

Fig. 13. *Robertsonites*? *reticuliforma* (Ishizaki, 1966)

Lateral view of left valve (CA 9852, sample 61 of Yajima, 1978, B Member of Kisarazu area). ×54.

Figs. 14–17. *Wichmannella bradyformis* (Ishizaki, 1968)

14. Lateral view of left immature (A-2) valve (CA 9849, sample 224–2, Kamiiwahashi Formation). ×85. 15. Lateral view of right male valve, (CA 9847, sample 279–4, Kamiiwahashi Formation). ×57. 16. Lateral view of left female valve (CA 9848, sample 279–4). ×57. 17. Lateral view of left immature (A–2) valve (CA 9850, sample 224–2). ×85.

Figs. 18, 19. *Australimoosella tomokoae* (Ishizaki, 1968)

18. Lateral view of right valve (CA 9853, sample 279–2, Kioroshi Member of Kioroshi Formation). ×60. 19. Lateral view of left valve (CA 9845, sample 279–2). ×60.

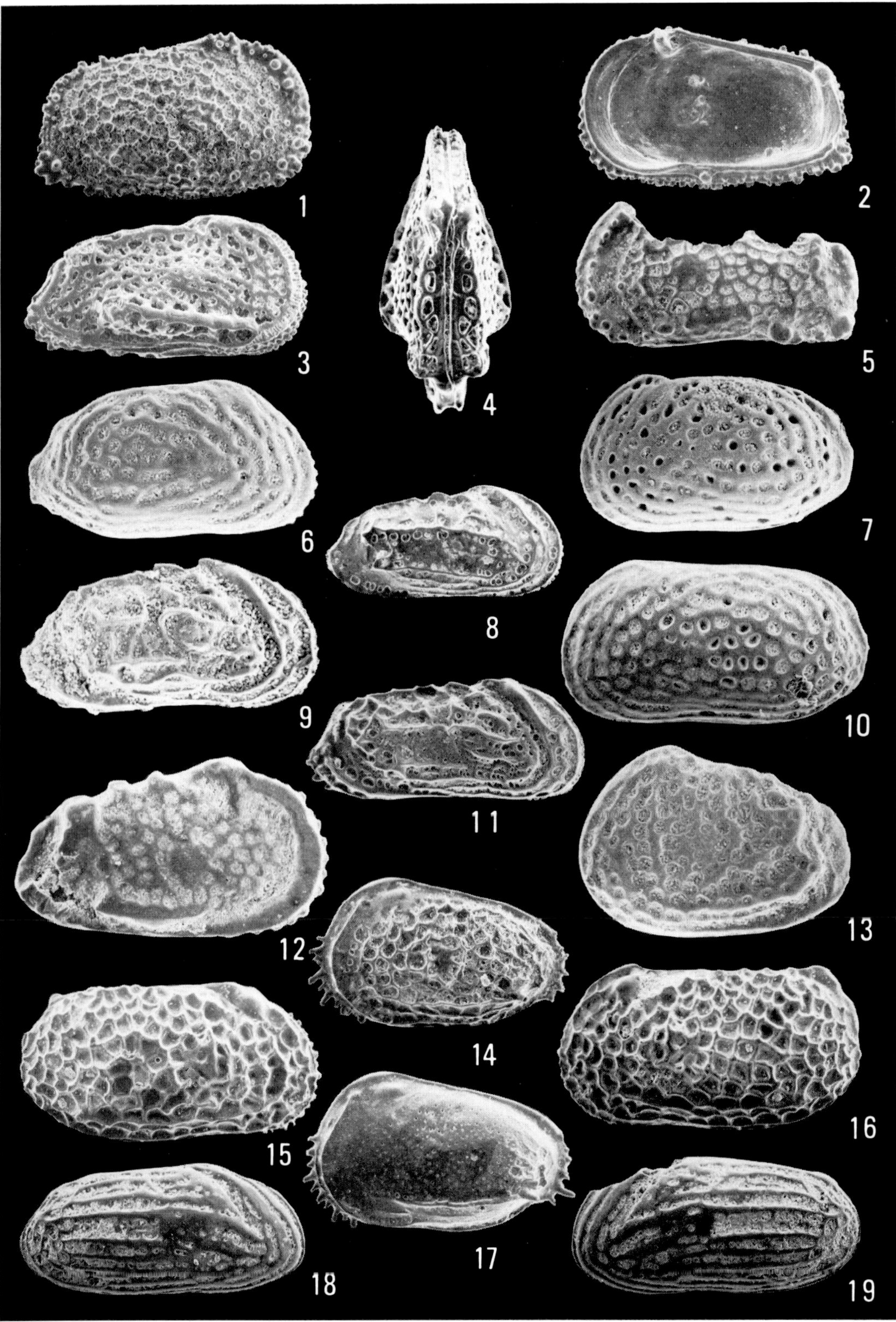

YAJIMA, Late Pleistocene Ostracoda

Explanation of Plate 13

Figs. 1–4. *Nipponocythere bicarinata* (Brady, 1880)

1. Lateral view of right female valve (CA 9911, sample 224–2, Kamiiwahashi Formation). ×85. 2. Lateral view of left female valve (CA 9912, sample 224–2). ×85. 3. Lateral view of right female valve (CA 9909, sample 251–1, Kioroshi Member of Kioroshi Formation). ×90. 4. Lateral view of left female valve (CA 9910, sample 251–1). ×85.

Fig. 5. *Aurila* sp.

Lateral view of left valve (CA 9866, sample 120–1, Yabu Member of Yabu Formation). ×70.

Figs. 6–8. *Robustaurila assimilis* (Kajiyama, 1913)

6. Lateral view of right male valve (CA 9868, sample 74–2, Kiyokawa Formation). ×71. 7. Lateral view of left male valve (CA 9869, sample 74–2). ×71. 8. Dorsal view of male carapace (CA 9867, sample 74–2). ×71.

Figs. 9–11. *Aurila kiritsubo* n. sp.

9. Lateral view of right valve (CA 9864, sample 49–1, Toyonari Member of Kioroshi Formation). ×65. 10. Lateral view of left valve (holotype, CA 9863, sample 49–1), ×65. 11. Dorsal view of carapace (CA 9865, sample 49–1). ×55.

Figs. 12, 13. *Caudites? posterocostatus* (Ishizaki, 1966)

12. Right valve view of carapace (CA 9871, sample 120–1, Yabu Member of Yabu Formation). ×65. 13. Dorsal view of carapace (CA 9871). ×55.

Fig. 14. *Urocythereis? gorokuensis* Ishizaki, 1966

Lateral view of left valve (CA 9870, sample 120–3, Yabu Member of Yabu Formation). ×55.

Fig. 15. *Pseudocythere frydli* n. sp.

Right valve view of carapace (holotype, CA 9876, sample 262–1, Kioroshi Member of Kioroshi Formation). ×70.

Fig. 16. *Bythoceratina* sp.

Lateral view of right valve (CA 9875, sample 305–4, Yabu Member of Yabu Formation). ×68.

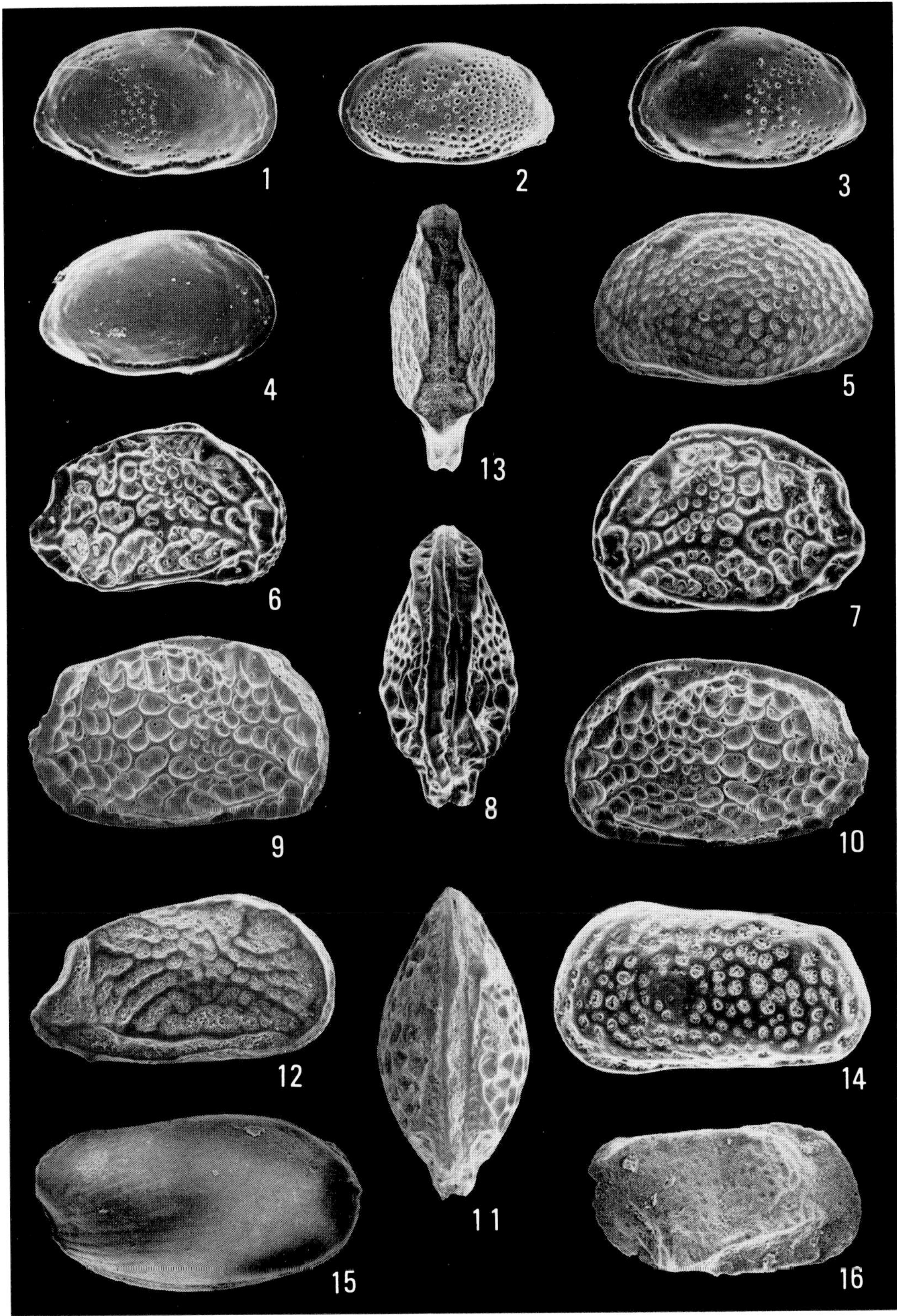
1
2
3
4
13
5
6
7
8
9
10
12
11
14
15
16

Explanation of Plate 14

Figs. 1–8, 17. *Semicytherura wakamurasaki* n. sp.

1. Lateral view of right male valve (CA 9884, sample 190–1, Kioroshi Member of Kioroshi Formation). ×110. 2. Lateral view of left male valve (CA 9887, sample 271–2, Kiroroshi Member of Kioroshi Formation). ×110. 3. Lateral view of right female valve (CA 9883, sample 190–1). ×110. 4. Dorsal view of male carapace (CA 9888, sample 271–2). ×110. 5. Dorsal view of female carapace (CA 9889, sample 271–2). ×110. 6. Lateral view of left female valve (holotype, CA 9882, sample 190–1). ×110. 7. Inner view of left female valve (CA 9885, sample 190–1). ×110. 8. Inner view of right female valve (CA 9886, sample 190–1). ×110. 17. Normal pore opening with lip on exterior surface of subcentral part of right female valve (CA 9883). ×4000.

Fig. 9. *Semicytherura* sp.

Lateral view of left valve (CA 9891, sample 305–1, Yabu Member of Yabu Formation). ×103.

Figs. 10, 11, 14. *Kangarina hayamii* n. sp.

10. Dorsal view of carapace (holotype, CA 9892, sample 120–1, Yabu Member of Yabu Formation). ×130. 11. Lateral view of right valve (CA 9894, sample 273–2, Kioroshi Member of Kioroshi Formation). ×124. 14. Lateral view of left valve (CA 9893, sample 120–1). × 120.

Figs. 12, 13, 15, 16. *Eucytherura utsusemi* n. sp.

12. Inner view of right valve (holotype, CA 9879, sample 120–3, Yabu Member of Yabu Formation). ×164. 13. Dorsal view of carapace (CA 9881, sample 120–2, Yabu Member of Yabu Formation). ×164. 15. Lateral view of right valve (holotype, CA 9879). ×180. 16. Lateral view of left valve (CA 9880, sample 305–2, Yabu Member of Yabu Formation). ×185.

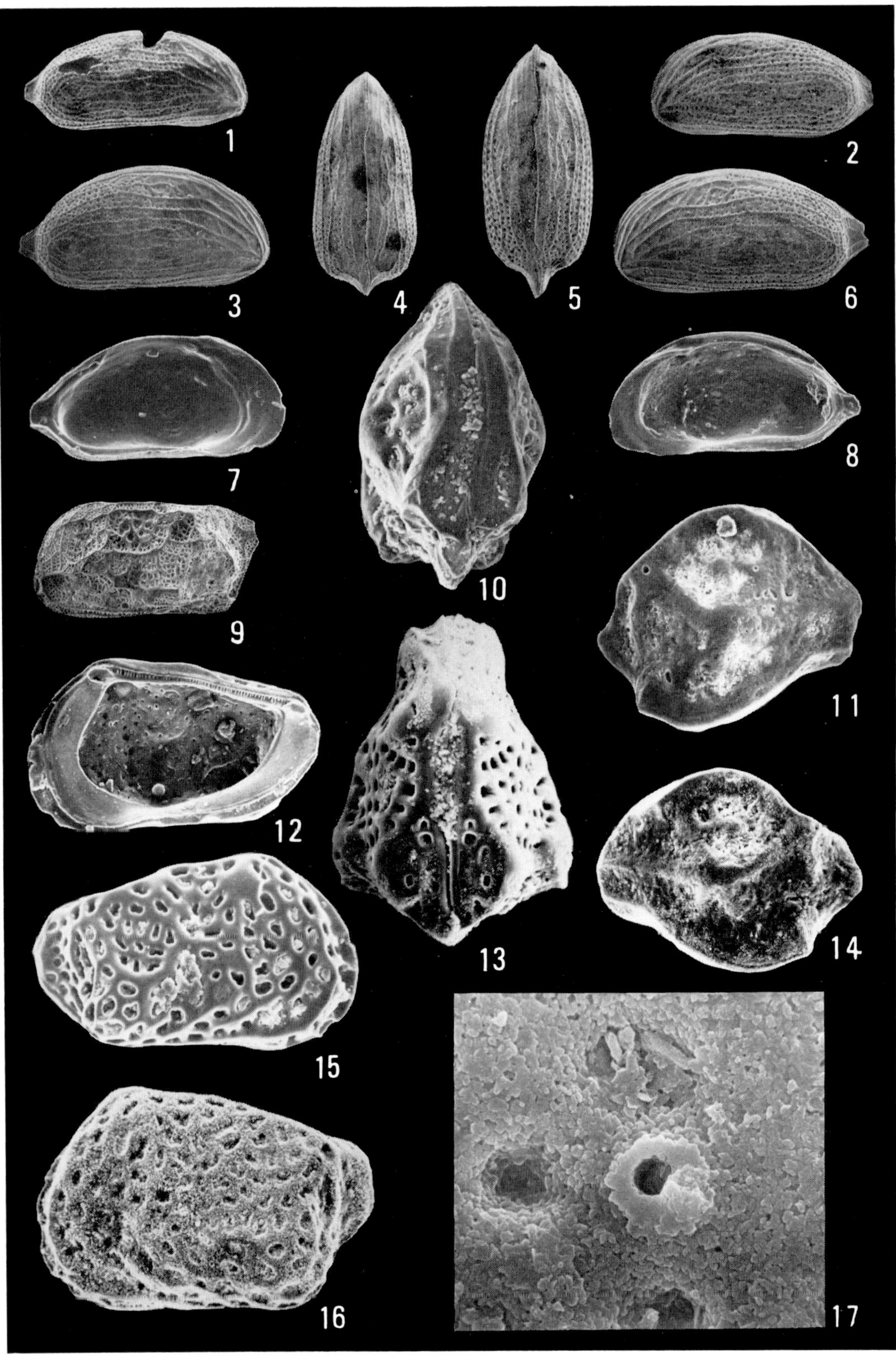
1
2
3
4
5
6
7
8
9
10
11
12
13
14
15
16
17

Explanation of Plate 15

Figs. 1, 2, 5. *Propontocypris* (*Propontocypris*) sp.

1. Lateral view of left immature valve (CA 9802, sample 273–2, Kioroshi Member of Kioroshi Formation). ×95. 2. Inner view of left immature valve (CA 9802). ×95. 5. Dorsal view of carapace (CA 9801, sample 237–2). ×95.

Figs. 6, 9. *Paracypris* sp.

6. Lateral view of left valve (CA 9804, sample 305–5, Yabu Member of Yabu Formation). ×48. 9. Lateral view of right valve (CA 9803, sample 80–2, Kiyokawa Formation). ×48.

Figs. 3, 4, 7, 8, 13, 15. *Paracytheroma* sp.

3. Inner view of left valve (CA 9921, sample 228–2, Kamiiwahashi Formation). ×84. 4. Dorsal view of carapace (CA 9922, sample 6–3, Toyonari Member of Kioroshi Formation). ×75. 7. Lateral view of right valve (CA 9923, sample 6–3). ×66. 8. Lateral view of left valve (CA 9924, sample 228–3). ×66. 13. Muscle scars on left valve (CA 9921). ×600. 15. Normal pore opening on exterior suface of subcentral part of right valve (CA 9923). ×4000.

Fig. 10. *Sclerochilus* sp. 1

Lateral view of left valve (CA 9919, sample, 308–1 Kiyokawa Formation). ×84.

Figs. 11, 12. *Xestoleberis suetsumuhana* n. sp.

11. Lateral view of right valve (CA 9914, sample 284–3, Yabu Member of Yabu Formation). ×56. 12. Lateral view of left valve (holotype, CA 9913, sample 138–5, Kamiizumi Member of Yabu Formation). ×56.

Figs. 14, 16. *Aurila kiritsubo* n. sp.

14. Sieve-type normal pore openings on exterior surface of subcentral part of left valve (holotype, CA 9863, sample 49–1, Toyonari Member of Kioroshi Formation). ×900. 16. Simple normal pore opening on muri at ventral part of left valve (holotype, CA 9863). ×900.

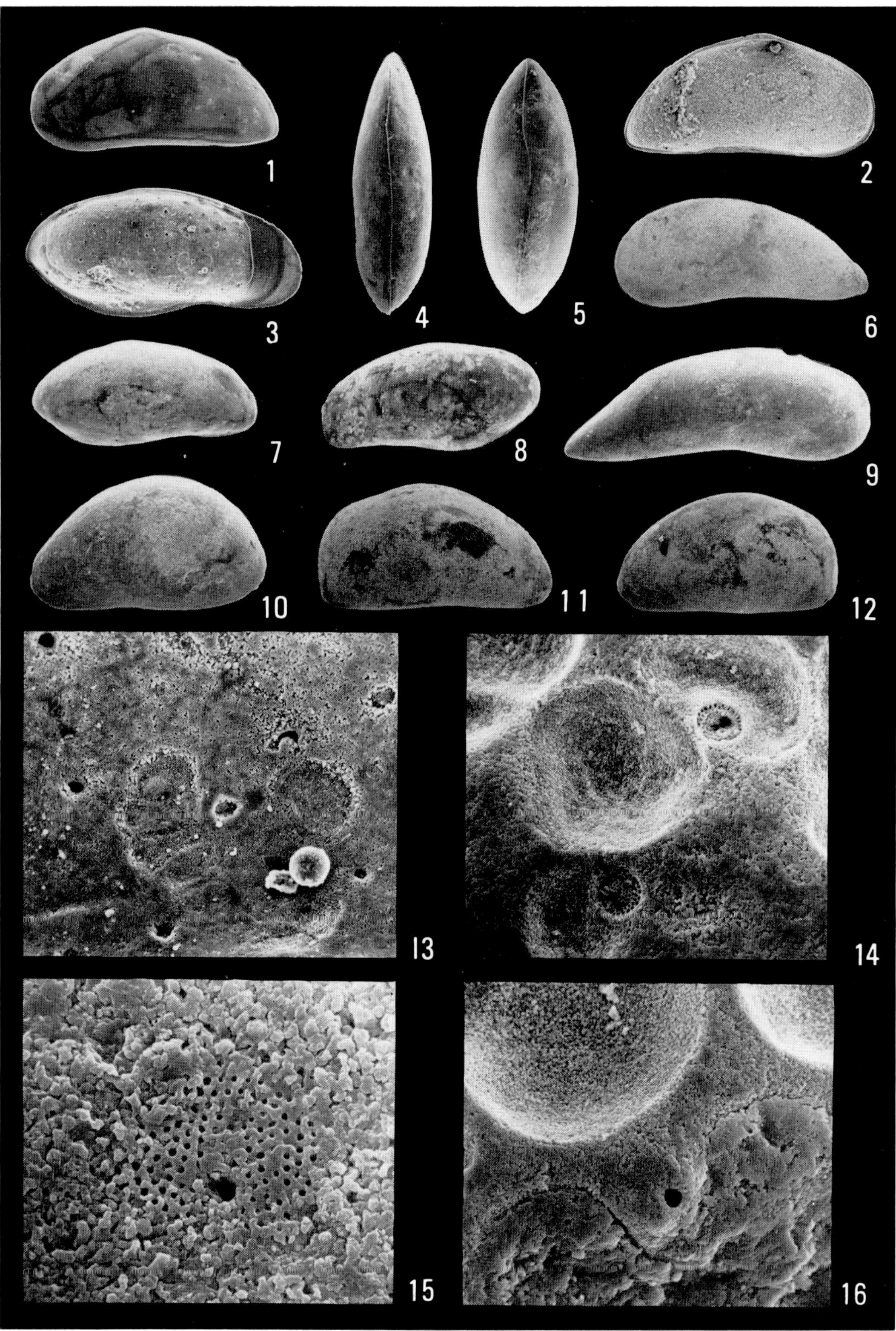
1
2
3
4
5
6
7
8
9
10
11
12
13
14
15
16

Explanation of Plate 16

Sections prepared after embedding in resin. Figs. a, b. Light-micrographs of a 1-μm section. Fig. c. A montage electron-micrograph. Arrows indicate the apical portions of the boundary of outer epidermal cells. e: epiculicle, h: hinge, m: membraneous layer, il: inner lamella, ol: outer lamella.

Scale bars in plates 16–30 and in text-figs. 1–17 indicate 1 μm (single bar), 10 μm (double bar), 100 μm (triple bar), and 0.1 μm (broken bar) unless otherwise stated.

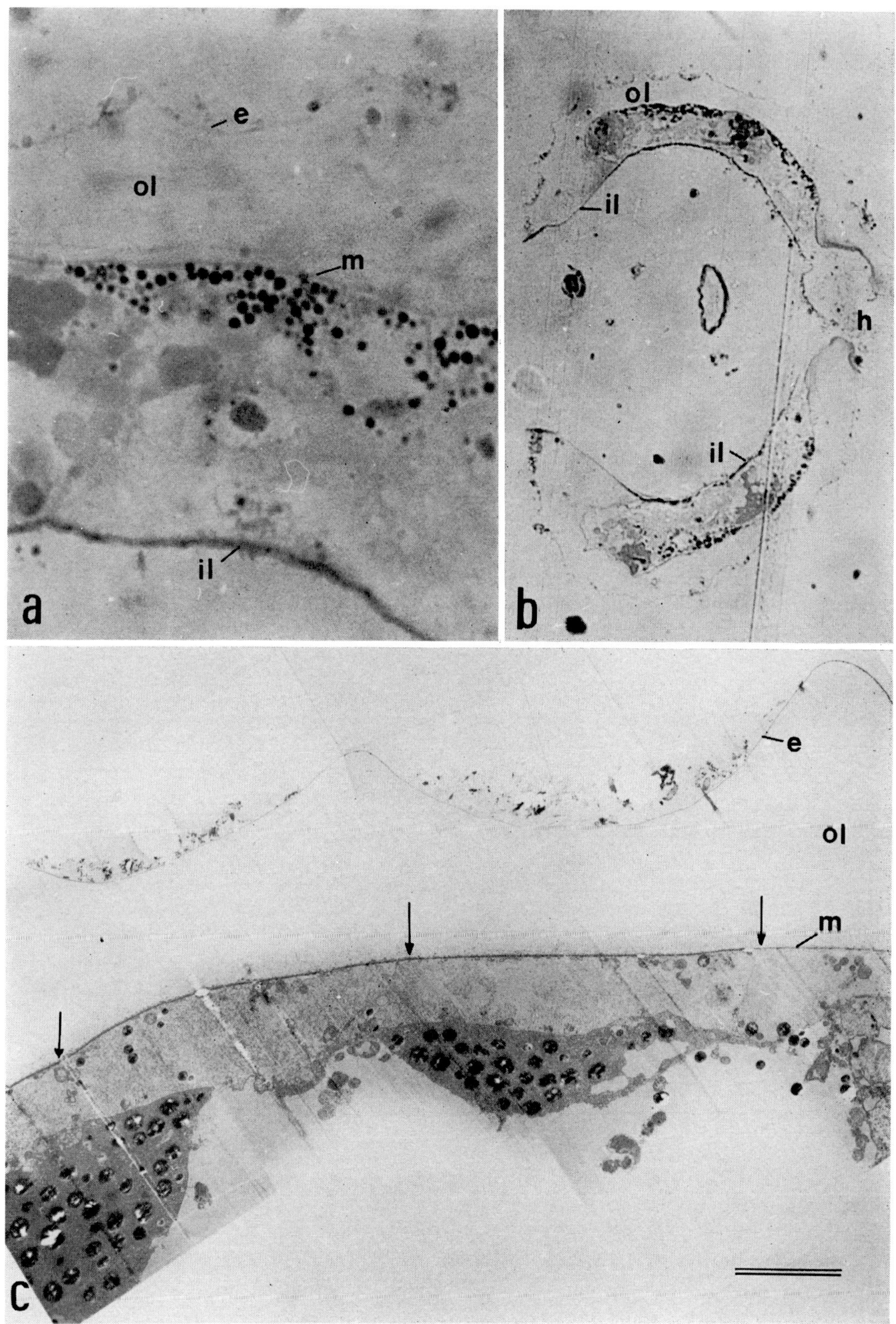
e
ol
m
il
a
ol
il
h
il
b
e
ol
m
c

Explanation of Plate 17

Figs, a, b. Apical part of outer lamella cuticle at molt stage C in fossa(a) and ridge (b). Figs. c-f. Epicuticle and adjacent procuticle at molt stage C in fossa (c) and ridge (d) of outer lamella, and in uncalcified part (e) and calcified part (f) of inner lamella. Figs. g, h. Uncalcified parts of inner lamella cuticle and apical boundaries between adjacent inner epidermal cells. d: desmosomes, e: epicuticle, g: gap-type junction, m: mitochondrion, p: procuticle. Fig. i: Calcified part of inner lamella cuticle where the piled membrane structure in procuticle is well developed.

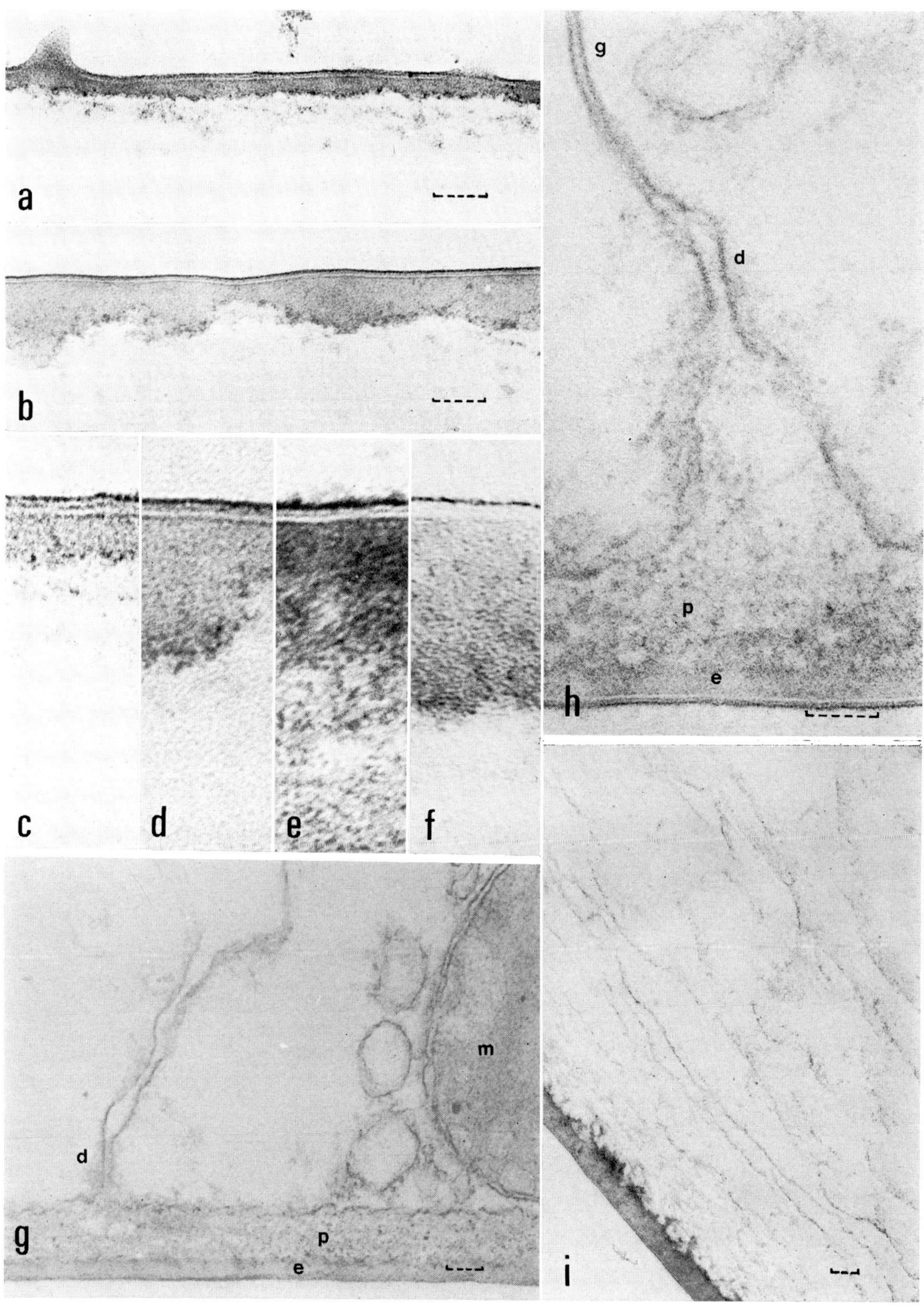

OKADA, Carapace Ultrastructures

Explanation of Plate 18

Subdermal cells characterized by abundant r-ER. Outher lamella cuticle (ol) and outer epidermal cell (oe) are also shown. g: granule, m: membranous layer, n: nucleus, sd: subdermal cells.

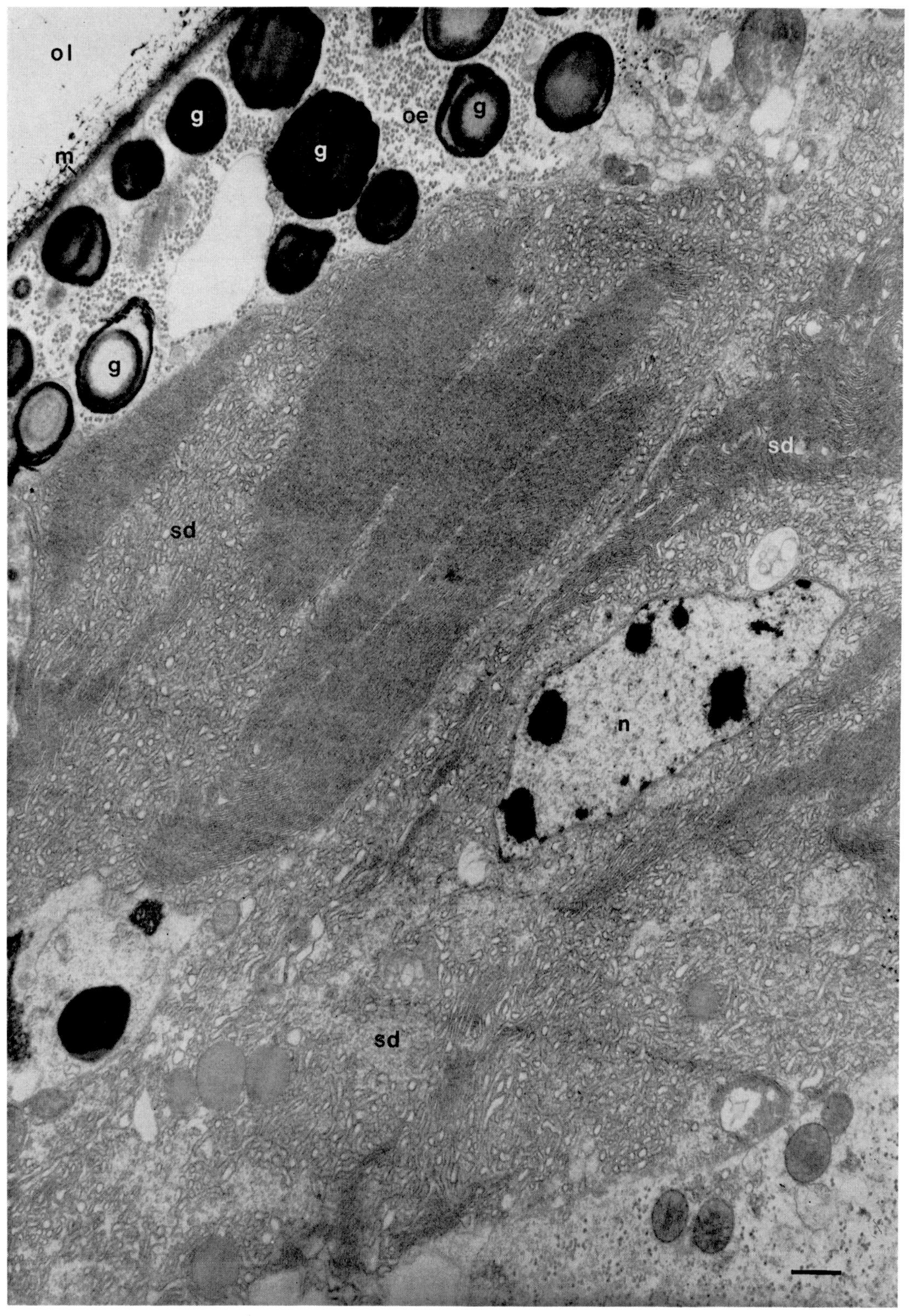
ol
m
g
g
oe
g
g
sd
sd
n
sd

Explanation of Plate 19

Granules (Figs. b-d) and layered r-ER (Figs. a, e) in subdermal cell.

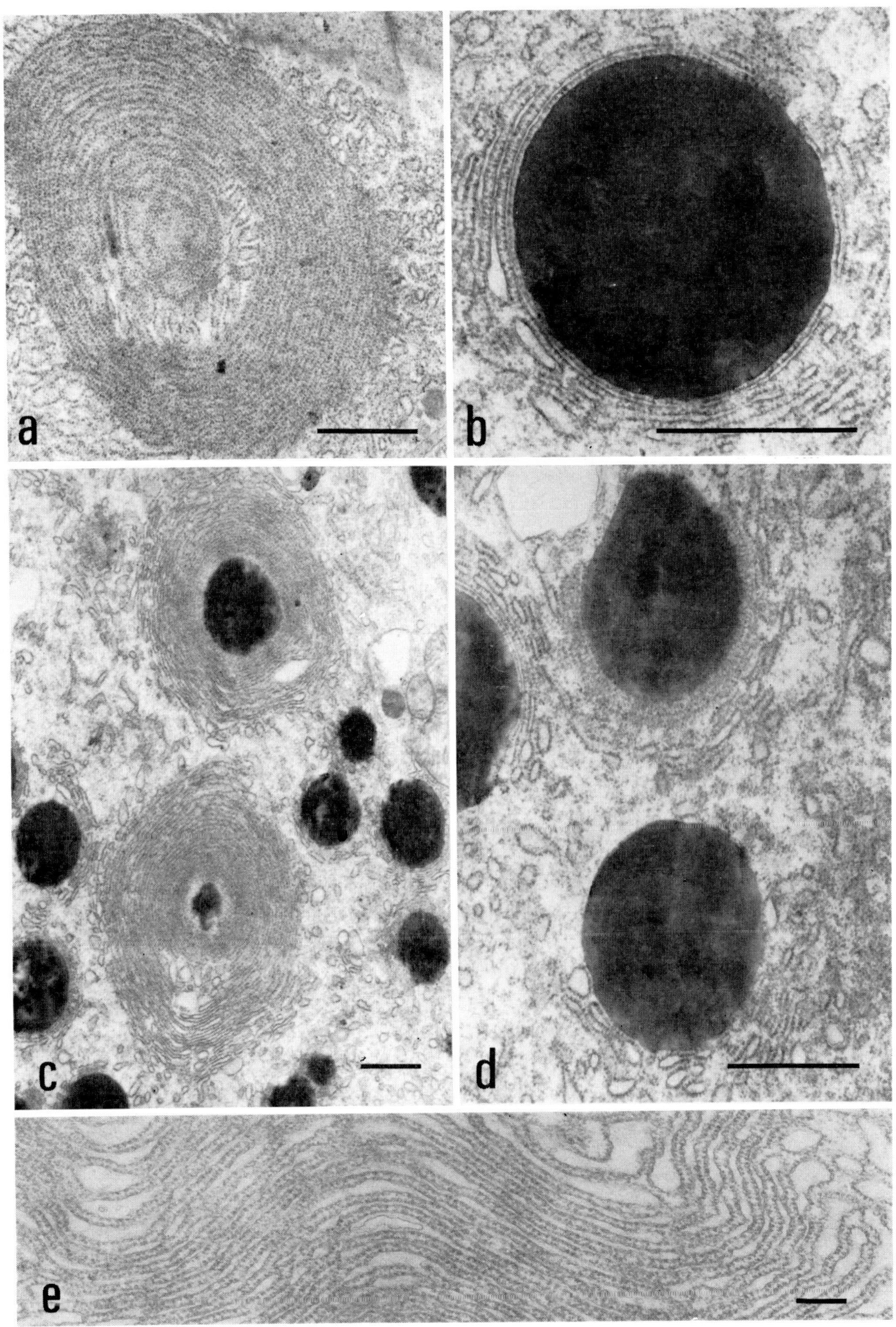
a
b
c
d
e

Explanation of Plate 20

Figs. a–d. Posterior spines. a: SEM micrograph of a spine, its distal end was cut off. b: TEM micrograph of a cross section of a spine. c-d. Newly formed spine just before ecdysis. Spines are completely covered with epicuticle. Fig. e. Longitudinal section of a sensillum. Bending of setae may affect the pitch of circular fold. Proximal end of setal cuticle is indicated with arrows. d: depression of the multiple pouch structure, e: epicuticle, f: circular fold.

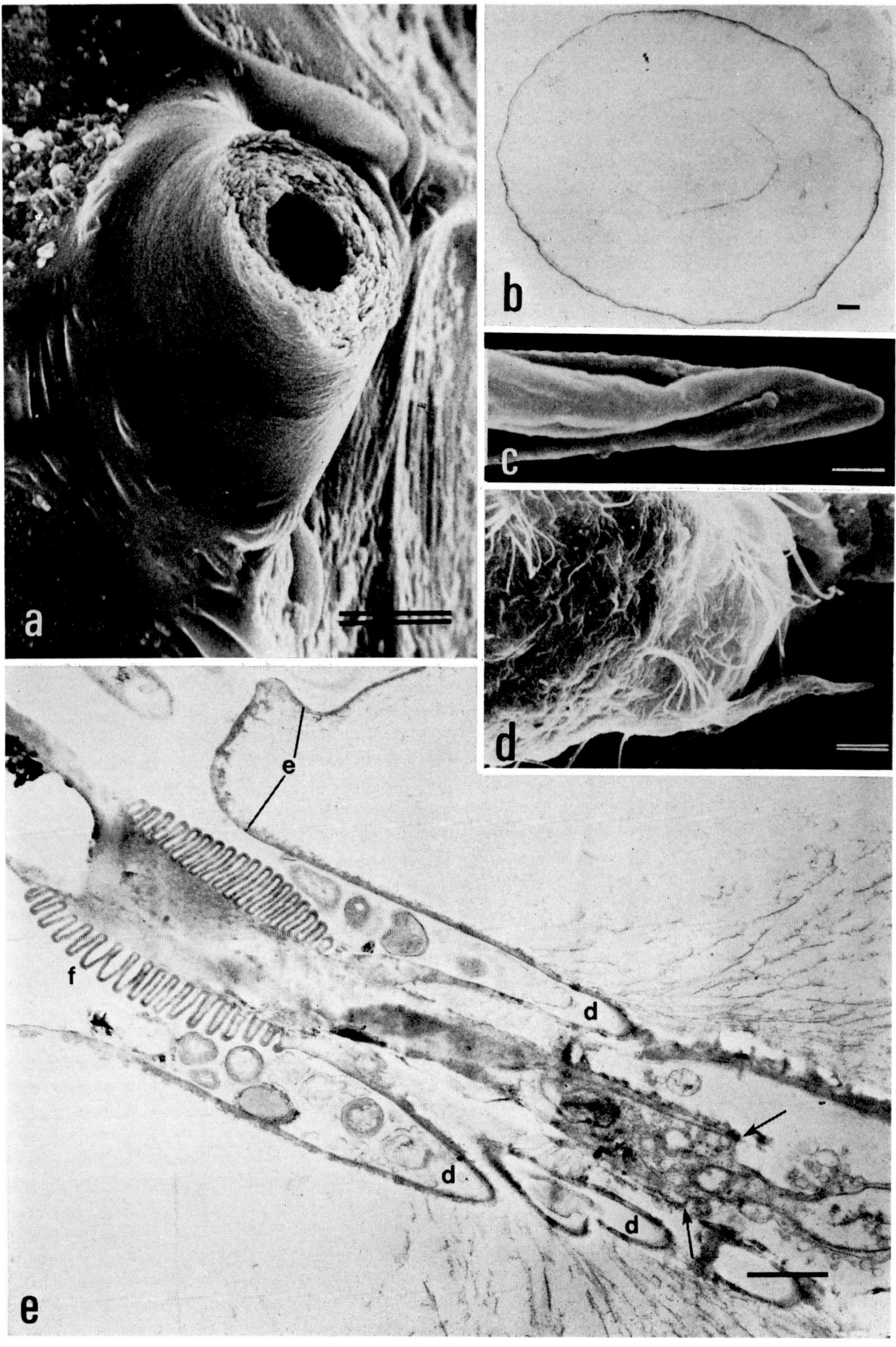
a
b
c
d
e
e
f
d
d
d

Explanation of Plate 21

Two long cilia surrounded with a dense tube. They seem to coil in setal cuticle as a double helix. Serial cross sections of a single sensillum are shown. Usc knife marks as an indicatior of the same direction (arrows).

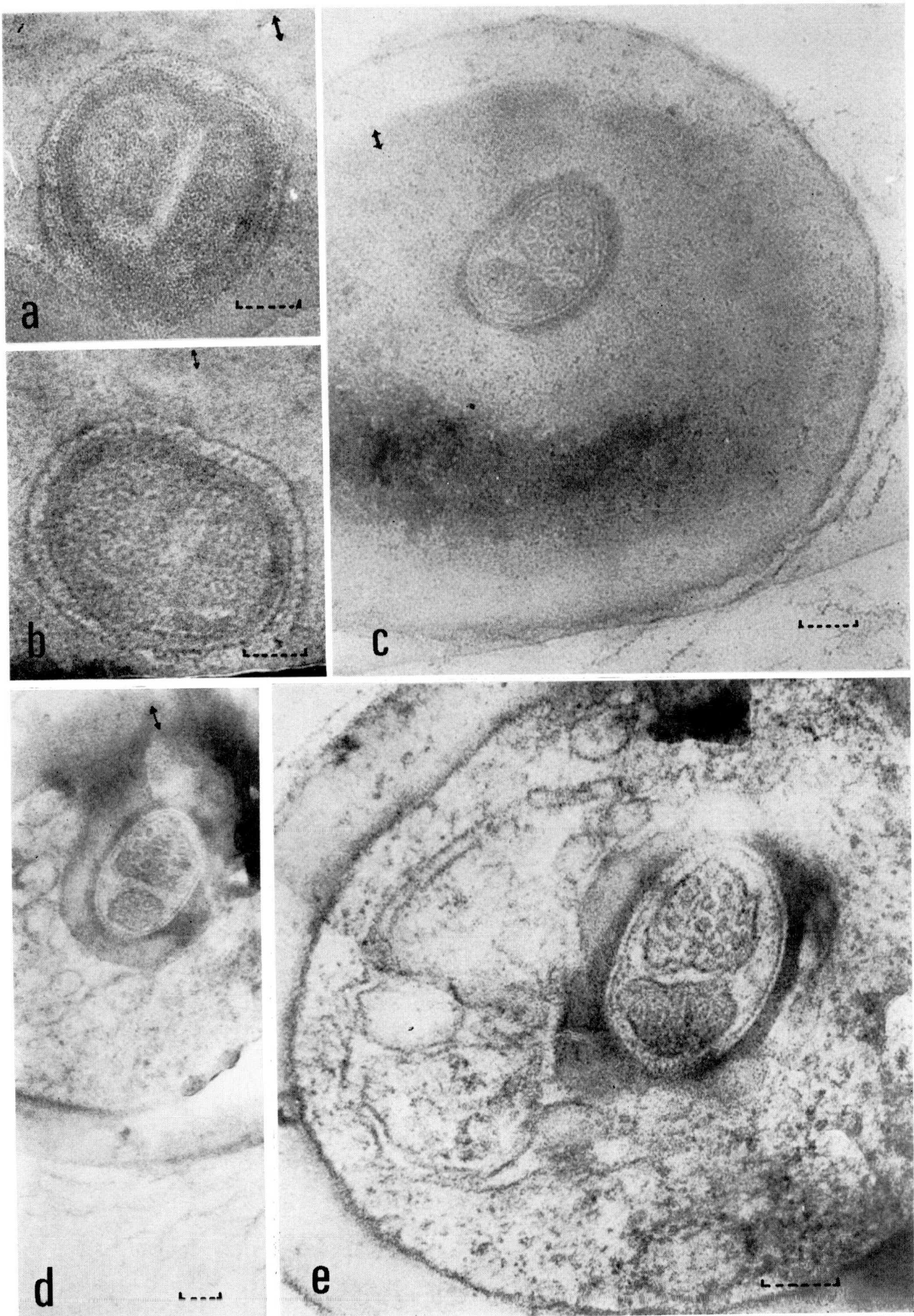

OKADA, Carapace Ultrastructures

Explanation of Plate 22

Figs. a, b. Median (a) to proximal (b) parts of multiple pouch structure. A layer of filamentous sheath, a dense tube and a sh-2 cell enclose two cilia in setal cuticle. Closed bottoms of tubular depressions (18, 21, 23 and 27) are shown (b). Fig. c. Proximal end of setal cuticle with open bottom. Distal part of a branched sh-1 cell also encloses the cilia.

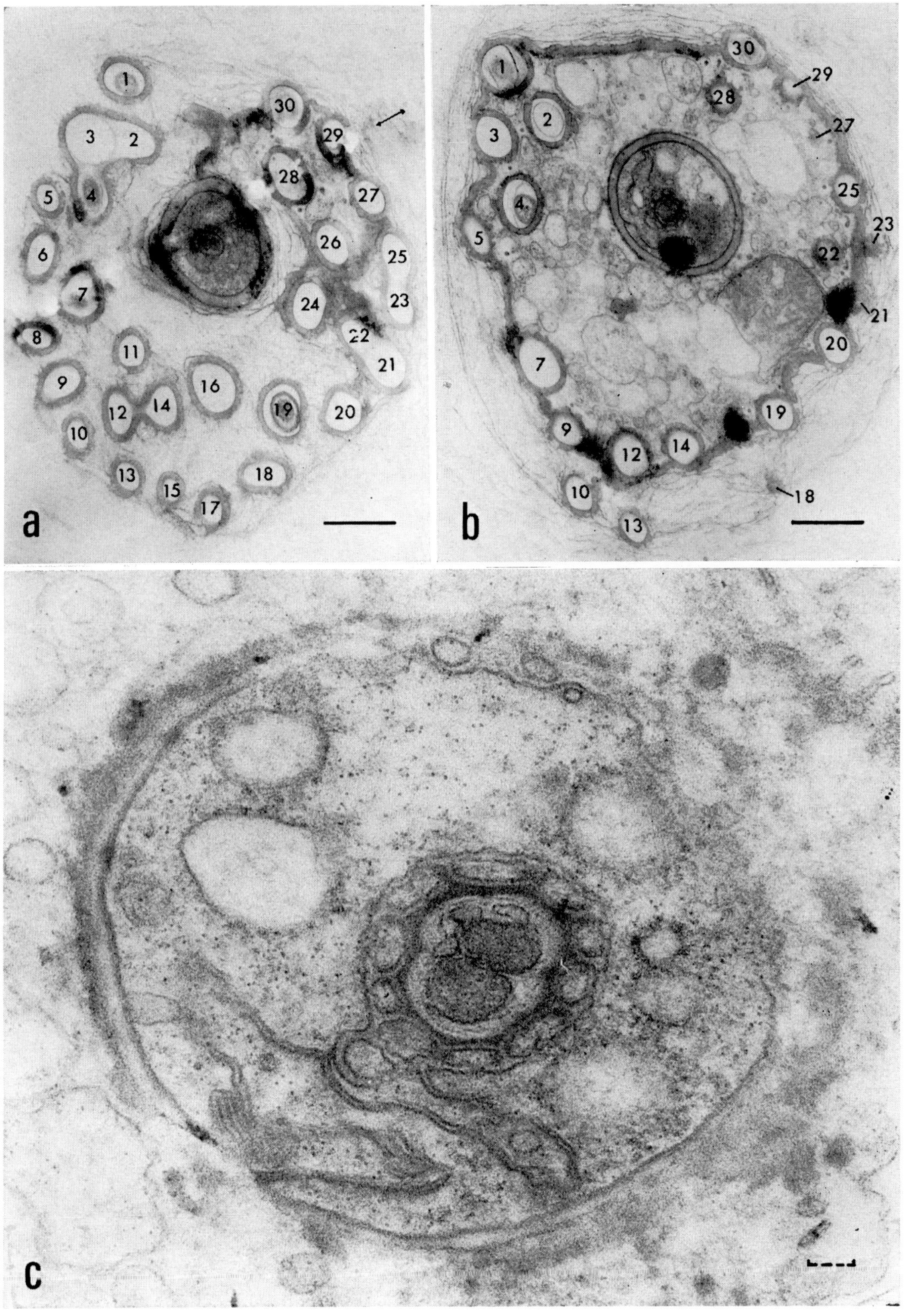
a
b
c
1
2
3
4
5
6
7
8
9
10
11
12
13
14
15
16
17
18
19
20
21
22
23
24
25
26
27
28
29
30

Explanation of Plate 23

Branched sh-1 cell. Two thicker cilia have dense matrix and microtubles which are traceable to micrographs of other sections. Two minor cilia are tips of the short cilia with electron-lucid matrix.

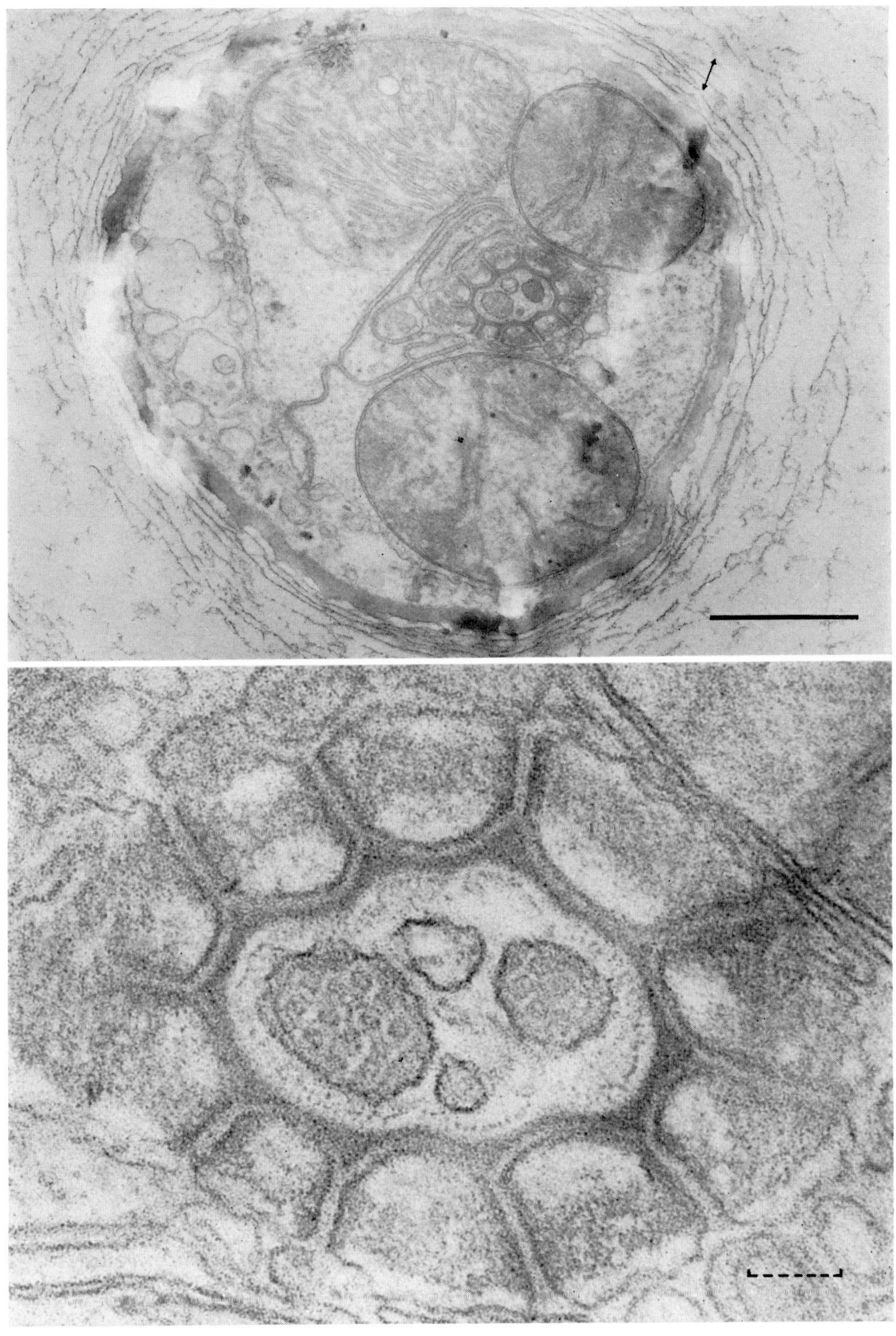

Explanation of Plate 24

Distal parts of the short cilia with electron-lucid matrix, one with a 9+0 pattern and the other without. The structure of distal ends of a cilium is shown in figs. b and c. A non-branching sh-1 cell forms a mesaxon. The dense tube does not enclose cilia. Bundles of filaments are observable instead of the filamentous sheath. These filaments and those in the sh-1 cell resemble actin filaments.

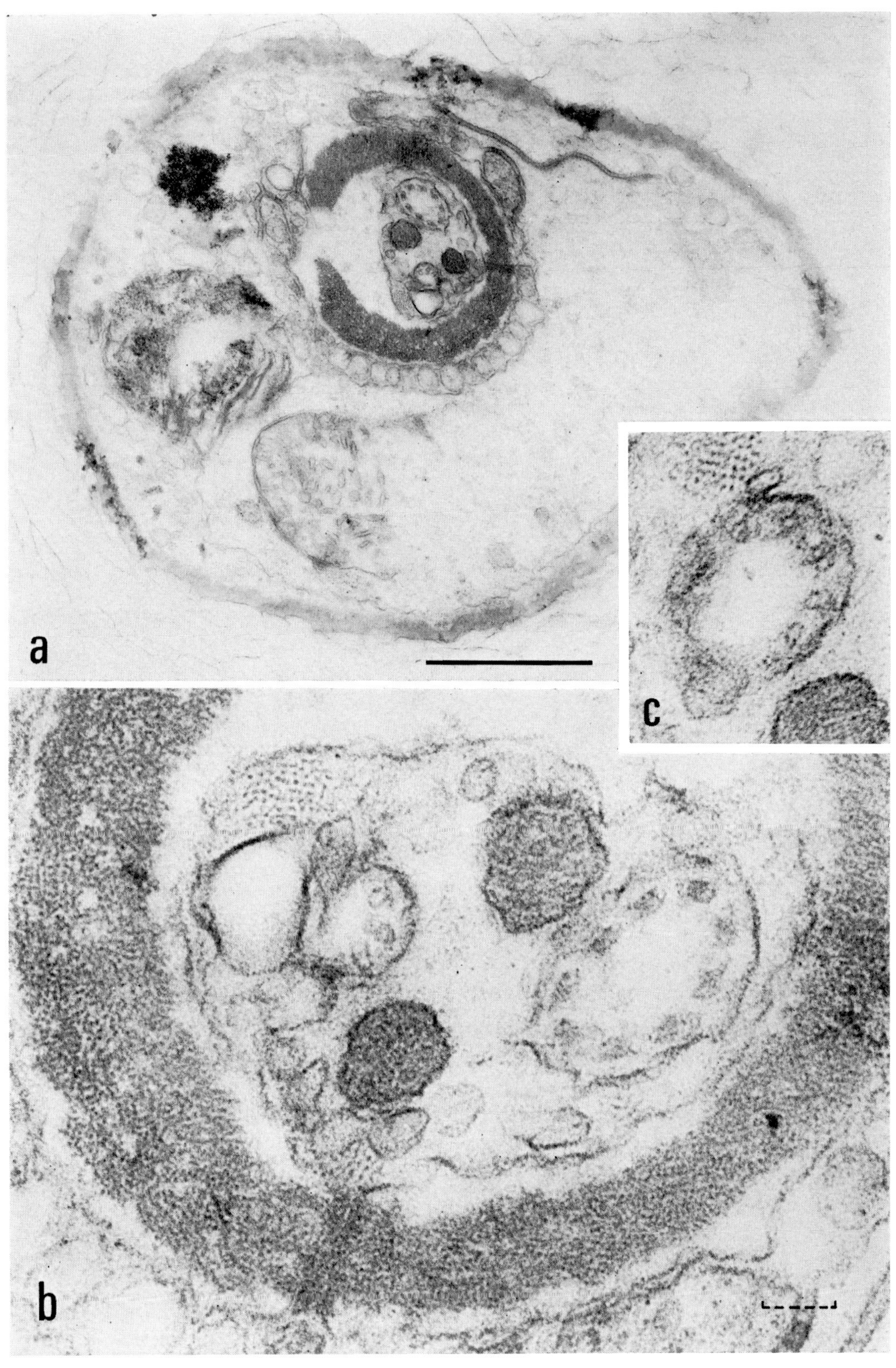
a
c
b

Explanation of Plate 25

9+0 pattern of two types of cilia. Two cilia have the well-known 9+0 pattern. The sh-1 cell does not form a complete tube. Fig. a. Enlargement of text-fig. 15a. Fig. c. Enlargement of fig. b.

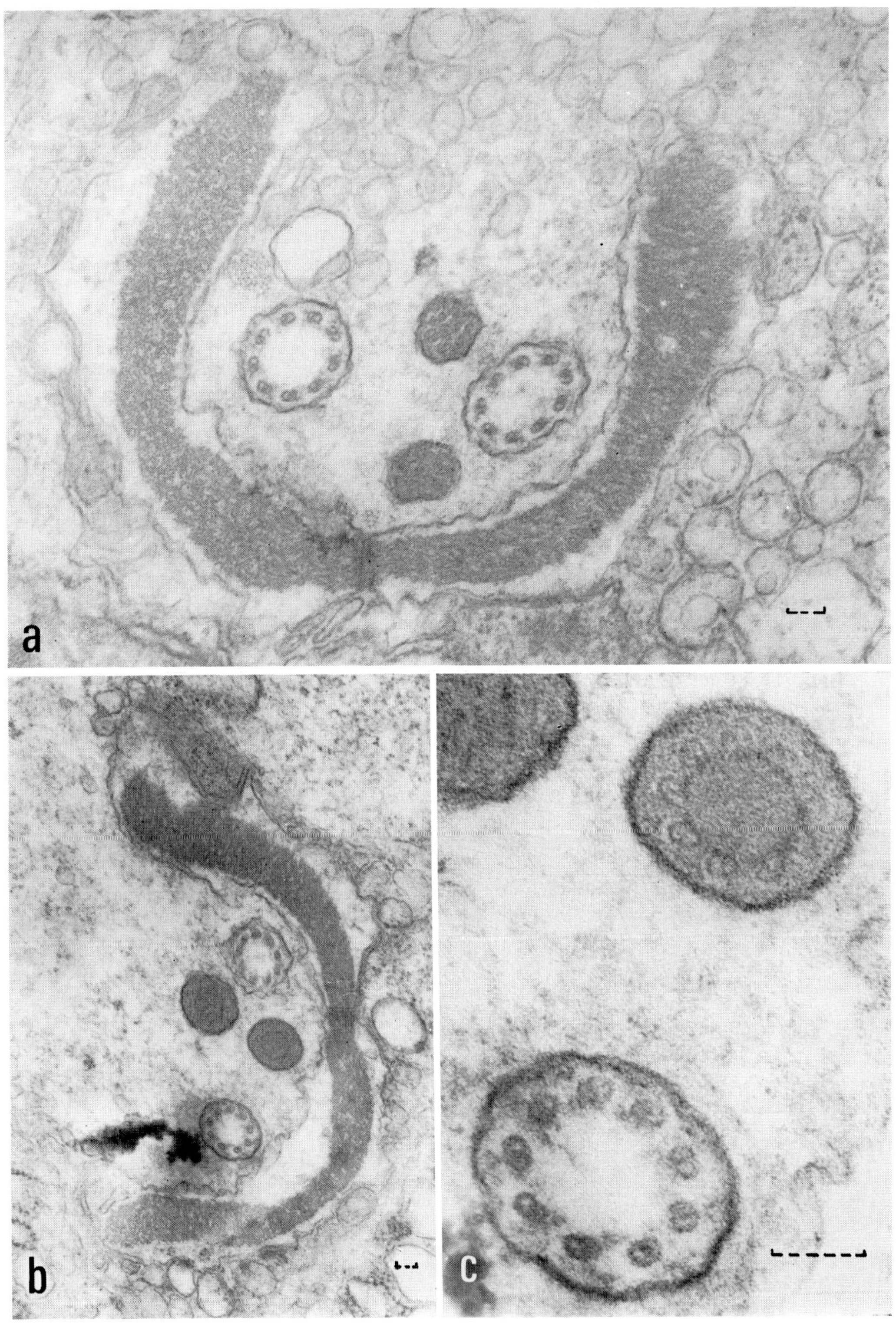
a
b
c

Explanation of Plate 26

Basal parts of cilia. Microtubules appear to be connected by electron-dense substance which tubularly encloses the center of a cilium. A center tubule exists in an electron-lucid cilium (b).

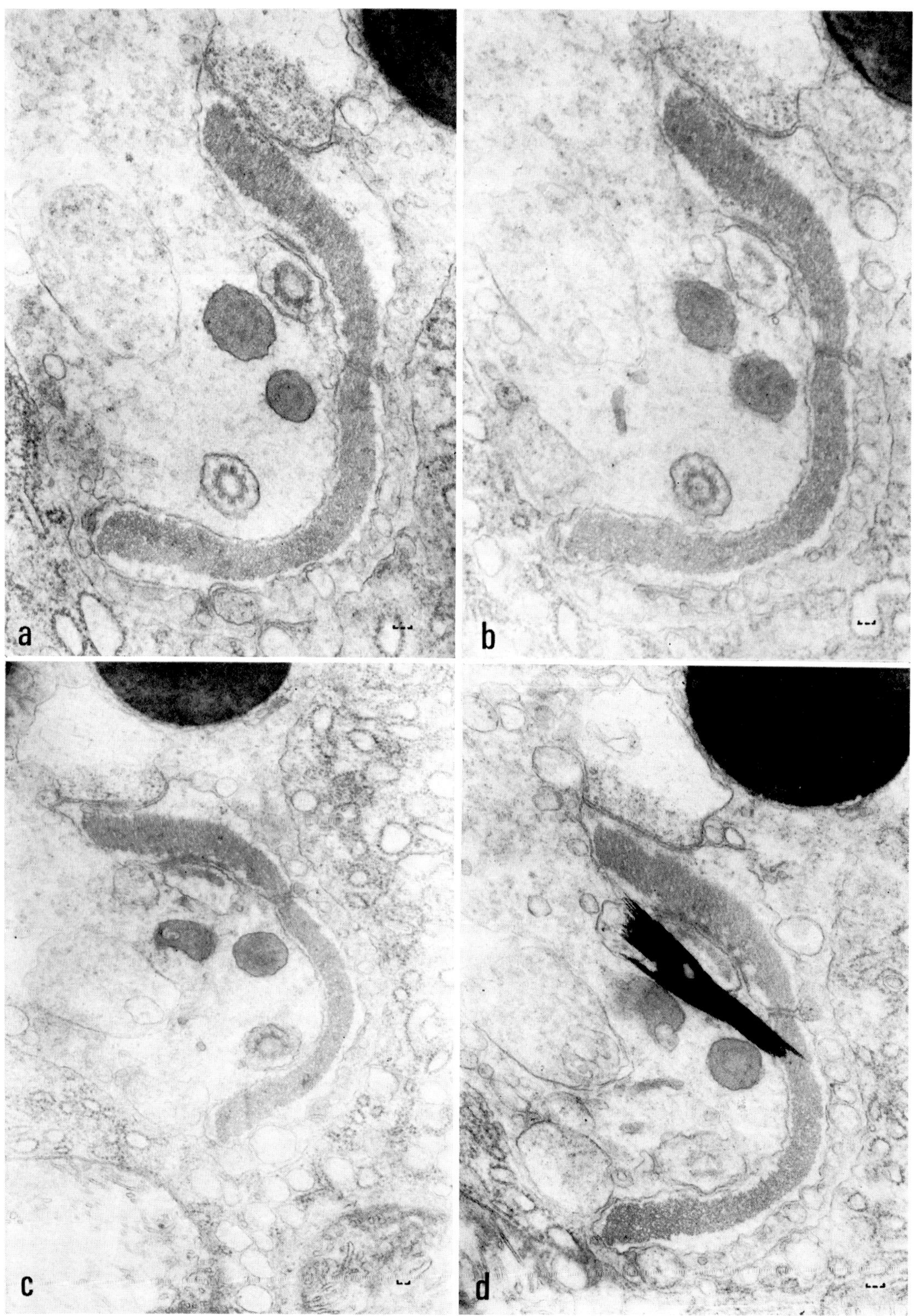

OKADA, Carapace Ultrastructures

Explanation of Plate 27

Dendrites under cilia. Fig. a. Rootlets in the electron-lucid cilia make desmosome-like connections with the sh-1 cell. Figs. b, c. Near the basal end of the sh-1 cell (1). Every dendrite contains mitocondria. The proximal part of an electron-lucid dendrite contains as many solitary microtubules as usual nerve fibres.

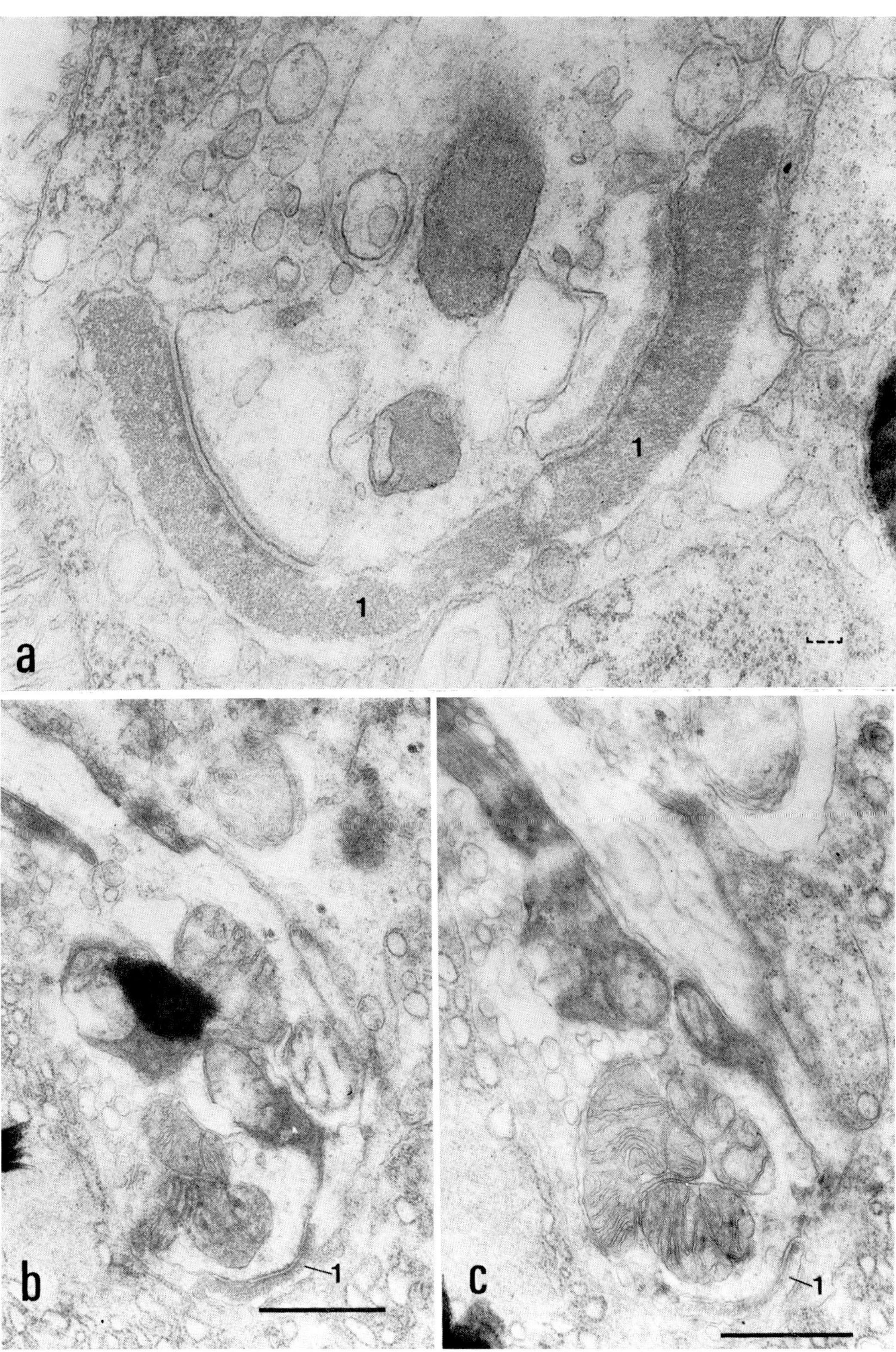
1
1
a
b
1
c
1

Explanation of Plate 28

Figs. a, b. Branching of a sh-1 cell. The cell branches into eight parts at a distal portion (a), and incomplete branching can be seen at a proximal portion (b), Fig. c. Longitudinal section of basal parts of cilia in the same section as text-fig. 15b. Two center tubules are shown in a cilium. c: cilium, d: dendrite, r: rootlet, sh-1: sh-1 cell.

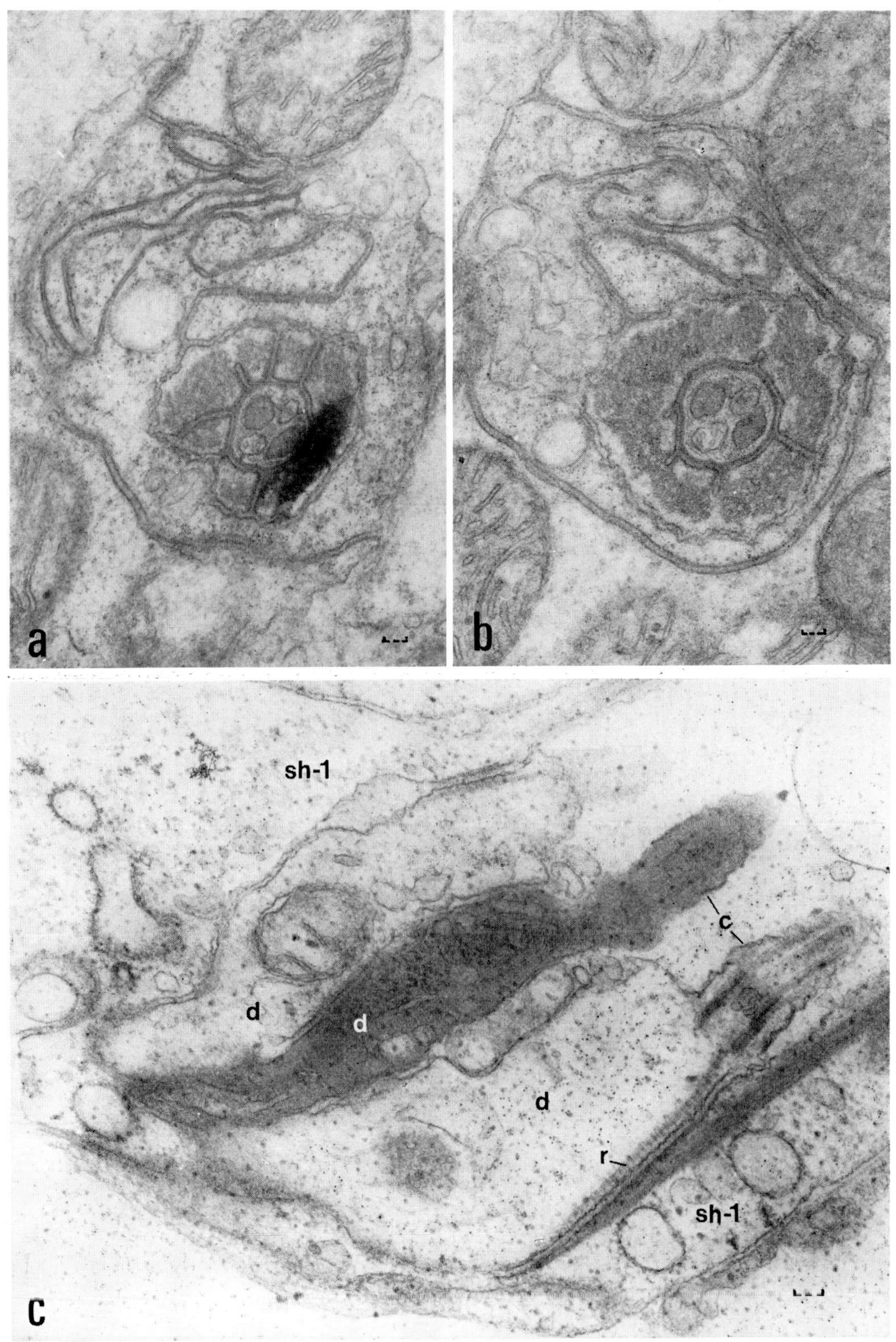
a
b
sh-1
c
d
d
d
r
sh-1
c

Explanation of Plate 29

Pores near posterior margin. Fig. a. Distal part. Fig. b. Median part. Fig. c. Proximal part.

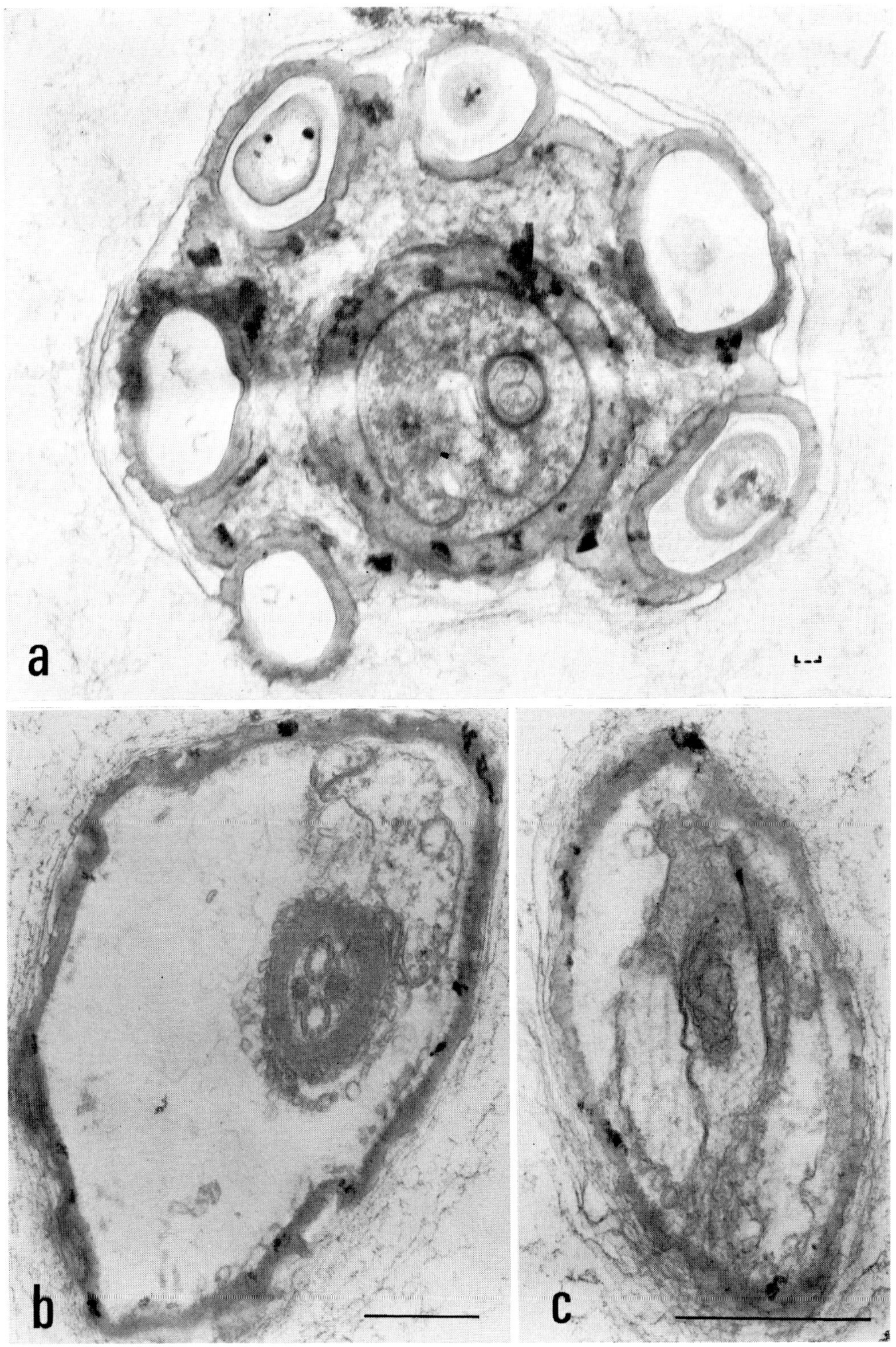
a
b
c

Explanation of Plate 30

Anterior marginal pores. Fig. a. Distal part. Fig. b. Median part. Fig. c. Proximal part.

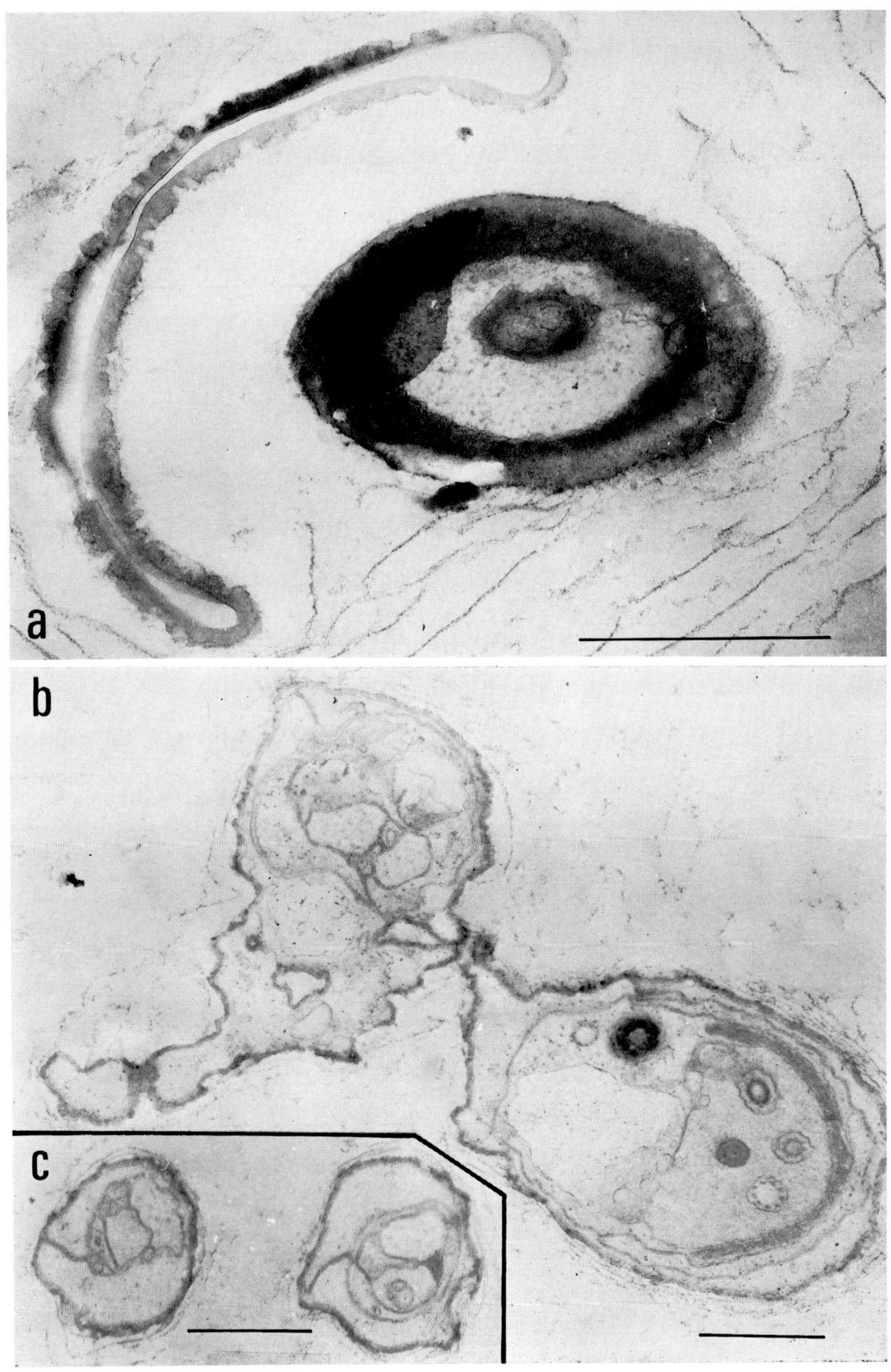
a
b
c